Advanced Sciences and Technologies for Security Applications

The series Advanced Sciences and Technologies for Security Applications comprises interdisciplinary research covering the theory, foundations and domain-specific topics pertaining to security. Publications within the series are peer-reviewed monographs and edited works in the areas of:

- biological and chemical threat recognition and detection (e.g., biosensors, aerosols, forensics)
- crisis and disaster management
- terrorism
- cyber security and secure information systems (e.g., encryption, optical and photonic systems)
- traditional and non-traditional security
- energy, food and resource security
- economic security and securitization (including associated infrastructures)
- transnational crime
- human security and health security
- social, political and psychological aspects of security
- recognition and identification (e.g., optical imaging, biometrics, authentication and verification)
- smart surveillance systems
- applications of theoretical frameworks and methodologies (e.g., grounded theory, complexity, network sciences, modelling and simulation)

Together, the high-quality contributions to this series provide a cross-disciplinary overview of forefront research endeavours aiming to make the world a safer place.

The editors encourage prospective authors to correspond with them in advance of submitting a manuscript. Submission of manuscripts should be made to the Editor-in-Chief or one of the Editors.

More information about this series at http://www.springer.com/series/5540

Igor Jovanovic • Anna S. Erickson
Editors

Active Interrogation in Nuclear Security

Science, Technology and Systems

Springer

Editors
Igor Jovanovic
Department of Nuclear Engineering
and Radiological Sciences
University of Michigan
Ann Arbor, MI, USA

Anna S. Erickson
Nuclear & Radiological Engineering
Program
G.W. Woodruff School of Mechanical
Engineering
Georgia Institute of Technology
Atlanta, GA, USA

ISSN 1613-5113 ISSN 2363-9466 (electronic)
Advanced Sciences and Technologies for Security Applications
ISBN 978-3-319-74466-7 ISBN 978-3-319-74467-4 (eBook)
https://doi.org/10.1007/978-3-319-74467-4

Library of Congress Control Number: 2018942899

Printed on acid-free paper

This Springer imprint is published by the registered company Springer International Publishing AG part of Springer Nature.
The registered company address is: Gewerbestrasse 11, 6330 Cham, Switzerland

To our families

Preface

Nuclear security represents one of the most pressing global challenges. Our societies are threatened asymmetrically by terrorist groups, either operated independently or sponsored by rogue states. The prospect of nuclear terrorism, especially one that would involve a nuclear explosive constructed using special nuclear material (SNM), motivates a concerted effort spanning technical and policy means to control the production, possession, and movement of SNM. Formidable physics obstacles stand in the way of addressing this challenge. SNM spontaneously produces only a low level of unique detectable signature in the form of penetrating radiation. This signature is immersed in an abundant natural background radiation. In addition, the radiation signatures of SNM can be shielded with relative ease, effectively eliminating the possibility to detect even substantial quantities of SNM that could be used to construct a functioning nuclear explosive device.

These fundamental physics challenges have motivated the development of alternative methods to detect, locate, and identify SNM. It is becoming accepted today in the physics and security community that the only promising way to detect SNM in movement, especially associated with burgeoning international trade, is using an active method, frequently referred to as active interrogation (AI). The objective of this book is to address the multiple technical facets of AI in an integrated and coherent form that has not been available to date. The available literature in this area is mostly in the form of relatively disjointed project reports and technical journal articles. The existing books usually treat only a single aspect of AI, such as radiation sources, detectors, algorithms, electronics, and physics of the detection process. To the best of our knowledge there is no volume that presents this subject in integrated form. Aggregating the knowledge in disparate technical areas of AI and connecting it in a logical fashion therefore holds a great educational and more general scientific merit, which is why we decided to undertake the project of writing this book. This book has been prepared in the format of an edited volume, since we believe that the best way to convey the state of the art in AI is to engage some of the leading researchers and educators who perform research on this subject, or have made important technical contributions to it.

We believe this is the right time to publish a book with this type of content. The community is highly engaged, the research activity is consistently high, and many students and researchers are entering the field. Graduate students and masters- and doctoral-level professionals working in the area of nuclear security or security in general could profit from a volume that will provide them with a concise physical basis of AI, along with in-depth current status of key issues in AI, as the technical method of choice to counter the threat of nuclear terrorism.

The goal of the book is to achieve a high level of alignment with the material taught in courses associated with a major growing area of nuclear engineering while integrating the fundamental physics principles with the current and future technology trends. The book should be an appropriate reference for both senior undergraduate and graduate courses in the area of nuclear security, nonproliferation, safeguards, and detection. It could even be a principal text for a special topic course on AI, or a required/optional text in the broader areas listed above. Parts of the book would also find audience among a broader group of engineers (mechanical and electrical) and scientists (physicist, materials scientists, and chemists), as well as policy analysts. At the same time, the book does not purport to replace the well-known texts that specialize in any of the technical areas covered here.

A fraction of this book draws upon the material developed for special topics courses taught at Purdue University and Penn State University for a number of years, including the ones that resulted from the collaborative effort of Penn State University, Massachusetts Institute of Technology, and Texas A&M University to spearhead a formal Nuclear Security Education Program. We wish to thank our contributors, who invested their valuable time and effort into making this book possible. Simultaneously, we would like to acknowledge the support we have received over the years from our colleagues at universities, national laboratories, and federal agencies, which further motivated the writing of this book.

Ann Arbor, MI, USA — Igor Jovanovic
Atlanta, GA, USA — Anna S. Erickson
November 2017

Contents

Contributors

Miltiadis Alamaniotis Purdue University, West Lafayette, IN, USA

Shaheen Dewji Oak Ridge National Laboratory, Oak Ridge, TN, USA

Mark Ellis Atomic Weapons Establishment, Aldermaston, Reading, Berkshire, UK

Anna S. Erickson Nuclear & Radiological Engineering Program, G.W. Woodruff School of Mechanical Engineering, Georgia Institute of Technology, Atlanta, GA, USA

Robert Garnett Los Alamos National Laboratory, Los Alamos, NM, USA

Cameron Geddes Lawrence Berkeley National Laboratory, Berkeley, CA, USA

Nolan Hertel Georgia Institute of Technology, Atlanta, GA, USA

Igor Jovanovic Department of Nuclear Engineering and Radiological Sciences, University of Michigan, Ann Arbor, MI, USA

Richard Kouzes Pacific Northwest National Laboratory, Richland, WA, USA

Mitaire Ojaruega Defense Threat Reduction Agency, Fort Belvoir, VA, USA

Sara A. Pozzi University of Michigan, Ann Arbor, MI, USA

Dennis Slaughter Lawrence Livermore National Laboratory, Livermore, CA, USA

Haori Yang Oregon State University, Corvallis, OR, USA

Chapter 1
Introduction

Igor Jovanovic

Abstract The context for the emergence of nuclear security challenges is discussed, including some of the milestones in the development of nuclear weapons and the international efforts aimed to direct the use of nuclear technology to peaceful applications. The key terminology pertaining to nonproliferation is introduced, after which the focus of the chapter shifts to the problem of nuclear terrorism. The effects of a hypothetical terrorist attack using a relatively small, improvised, nuclear device are discussed. Next, the policy and the relevant treaties that have shaped the arms control regime over the past decades are introduced. The national and international efforts are subsequently discussed, which aim to provide a framework for prevention of nuclear terrorism. Finally, the general characteristics of the active interrogation technique are discussed, which will be expanded upon in the remainder of the book.

1.1 Historical Perspective on Nuclear Security

While the process of laying the foundations of the nuclear science can be traced to the early 1900s, in this context the discovery of the nuclear fission in 1938 by Meitner [1] can be singled out as a truly disruptive event. The potential for nuclear fission to lead to a new class of weapon that eclipses all the others in terms of its destructive power was quickly recognized (see, for example, the 1939 letter by Albert Einstein to US President Franklin Delano Roosevelt [2]). The discovery and exploration of the prospects for application of nuclear fission largely coincided with the Second World War, the most extensive and lethal conflict to date. In large part due to these circumstances, the effort to employ fission to develop a nuclear explosive device and to engineer its size and packaging such that is compatible with the state-of-the-art weapons delivery systems gained the highest priority in the U.S.

I. Jovanovic (✉)
Department of Nuclear Engineering and Radiological Sciences, University of Michigan, Ann Arbor, MI, USA
e-mail: ijov@umich.edu

I. Jovanovic, A. S. Erickson (eds.), *Active Interrogation in Nuclear Security*, Advanced Sciences and Technologies for Security Applications,
https://doi.org/10.1007/978-3-319-74467-4_1

war effort, and is known as the *Manhattan Project* [3]. This impressive scientific and technological development was fueled by an unprecedented combination of high concentration of scientific talent and essentially unlimited government support. At a final cost of approximately $27 billion (adjusted to 2016 dollars) [4], by 1945 a large team of scientists and engineers has developed two major embodiments of a nuclear bomb, based on highly enriched uranium and plutonium. The regrettable historical episode in which both of those designs were employed to accelerate the end of the war with Japan also led to an immediate and universal recognition of the urgent need to control the development and use of such weapons and limit the proliferation of nuclear weapons technology.

United States held the monopoly to nuclear weapons only for several years, until the demonstration of the Soviet plutonium implosion weapon in 1949. This quickly escalated into a nuclear weapons race, initially only between the United States and Soviet Union, but by 1960s the circle of nuclear weapons states has expanded to include the United Kingdom, France, and China. The nuclear weapons race came at a substantial cost, which required public support. In United States, the *Project Candor* was initiated in 1953 as a media campaign aimed to gather support for extensive nuclear armament expenditures by regularly informing the public about the perils of falling behind the Soviet nuclear weapons capabilities. As a part of this campaign, President D. Eisenhower delivered the famous *Atoms for Peace* speech to the United Nations General Assembly in December 1953, in which he urged the worldwide effort to employ nuclear energy for peaceful purposes. Within the Atoms for Peace program (Fig. 1.1), the United States government spearheaded an effort to pursue research into peaceful applications of nuclear science by providing training, equipment, and help to construct facilities such as research reactors in many countries that lacked nuclear capabilities at that time. Regrettably, the origin of the indigenous efforts to develop nuclear weapons programs in countries such as Israel, Iran, and Pakistan can be traced to the U.S. assistance provided under the Atoms for Peace program.

In the same address to the United Nations General Assembly in which he announced the establishment of the Atoms for Peace program, President Eisenhower also proposed that an international agency be established under the aegis of United Nations, with a dual mission: (1) to promote the peaceful uses of nuclear

Fig. 1.1 A 1955 U.S. stamp highlighting the Atoms for Peace program

energy, and (2) to inhibit its use for any military purpose, including nuclear weapons (https://www.iaea.org/about/mission). The agency was established in 1957 as the *International Atomic Energy Agency* (IAEA) and is headquartered in Vienna. It membership (as of 2017) includes 168 states.

With the formation of IAEA, the basis of the international framework was established such that the future use of nuclear science and engineering for weapons purposes can be curtailed. This philosophy has been implemented through the introduction and increased acceptance of various international treaties. This formal framework is described in more detail in Sect. 1.3.

It is in order here to introduce the key terminology relevant to the discussion of nuclear security in the remainder of this book.

- *Nuclear weapons states* are the five countries that have acquired nuclear weapons before January 1, 1967, and they include United States, Russia (as a successor to Soviet Union), United Kingdom, France, and China.
- *Nuclear proliferation* refers to the spread of nuclear weapons, fissile material, and the requisite technology and information that could be used to construct them, to the states that are not among the five nuclear weapons states. *Nonproliferation* then refers to efforts to curtail such proliferation activities through diplomatic, legal, and administrative methods.
- *Nuclear counterproliferation* is the effort to combat nuclear proliferation, but focusing on intelligence and military methods.
- *Nuclear safeguards* are measures that can be used to verify that countries comply with their international obligations not to use nuclear materials as nuclear explosives.
- *Nuclear treaty verification* refers to development and deployment of measures to ensure verifiable compliance with treaties and other international agreements, implementation of regimes to reduce nuclear weapons, and detection and dismantlement of undeclared nuclear programs.
- *Nuclear forensics* refers to the development of methods and practices that can be used to determine the provenance (origin) of the nuclear material, whether after its use (post-detonation) or in the cases when it is intercepted (pre-detonation).
- *Weapons of mass destruction* include weapons that operate on the basis of nuclear, radiological, chemical, biological, or other principles that can lead to large loss of life, damage to infrastructure, or biosphere.

The list of states that developed their own nuclear weapons has since expanded to India (1974), Pakistan (1998), and North Korea (2006); it is widely accepted that Israel has also acquired nuclear weapons capability as early as in 1960s. There are also several nuclear weapons sharing states (Belgium, Germany, Italy, the Netherlands, and Turkey), which host U.S. nuclear weapons under the NATO's nuclear sharing policy. While South Africa developed nuclear weapons, it has voluntarily given up this capability, along with the former Soviet states of Belarus, Ukraine, and Kazakhstan (Fig. 1.2).

The nuclear weapons states have found the cost of developing and maintaining a nuclear stockpile to be substantial, and therefore a large burden on their

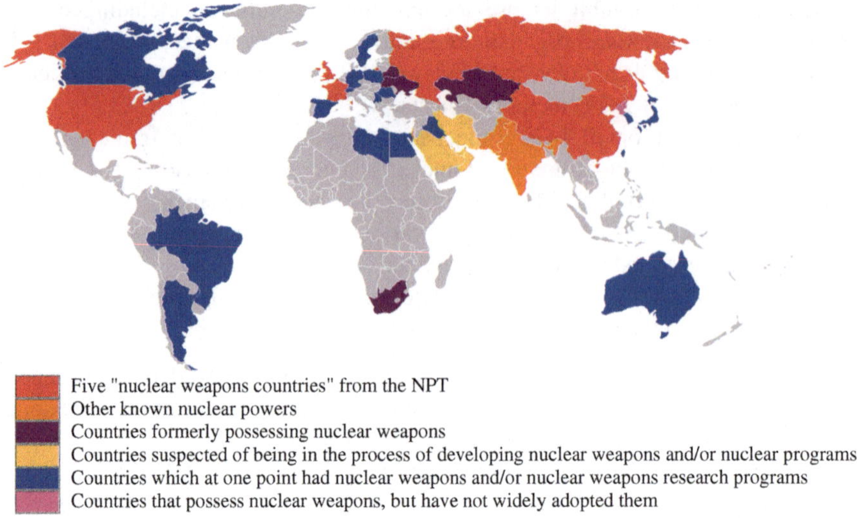

Fig. 1.2 Current status of nuclear weapons possession worldwide (2017)

economies. This extraordinary and recurring cost of maintaining a large nuclear weapons capability has contributed to the collapse of the Soviet economy and the subsequent dissolution of Soviet Union in 1991. The U.S. National Nuclear Security Administration has played an important role in ensuring the security of the former Soviet nuclear stockpile and the consolidation of Soviet nuclear material in Russia.

With the demise of the Soviet Union and the subsequent large (multifold) reductions of nuclear weapons stockpiles, the risk of nuclear conflict and the magnitude of its aftereffects was significantly reduced, despite the increasing threat of a nuclear conflict surrounding the recent development of nuclear capability and rapid advancement towards intercontinental delivery systems by North Korea. As a result, since the 1990s, the international community has shifted its focus from the threat of full-scale nuclear conflict to the threat of nuclear terrorism. The recognition of this important threat has gained momentum after the series of coordinated terrorist attacks on September 11, 2001 and the continued political instability, conflict, and nuclear ambitions in the Middle East and in North Korea.

1.2 The Problem of Nuclear Terrorism

Nuclear terrorism refers to an act of terrorism in which a person or people belonging to a terrorist organization detonates a nuclear or radiological device. In this context, a *radiological device* refers to an assembly containing a significant amount of radioactive material that could be used to inflict damage by delivering large radiation doses to humans or contaminating areas important to economic activities, such as city centers or transport hubs.

While the threat of nuclear terrorism has gathered more attention since the 1990s, it has been anticipated in the early days of development of nuclear weapons. In 1945, the Manhattan Project director J. Robert Oppenheimer attended a congressional hearing, in which Senator William Milliken famously asked:

> "We... have mine-detecting devices, which are rather effective... I was wondering if anything of that kind might be available to use as a defense against *that particular type* of use of atomic bombs."

Here, Senator Milliken was referring to the possibility for a nuclear bomb to be delivered not by a standard military vehicle (at that time, a heavy bomber), but rather as a clandestine shipment, owing to the weapon's relatively small size. The Oppenheimer's response was:

> "If you hired me to walk through the cellars of Washington to see whether there were atomic bombs, I think my most important tool would be a screwdriver to open the crates and look. I think that just walking by, swinging a little gadget would not give me the information."

As the lead developer of the atomic bomb, Oppenheimer had a deep understanding of the physical principles governing weapons design and the properties of fissile materials such as highly enriched uranium and plutonium, including the nature and strength of their characteristic signatures. This mastery of the physical problem, along with the knowledge of the state-of-the-art detection technology, led him to immediately conclude that the task of detecting clandestine nuclear device would be very difficult and beyond the reasonable practical reach of the then-available technology. Importantly, Oppenheimer recognized that there was no obvious path to the development of advanced technological means to detect special nuclear material, especially in shielded configurations.

The possibility of nuclear terrorism has been identified as one of the important threats of our time by many leaders. For example, in the inaugural Nuclear Security Summit in 2010 President Obama remarked [5]:

> "The single biggest threat to U.S. security, both short-term, medium-term and long-term, would be the possibility of a terrorist organization obtaining a nuclear weapon."

He further summarized the potential impact of a nuclear terrorist attack at the Nuclear Security Summit held 6 years later [6]:

> "At hundreds of military and civilian facilities around the world, there's still roughly 2000 tons of nuclear material, and not all of this is properly secured. And just the smallest amount of plutonium – about the size of an apple – could kill and injure hundreds of thousands of innocent people. It would be a humanitarian, political, economic, and environmental catastrophe with global ramifications for decades. It would change our world."

It has been long recognized that the single greatest obstacle to developing a nuclear weapon is obtaining the necessary special nuclear material (highly enriched uranium or plutonium), since the production of a suitable quantity and quality of material is beyond the capability of any non-state actor. With highly enriched material in possession, even a moderately technically advanced group would likely construct an operational weapon relatively quickly. In the case of plutonium, the construction of a nuclear weapon would prove more challenging, but the required

Fig. 1.3 A model of special atomic demolition ammunition exhibited in the National Museum of Nuclear Science and History, Albuquerque, New Mexico

quantity of material is significantly smaller. The possible pathways for a terrorist group to obtain special nuclear material include theft and smuggling, unauthorized receipt of any unaccounted material from major weapons programs (such as the former Soviet Union's nuclear weapons enterprise), or a voluntary supply by a rogue state that has the capability and resources to enrich uranium or nuclear reactors to produce plutonium.

There are many possible methods of delivery that are available to a terrorist group, but it is important to recognize that the accuracy of delivery is not as important as in the case of conventional explosives to achieve the desired effect. To illustrate the ease of delivery, one can consider the case of *special atomic demolition ammunition*, a class of man-portable nuclear devices that have been developed for the purpose of destruction of infrastructure or port facilities. For example, the device shown in Fig. 1.3 has been designed to be delivered as a "backpack" by two members of Special Forces. It is therefore conceivable that a nuclear device could be readily delivered in a personal vehicle, truck, boat, or even a small private aircraft.

By far the most studied mode of delivery has been the cargo traffic. There are millions of cargo containers that cross borders worldwide every year, and they are exceptionally well suited for transporting clandestine nuclear materials due to the relative ease with which they can be shielded and the absence of an effective inspection method that is compatible with the large flow of container traffic. Research on how to detect special nuclear material in cargo containers has therefore been the centerpiece of the research portfolio by agencies such as the U.S. Department of Homeland Security. There has been also been a significant interest in *standoff* detection, which could help address scenarios such as small boats approaching ports from international waters. The problem of true standoff detection is daunting and in its most literal interpretation, where no detector or probe can be located near the object, no technological solutions have been demonstrated even at a rudimentary level.

The effects of nuclear explosions are relatively well understood and can be used to predict the consequences of a nuclear detonation carried out by terrorists in a major population center. Besides blast damage and thermal radiation, the population and structures would be exposed to powerful ionizing radiation and an electromagnetic pulse. Since one major effect of a nuclear explosion (especially when conducted at higher altitudes) is the occurrence of a powerful electromagnetic pulse capable of destroying critical electrical infrastructure with severe long-term consequences, the possibility of such terrorist attack must be carefully considered as well. Models of impact of a nuclear terrorist attack have been developed, and here we first provide an example of study conducted by the RAND Corporation for Department of Homeland Security [7], where a relatively modest 10-kt bomb delivered in a shipping container is detonated in a major port (Long Beach, California). The RAND model predicts 60,000 short-term fatalities, 150,000 people requiring emergency medical treatment, and an economic impact on the order of $1 trillion. In another study led by Lawrence Livermore National Laboratory [8], extensive modeling has been done to predict a devastating aftermath of a hypothetical nuclear terrorist attack carried out in downtown Washington, DC (Fig. 1.4). One may further speculate about the wide-ranging effects that such an event would have on the society, including the possible global retreat to more authoritarian forms of government.

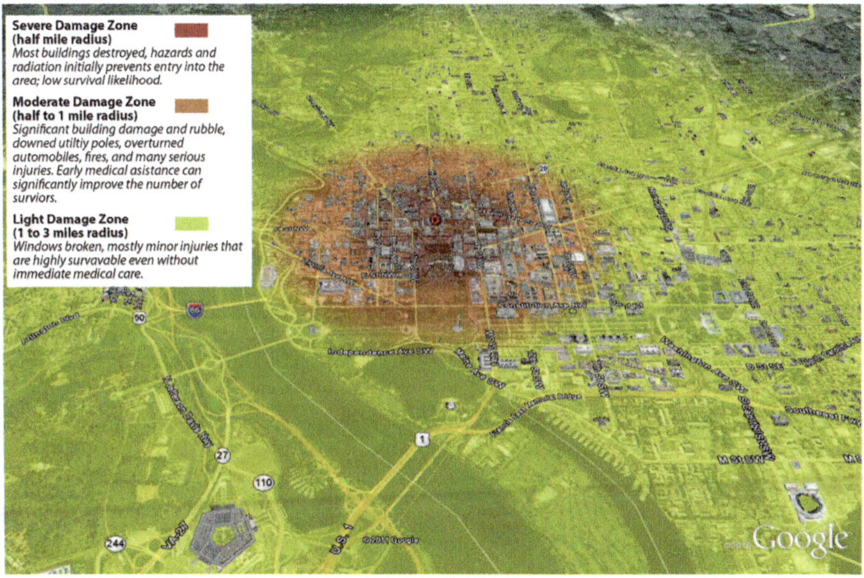

Fig. 1.4 Summary of prompt damage expected from a hypothetical 10-kt nuclear explosion in Washington, DC. Reproduced from Ref. [8]

1.3 The Role of Policy in Nuclear Security and Arms Control

It is a well-known fact that global nuclear security, especially its nonproliferation aspect, has relied on an evolving policy framework established by a growing international consensus. The basis for this framework are several key international treaties, which are succinctly introduced here.

The *Nonproliferation Treaty* (NPT) is an abbreviated form of the *Treaty on the Non-Proliferation of Nuclear Weapons*. NPT is first of the major treaties put into place in an effort to prevent the spread of nuclear weapons and the associated technology, as well as promote cooperation on the peaceful uses of nuclear energy. NPT has another ambitious and synergistic goal, and that is to promote nuclear disarmament and general and complete disarmament. The treaty was opened for signature in 1968 and became effective in 1970, and was ultimately signed by 191 states as of 1991. Notably, the known nuclear powers of India, Pakistan, and Israel have not joined the NPT, along with South Sudan. The case of North Korea is special in that is joined in 1985, but then withdrew in 2003 after conducting its first nuclear test. The NPT formally defines the nuclear-weapons states (United States, Russia, United Kingdom, France, and China). It further requires all other signatory states to agree never to acquire nuclear weapons; in exchange, the NPT nuclear-weapon states pledge to share the benefits of peaceful nuclear technology and to pursue nuclear disarmament aimed at the ultimate elimination of their nuclear arsenals (http://disarmament.un.org/treaties/t/npt/text, 1968). A common interpretation of the NPT is to consist of three major pillars: nonproliferation, disarmament, and the right to use nuclear technology for peaceful purposes.

The treaty requires ratification by the respective legislators and this has been the case with only 40 signatories to date, albeit they include the important powers of United States, Russia (ratified as Soviet Union), and United Kingdom. The NPT has been considered a wide success, not only because of being the most widely embraced arms control treaty in history, but also because of its quantifiable performance. Namely, despite the predictions of the wide adoption of nuclear weapons at the time the treaty was put forward, the list of additional states that acquired nuclear weapons since is relatively short and can in large part be credited to global efforts articulated through the NPT.

The NPT was augmented by two other major instruments that can be considered highly related to the original treaty. First, in the aftermath of the Indian nuclear test in 1974, the *Nuclear Suppliers Group* (NSG) was founded to further curtail the export of material, equipment, and technology that could find use in construction of nuclear weapons (http://www.nuclearsuppliersgroup.org/en/, 1975). The group is sometimes referred to as the *London Club* because it held a series of meetings in London. Today, NSG has 48 signatory states. NSG implements two sets of additional guidelines for nuclear and nuclear-related exports, with sensitive items constituting the so-called *Trigger List*.

The second major supplement to the NPT was introduced by the IAEA in 1993 and is referred to as the *Additional Protocol*. The objective of this program is to extend and strengthen the nuclear safeguards, which improves the ability of IAEA to uncover undeclared nuclear activities, especially those that have no clear connection to civilian applications. Under the Additional Protocol, more information on nuclear and nuclear-related activities is provided to the IAEA, the right of access of IAEA inspectors are greatly enhanced (for example, by shortening the inspection notice to 2 h and granting automatic visa renewals), and development of safeguards is deemed to be state-specific. The Additional Protocol was signed and brought into force by the vast majority of NPT signatories.

A major recent development related to NPT is the international agreement termed the *Joint Comprehensive Plan of Action* or the *Iran deal*, which was adopted in 2015 by Iran, the permanent members of the United Nations Security Council plus Germany, and the European Union. In this landmark recent agreement, Iran will take steps to limit the capacity of its uranium enrichment facilities, eliminate its medium-enriched uranium, reduce the stockpile of low-enriched uranium, and delay the construction of any heavy-water facilities, in return receiving a relief of the economic sanctions imposed in the wake of Iran's nuclear activities. The Iran deal has seen significant criticism and remains the subject of intense international attention, one of the critical components being the mechanism of its verification.

There are several major arms control and disarmament efforts treaties that went into effect over the past several decades, significantly reducing the deployed nuclear stockpiles. The *Strategic Arms Limitation Talks* (SALT) started in 1969 and consisted on two rounds (SALT I and SALT II). As a result of SALT I, the *Anti-Ballistic Missile Treaty* was signed in 1973 and terminated in 2002. While SALT II aimed to reduce the number of strategic nuclear forces to 2250 delivery vehicles and was agreed upon in 1979, it was not ratified amidst international tensions surrounding Afghanistan and expired in 1985. The *Strategic Arms Reduction Treaty* (START, also START I) between Soviet Union and the United States was signed on 1991 and went into force (with Russia) in 1994, limiting the signatories to 6000 deployed nuclear warheads on 1600 inter-continental ballistic missiles and bombers. The START I expired in the 2009, and in 2010 a follow-up treaty was signed (*New START* implementing further reductions by a factor of two. While the START II was negotiated, it has not come into force, and the START III treaty has not been negotiated yet.

In the area pertaining to nuclear tests, the landmark multilateral agreement has been the *Comprehensive Nuclear-Test-Ban Treaty* (CTBT), which prohibit all nuclear explosions regardless of purported application. CTBT was adopted in 1996 and signed by 183 countries and ratified by 166 of them. Notably, eight major countries with nuclear capability, including China, India, Pakistan, North Korea, and the United States, have not signed or ratified the treaty.

A key component of any international treaty related to nonproliferation and arms control is the implementation of a solid verification regime. The goal of a verification regime is to monitor and verify nuclear reduction agreements and detect violations of treaties and other nuclear nonproliferation commitments. To do, measures are developed and deployed to ensure verifiable compliance with treaties and other international agreements, implement regimes to reduce nuclear weapons, and detect and dismantle undeclared nuclear programs (https://nnsa.energy.gov/aboutus/ourprograms/nonproliferation-0/npac/verification, 2017).

There are many technologies that have been developed and put into use over the years to support the objectives of treaty verification, and they follow the principle of *trust, but verify* attributed to a Russian proverb and used by President Ronald Reagan in 1987, and now also used as a motto of the Defense Threat Reduction Agency's On-Site Inspection Agency (https://www.legistorm.com/stormfeed/view_rss/254149/organization/31751.html, 1995). The technologies used for verification measurements are diverse and include satellite imagery, telemetry, electro-optical and radar senors, space-based sensors, seismic and infrasound monitoring for suspected nuclear explosions, on-site inspection, and air sampling.

In the recent period, the U.S. Department of Energy's National Nuclear Security Administration (NNSA) has stepped up effort to integrate the U.S. academic community into its effort to maintain and improve on the treaty verification regime. Since 2014, NNSA has provided a substantial and coordinated support to a group of universities and national laboratories to engage in a combination of education and research at the intersection of science, technology, and policy. The goal of this effort is not only to provide future workforce proficient in the current methods of verification, but also to advance the fundamental research into science and technology that could support current and future verification challenges. This framework has been led by the University of Michigan under the *Consortium for Verification Technology* (https://cvt.engin.umich.edu, 2014).

1.4 Institutionalized Efforts to Curb Nuclear Terrorism

While the extensive policy framework has evolved over decades in response to a major buildup of nuclear weapons and means of their delivery by nation-states, the problem of nuclear terrorism requires a different set of measures and programs. It is certain that the arms control reductions have had a positive impact on the efforts to prevent nuclear terrorism by the virtue of material reductions and implementation of better security and verification measures. The complementary measures to prevent nuclear terrorism are arguably significantly easier to agree upon in the broad international community, since they are seen to pertain to non-state actors.

The problem of nuclear terrorism is seen to be of persistent nature and of limited, albeit significant local impact (see Sect. 1.2). It has also been seen to be an example of a *rare event*, associated with a low probability of occurrence and a sparse historical record from which to develop predictive models based on past statistics [9]. It has been a conclusion of numerous assessments that terrorist groups have the ambition to acquire, and possibly use, unconventional weapons such as improvised nuclear explosive devices and radiological dispersal devices.

The problem of domestic nuclear terrorism in the United States has been addressed primarily by the Domestic Nuclear Detection Office (DNDO), an organi-

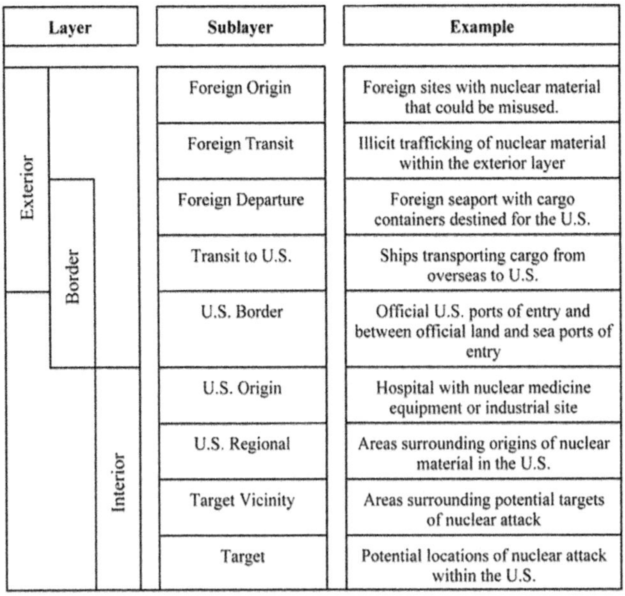

Fig. 1.5 Layers of the global nuclear detection architecture. Reproduced from Ref. [10]

zation established in 2005 within the department of Homeland Security to centralize the coordination of the federal government's response and to prevent nuclear terrorism by continuously improving capabilities to deter, detect, respond to, and attribute attacks, in coordination with domestic and international partners (http://www.dhs.gov/domestic-nuclear-detection-office, 2017). The SAFE Port Act (https://www.congress.gov/bill/109th-congress/house-bill/4954, 2006) established DNDO and passed to it the specific statutory responsibilities to protect the United States against radiological and nuclear attack, including the responsibility to develop a *global nuclear detection architecture* (GNDA) [10].

The GDNA (Fig. 1.5) is envisioned to consist of three partially overlapping areas and nine sublayers, and therefore has a broad scope and international character. The *exterior layer* comprises the foreign origin, foreign transit, and foreign departure sub-layers. DNDO improves upon radiological and nuclear material detection abroad through efforts that encourage foreign nations or regions to develop and enhance their nuclear detection architectures. The *interior layer* of the GNDA includes all areas within and up to, but not including, the U.S. border. The interior layer focuses on increasing nuclear detection capabilities across the maritime, air, and land pathways and addresses a wide array of potential threats. Finally, the *transit and border layer* is composed of transit to the U.S. from a foreign port of departure or non-port of departure, as well as passing through the U.S. border prior to entering the U.S. interior. This represents the last opportunity to detect radiological or nuclear materials prior to their arrival onto U.S. territory, and initiatives in this layer emphasize maritime domain awareness related to preventive radiological/nuclear detection (https://www.dhs.gov/global-nuclear-detection-architecture, 2017).

GNDA architecture encompasses other programs, such as the *DNDO GRaDER Guidance for Users*, which provides a continuous means of independently testing and evaluating commercially available radiological and nuclear (Rad/Nuc) detection equipment against ANSI N42 performance standards to ensure that only the best radiation detector capabilities are funded by government procurement and grant programs. GRaDER provides performance and operationally relevant technical information on these systems to Department components, other federal agencies, and state, local and tribal governments and first responders. Also, as a part of GNDA, DNDO support cross-cutting efforts, which focus on programs and capabilities spanning multiple layers and pathways of the GNDA. Efforts undertaken in this layer provide the basis for time-phased deterrence and detection strategies. These elements streamline existing capabilities, improve overall coordination, and ultimately seek to enhance radiological and nuclear detection at the federal, state, territorial, tribal and local levels (https://www.dhs.gov/global-nuclear-detection-architecture, 2017).

DNDO work is supported by the activities of the Department of Defense, Department of Energy, Department of Justice, and Department of State, as well as the U.S. Nuclear Regulatory Commission and the Office of the Director of National Intelligence, in coordination with state, local, and tribal authorities, international partners, and private entities. Together, programs supporting the GNDA create a multi-layered defensive network to detect and assist interdiction of radiological and nuclear materials out of regulatory control.

Examples of efforts supporting the GNDA include the radiation portal monitors, which scan for radiological and nuclear materials at international border crossings, employment of radiation detectors by law enforcement and public safety personnel to protect special events, and the use of radiation detection equipment by the U.S. Coast Guard teams when boarding vessels (https://www.dhs.gov/global-nuclear-detection-architecture, 2017).

Similar to the NNSA's efforts in verification, DNDO has been making a significant investment to engage the U.S. academic community in the basic and applied research in the area of detection for homeland security applications, such as prevention of nuclear terrorism. To support this engagement, an Academic Research Initiative program has been established, which provides the necessary education and training to graduate students in the areas of advanced radiation detection and measurement while engaging them in basic, high-risk and potentially high-payoff research in this area.

A *Second Line of Defense* initiative has also been developed under the name *Megaports Initiative*, which is of international character. The program is the part of the Office of International Material Protection and Cooperation within the NNSA. The main idea behind the Megaports Initiative is to enhance the security of the international maritime shipping network (containerized traffic) by equipping the major international seaports with radiation detection equipment and alarm communication systems. Also, the initiative provides comprehensive training for foreign personnel, short-term maintenance coverage, and technical support to ensure the long-term viability and sustainability of installed radiation detection systems [11]. The Megaports Initiative is a joint effort of NNSA, DHS, and the Department of State.

International efforts to prevent nuclear terrorism and ultimately led by the IAEA, but a number of national entities and organizations has emerged to support this mission. For example, the Stanton Foundation has provided support to spearhead the US-Russia Initiative to Prevent Nuclear Terrorism and includes prominent institutions such as the Center for International Security, Institute for World Economy and International Relations, Russian Academy of Sciences in Russia and the Nuclear Threat Initiative (US).

The umbrella program of the European Union in the area of nuclear power is the *Euratom* (https://ec.europa.eu/programmes/horizon2020/en/h2020-section/euratom, 2017). Within the Euratom program, the Joint Research Center of the European Commission supports the development and qualification of nuclear forensics methods and techniques to fight against illicit trafficking and provide operational support to member states and international organizations. The European Commission has further established the European *Nuclear Security Training Centre* (EUSECTRA), which aims to improve member states' capabilities to address the threats associated with illicit incidents involving nuclear or other radioactive materials by providing hands-on training using real nuclear materials to front line officers, their management, trainers, and other experts in the field (https://ec.europa.eu/jrc/en/european-nuclear-security-training-centre-eusectra/about, 2017).

In Japan, the Japan Atomic Energy Agency has established the *Integrated Support Center for Nuclear Nonproliferation and Nuclear Security* (https://www.jaea.go.jp/04/iscn/index_en.html, 2017) to contribute to the improvement of the nuclear material management and strengthening of international nuclear nonproliferation. The mission of this organization includes the development of technologies for nuclear nonproliferation, measurement, detection, and forensics of nuclear material, and support to capacity building and infrastructure development.

1.5 Overview of the Active Interrogation Method

Active interrogation (AI) has become one of the most active areas of research and development in nuclear security and nuclear detection worldwide [12]. The intense focus on this topic by the international scientific community has been motivated not only by the increased level of attention and support by the main stakeholders, such the U.S. federal agencies, but also by the realization that the recent advances in the component technologies such as sources, detectors, and algorithms may allow a significant rate of progress to be realized. It is worth mentioning that, since many of the AI approaches employ advanced concepts, the AI research has been able to attract an impressive cadre of researchers with experience in fundamental sciences to work on the practical problem of detection of illicit nuclear materials.

The most challenging and urgent problem that demands the development of AI technology is the detection of special nuclear material, i.e. material that could be employed to construct nuclear weapons, with focus on ^{235}U and ^{239}Pu. The problems here include rapid clearing of objects such as containers in transit from nuclear threats, standoff detection of nuclear materials at kilometer-scale distances, and improved methods to material accountancy, quantification, safeguards, and verification. The main physics limitations associated with the use of conventional methods of detection are the very low spontaneous rate of characteristic signature generation (γ rays and neutrons), especially in the case of uranium, and the relative ease with which those spontaneous signatures can be shielded. To add to the complexity of this problem, almost all scenarios include the presence of relatively intense, complex, and variable radiation backgrounds.

The simplest method to address the problem of detection and characterization of nuclear material is *passive detection*. In this method, the objective is to reduce the minimum detectable amount of material or improve upon quantitative measurements by the use of sufficiently sensitive and selective radiation detectors, systems, measurement approaches, and algorithms. Some of the key technologies and performance metrics include radiation imaging, high-resolution spectroscopy, and background rejection. Unfortunately, it has been found that the passive detection methods frequently fail due to the aforementioned physics limitations.

The second widely employed possibility that could enable detection of special nuclear material is the use of probing radiation and transmission imaging. This type of measurement is usually referred to as *radiography* or *transmission radiography*. This method is well-developed and used in a wide range of applications, including security, medical, industrial, and basic research. Regrettably, nearly all radiographic methods (with the exception of specialized techniques such as nuclear resonance fluorescence radiography, fast neutron resonance radiography, and associated particle imaging) are only sensitive to the atomic structure along the propagation of the probe radiation. As a result, the radiographic techniques have been limited to imaging the gross distribution of material, the approximate atomic number, and, at best, identifying the locations where high-Z (high atomic number) materials such as uranium may be found. This is complementary to detection of special nuclear material, but in itself has insufficient selectivity.

The third, and the most promising, method that could enable detection of special nuclear material in challenging configurations, such as shielded highly enriched uranium, is referred as *active interrogation*. The key aspect of AI is that the probing radiation strongly interacts with the atomic nuclei to provide characteristic and significantly more intense signatures than in passive detection. To this end, probing radiation in the form of neutrons and energetic photons (X rays or γ rays) holds the greatest promise, although more exotic probes such as protons and muons have also been considered. The characteristic radiation emitted and detectable is in the form of γ rays and neutrons, and suitable detector systems need to be developed and optimally integrated to detect characteristic radiation in an AI environment, which is frequently associated with high radiation field.

One important characteristic of AI is that it establishes a definitive time structure of the expected signature signal. If the probing radiation can be pulsed, the time evolution of the emitted radiation may serve as an additional discriminant for special nuclear material and can help to further reduce the component of the background that overlaps with the signal in time. Such pulsed techniques thus rely on measurements within some coincidence window, or may apply the time-of-flight technique.

Directionality of the probing radiation may also be used to improve the detection sensitivity in AI. If the probing radiation is directed towards a small area, it may be possible to localize the threat by distinguish its characteristic response from the surrounding areas. While only few types of sources may offer inherent directionality, collimator arrangements may be used in conjunction with virtually all of them.

AI systems thus comprise several key components, all of which are discussed in more detail in the remainder of this book: sources, detector systems, and detection algorithms. One special consideration is the radiation dose delivered in AI, which has to be kept to a minimum to minimize the effects on stowaways and sensitive materials in the path of the interrogation beams.

It is tempting to consider the AI as a solution to the standoff detection problem. While AI may be able to provide a future path to gaining such capability, several remaining challenges must be addressed. First, there is a stringent requirement on source directionality (a component of source *brightness*). Second, the attenuation of the interrogating radiation on the way to the interrogated object and of the signature signal returning back to detector (preferably located near the source) must be considered. Finally, the characteristic signature radiation emitted by the interrogated object is emitted nearly isotropically, which limits the efficiency with which it can be collected at a standoff. Regardless of these limitations, it may be possible to establish ambitious (but realistic) goals for the use AI in standoff detection.

1.6 Synopsis of the Book

The organization of this book is as follows. Following the introductory remarks in this chapter, in Chap. 2 we turn our attention to the general characteristics of measurements in nuclear security. We introduce the material of interest, identify the characteristic signatures for detection and their backgrounds, and focus on their origins and the physics of their production. We also consider the methods that allow the statistical limits for their detection to be established.

In Chap. 3 we turn our focus on passive measurements as a baseline for subsequent discussion of the AI techniques, discussing issues such as strength of characteristic signatures, the general method with which passive measurements are performed, we introduce the associated technology, and consider the limitations that are encountered in passive measurements.

Chapter 4 provides the foundations of the detection approach via AI. A considerable attention is given to the impact of active measurements on the detectability of SNM and a synopsis of the AI technology is provided. We further discuss the modeling and simulation in the context of AI and which limitations arise in implementation of actual AI measurements.

Chapter 5 is focused on one of the key technological components of any AI system—the generation of the probing radiation. The principle of operation of linear accelerators, cyclotrons, and novel laser-driven sources is discussed, along with the review of radioisotope sources and the use of natural radiation for AI.

The next important component of an AI system are detectors, which are discussed in detail in Chap. 6. After discussing the general detector characteristics we devote special attention to photon and neutron detectors that may find use in AI as well is in any associated radiographic measurements.

In Chap. 7 we consider the various aspects of data acquisition and processing systems. In addition to introducing the more general aspects of those systems, we provide some discussion of the enabling techniques which may enable high rates in AI measurements, as well as methods that can allow the neutron from photon interactions with detectors to be distinguished from each other.

Chapter 8 discusses how the data generated by an AI system may be interpreted. Here we introduce the associated planar and tomographic radiography, principal component analysis, methods to unfold the signatures from measurements performed with detectors, and the use of algorithms to extract useful information from distributed detection systems.

We highlight some of the prototype systems that have reached a reasonable level of integration and testing readiness in Chap. 9. This includes the Nuclear Carwash, the Pulsed Neutron Fast Analysis system, the advanced scanner developed by Passport Systems, and the prototype laser-based scanner based on inverse Compton scattering.

Radiation dose is a major concern in all AI measurements, and the relevant aspects of dose deposition, measurement, and mitigation are discussed in Chap. 10. In Chap. 11 we introduce an effort to create a technical standard for AI systems, and summarize the main conclusions in Chap. 12.

References

1. L. Meitner, Nature **224**, 466 (1969)
2. Wikisource, Albert Einstein to Franklin D. Roosevelt - August 2, 1939 — Wikisource (2016). https://en.wikisource.org/w/index.php?title=Albert_Einstein_to_Franklin_D._Roosevelt_-_August_2,_1939&oldid=6288471. Online. Accessed 14 August 2017
3. R. Rhodes, *Making of the Atomic Bomb* (Simon & Schuster, New York, 2012). https://books.google.com/books?id=aSgFMMNQ6G4C
4. R.G. Hewlett, J. Oscar E. Anderson, *The New World, 1939/1946, A History of The United States Atomic Energy Commission*, vol. I (The Pennsylvania State University Press, University Park, PA, 1962)

5. Remarks by President Obama and President Zuma of South Africa before Bilateral Meeting. obamawhitehouse.archives.gov
6. Remarks by President Obama and Prime Minister Rutte at Opening Session of the Nuclear Security Summit. obamawhitehouse.archives.gov
7. C. Meade, R.C. Molander, Considering the effects of a catastrophic terrorist attack. Technical Report, Rand Center for Terrorism Risk Management Policy (2006)
8. B. Buddemeier, J.E. Valentine, K.K. Millage, L.D. Brandt, National capital region key response planning factors for the aftermath of nuclear terrorism. Technical Report (2011)
9. D. McMorrow, Rare events. Technical Report JSR-09-108, JASON, The MITRE Corporation (2009)
10. D.A. Shea, The global nuclear detection architecture: issues for congress. Technical Report, Library of Congress, Washington, DC (2009)
11. Office of the second line of defense. Technical Report PNNL-SA-74476, National Nuclear Security Administration (2010)
12. R.C. Runkle, D.L. Chichester, S.J. Thompson, Nucl. Instrum. Methods Phys. Res. Sect. A **663**(1), 75 (2012)

Chapter 2
Overview of Signatures and Measurement Needs

Mitaire Ojaruega and Anna S. Erickson

Abstract The Atomic Energy Act of 1954 defined special nuclear material (SNM) as plutonium or uranium enriched in the isotopes ^{233}U or ^{235}U, or any materials the U.S. Atomic Energy Commission determines to be SNM. With the current state of affairs, a key technical challenge on a quest to combat illegal use of SNM is detecting heavily shielded highly enriched uranium. For the purpose of this anthology the above definition holds. Active interrogation is the intentional bombardment of an object using ionizing radiation in order to induce nuclear reactions producing distinct, energetic emissions, hence making the target more readily detectable. This method further allows the detection of SNM quantity and other material properties. Traditionally, this technique uses ionizing radiation such as neutrons, muons, and γ rays as a probe. Combating this challenge has resulted in various government entities investing their substantial resources.

2.1 Overview of Signatures of SNM

For the past two decades, detection of SNM in all its forms has been an issue of significant importance. A number of techniques have been developed over the years to help address this challenge. The most promising approach for identifying the mass and other features of shielded SNM is active interrogation (AI), which involves introducing ionizing radiation to penetrate a cargo and other shielding. The interaction of different types of radiation with nuclear materials produces unique signatures. The main signatures used to identify the SNM within a cargo are delayed

M. Ojaruega
Defense Threat Reduction Agency, Fort Belvoir, VA, USA
e-mail: mitaire.i.ojaruega.ctr@mail.mil

A. S. Erickson (✉)
Georgia Institute of Technology, Atlanta, GA, USA
e-mail: erickson@gatech.edu

I. Jovanovic, A. S. Erickson (eds.), *Active Interrogation in Nuclear Security*, Advanced Sciences and Technologies for Security Applications,
https://doi.org/10.1007/978-3-319-74467-4_2

neutrons, delayed γ rays, prompt neutrons, and prompt γ rays, because of their penetrating power [1].

For the detection and characterization of shielded SNM, fission is the most important and most studied nuclear reaction process in AI detection techniques to date. In AI applications, the neutrons and γ rays emitted as part of the fission process can yield information about the following characteristics of SNM: yield from induced fission, energy, and half-life of the decaying nuclides produced in fission.

Fission fragments do not need another name. Usually the daughter products are the results of the fission fragments radioactive decay mostly beta (then gamma) and in very few cases via neutron emission. In addition to fission fragments, two to three prompt neutrons and about eight prompt γ rays are emitted within a very short time (on the order of 10^{-14}–10^{-12} s). Fission fragments are highly unstable and subsequently tend to undergo β decay, introducing comparatively long decay times (on the order of a second or longer), which is the origin of the delayed signature for on average fission. These delayed signatures are unique for SNM when using AI and yield six to seven γ rays and 0.01–0.02 neutrons per fission. The process of fission and the emission of associated particles is illustrated in Fig. 2.1.

The cross section is an important consideration when discussing fission-based SNM detection and characterization since it is a representation of probability of an interrogating radiation to cause fission or photofission in the given nuclear material. In general, the microscopic cross section, which is an inherent property of the isotope, increases with decreasing neutron energy. Cross sections for ^{235}U are shown in Fig. 2.2 and range from 600 b for thermal neutrons to about 1.2 b for 2.2-MeV neutrons [2], making thermal neutrons ideal candidates to cause fission in ^{235}U. As the energy of the neutron increases, the region of resonances can be observed. Resonances correspond to neutron energies that produce final states that coincide with compound nuclear energy levels. Since the level widths are rather small, the resonances are sharp and grow closer together as the kinetic energy of the neutron increases. At some point, resonances can no longer be resolved, and the cross section drops off sharply. For low-energy neutrons, the fission and radiative capture cross sections follow a $1/\sqrt{E}$ scaling with neutron energy (E), indicating that the probability of reaction is proportional to the time the neutron spends near the nucleus. Based on fission cross sections, one might predict that thermalized neutrons are best for AI. Unfortunately, interrogation with low-energy neutrons is by itself, in many cases not practical due to the high probability of neutron absorption in shielding and other cargo materials before they have a chance to reach the SNM.

As shown in Fig. 2.2, the cross section also heavily depends on the type of probing ionizing radiation. Interrogation sources used in AI produce neutrons or x-rays of appropriate energies to induce fission. For photon interrogating beams, the photofission threshold is 5–6 MeV. The fission process gives rise to a range of signatures, which are summarized in Table 2.1.

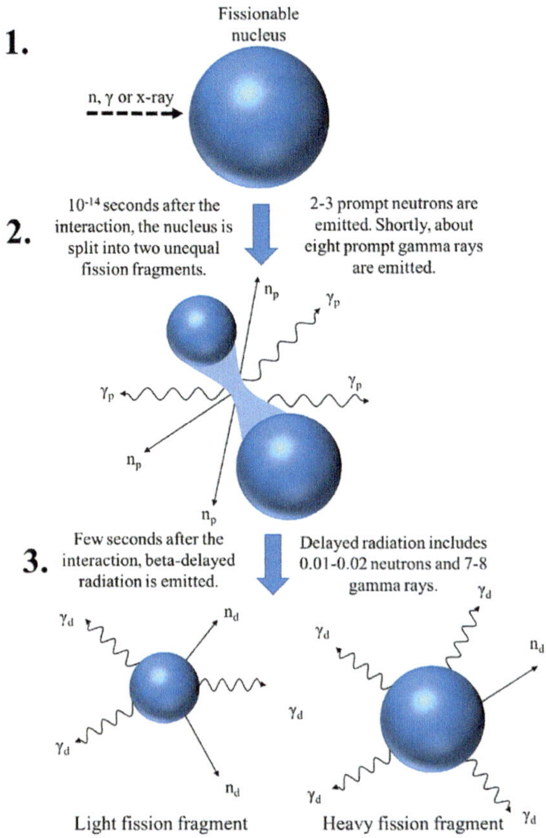

Fig. 2.1 Illustration of the time-dependent fission process

2.2 Prompt Signatures

Prompt signatures are emitted during the fission event or within a very short time afterwards, typically within a small fraction of a nanosecond, generally within a few nanoseconds. It has been argued that prompt fission emissions from an experimental standpoint are those that occur within 10 ns of probing [3]. They include fission γ rays and neutrons and very early isomeric γ rays preceding the β decay. For the purpose of AI, one must consider the cargo material when determining prompt signatures. For example, in situations where the container has many moderating or shielding materials surrounding the SNM, the prompt signatures can be affected by neutrons and photons absorption or photonuclear reactions, requiring more than simple timing techniques to characterize the signatures.

Prompt signatures of relevance to SNM detection are γ rays and neutrons.

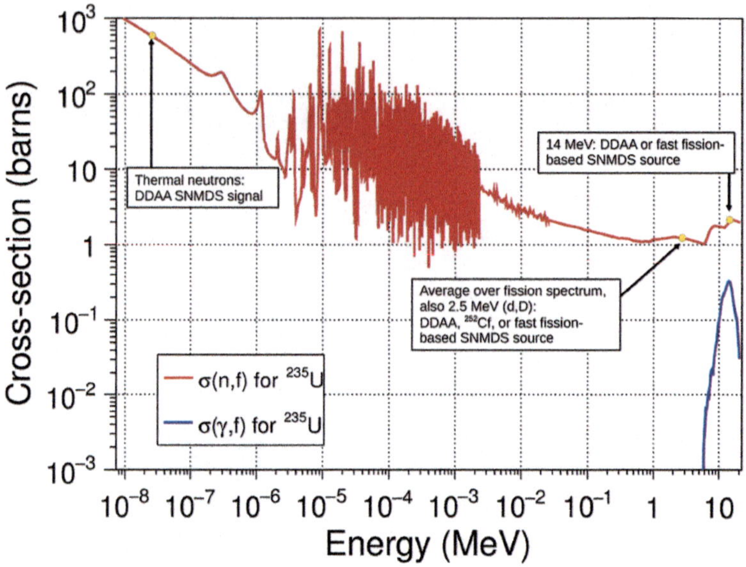

Fig. 2.2 Neutron and photon microscopic fission cross section of ^{235}U. Adapted from Ref. [2]

Table 2.1 Fission signatures relevant in AI applications [1]

Signature	Average number per fission	Multiplicity (2-σ range)	Detectable in real life	Detection modality	Detection efficiency in large objects
Delayed neutrons	~0.03	N/A	0.01	Moderated thermal neutrons	Low to medium
Delayed γ rays	~7	N/A	~0.60	Organic and inorganic scintillators	High
Prompt neutrons	2–5	2	~0.2	Moderated thermal neutron detectors liquid scintillators (LS) threshold activation (TAD)	High for thermal detectors (no energy threshold), low for LS and TAD
Prompt γ rays	~7	6	3	Fast coincidence	Very low

2.2.1 Prompt Gamma Rays

For all AI purposes, the strongest signature is the emission of prompt γ rays. On average about 7 photons γ rays are emitted per fission event with average energy of about 1 MeV per photon.

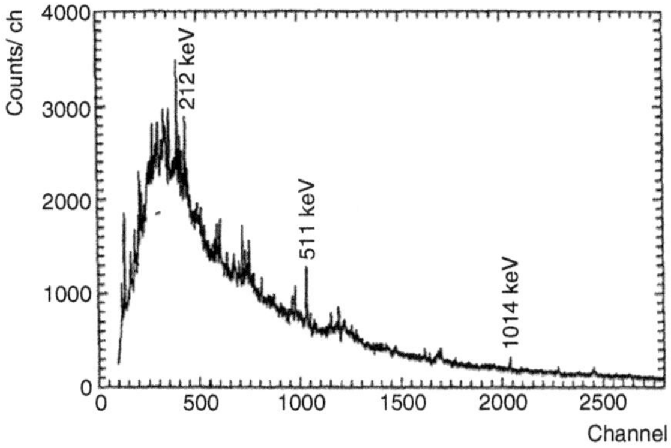

Fig. 2.3 Prompt γ-ray spectrum of ^{239}Pu interrogated with neutrons. Reproduced from Ref. [5]

$$N_\gamma(E) = \begin{cases} 6.6\ \gamma/\text{MeV} & \text{for } 0.1\ \text{MeV} < E < 0.6\ \text{MeV} \\ 20.2\exp(-1.78E)\ \gamma/\text{MeV} & \text{for } 0.6\ \text{MeV} < E < 1.5\ \text{MeV} \\ 7.2\exp(-1.09E)\ \gamma/\text{MeV} & \text{for } 1.5\ \text{MeV} < E < 10.5\ \text{MeV} \end{cases} \quad (2.1)$$

The shape of the prompt fission gamma spectrum can be approximated by analytical equations [4]. The spectral shape based on these equations is shown in Fig. 2.3. The prompt fission gamma ray exhibits a soft spectrum, often softer than the background radiations resulting from the interrogating neutrons inelastic and capture reactions. This fact and the presence of the interrogation radiation during the measurement of the prompt gamma rays, make this signature hardly useful in practical applications of active interrogation.

2.2.2 *Prompt Neutrons*

Prompt neutrons are also abundant during the AI procedure. The challenge with prompt neutrons is that meaningful measurements require detection of neutrons with 3 MeV or above. This is crucial as the background during prompt neutron detection is extremely high and can rise up to 2.5 MeV. The signature is still intense, and the prompt neutron yield is 2–5 per fission, with the average yield of 2.5. There is very limited data available for AI prompt neutron detection. However, it is believed that there is a difference in the prompt neutron detection spectrum for highly enriched, natural, and depleted uranium [6]. Interrogation of cargo is generally done with fast neutrons to ensure high penetration. If SNM is present, it fissions and produces prompt signatures. However, following the fission process some of the neutrons are thermalized and may linger for a longer period in a cargo container. This can make detection challenging because the thermal neutrons have a high fission cross section, as shown the Fig. 2.2.

Measurements of fission neutron energy spectra are very difficult and time consuming. The prompt neutron energies are quite similar for most fissionable

Table 2.2 Prompt neutron multiplicities for ^{235}U and ^{239}Pu undergoing neutron fission

Isotope	Neutron energy	Mean multiplicity
^{233}U	Thermal	2.49 [7]
^{235}U	Thermal	2.43 [8]
^{235}U	5.55 MeV	3.19 [9]
^{239}Pu	Thermal	2.89 [7]
^{241}Pu	Thermal	2.94 [7]

materials. However, one can characteristically describe the neutron energy by using the Watt spectrum, with an example for ^{235}U shown in Eq. (2.2), where E is the energy of emitted neutron in MeV.

$$\chi(E) = 0.4527 \exp(-E/0.965) \sinh(2.29E)^{1/2} \tag{2.2}$$

Neutron multiplicity can also be used as a signature of a particular isotope fissioning after neutron absorption. Note a significant difference between ^{235}U and ^{239}Pu exists, as outlined in Table 2.2. Moreover, the energy of interrogating neutron plays a role in measurements. A summary of prompt neutron multiplicity as a result of thermal fission can be found in Ref. [7].

2.3 Delayed Signatures

Delayed signatures are unique to SNM as their generation follows the β-decay of the fission products. The limitation of delayed neutron signatures is their relatively low abundance as compared to prompt emissions, amounting to about 0.6–5.4% of fission reactions [10]. On the other hand, delayed γ rays, while relatively abundant in their emissions, are hard to distinguish from the products of photonuclear reactions that can occur in non-fissionable materials.

2.3.1 Delayed Neutrons

Delayed neutrons are peculiar to the fission reaction. One can successfully argue that delayed neutrons appear almost exclusively in the fission reaction since they are a result of β decay of the fission products. The fission product undergoes β decay that leaves the nuclides in an excited state, most of which undergo γ decay. However, a small fraction of the excited nuclides emit neutrons. Delayed signatures are usually only measured after the probing ionizing radiation particle source is turned off, reducing background that often populates the detection spectrum, thus allowing clean measurements to be made. As indicated in Table 2.1, the delayed neutron signature is not very strong as compared to prompt signature. The average

Table 2.3 Delayed neutrons from ^{235}U and ^{239}Pu undergoing thermal fission

Group	Yield fraction (β), ^{235}U	Decay constant (λ), ^{235}U	Yield fraction (β), ^{239}Pu	Decay constant (λ), ^{239}Pu
1	0.000266	0.0127	0.000086	0.0129
2	0.001492	0.0317	0.000637	0.0311
3	0.001317	0.115	0.000491	0.134
4	0.002851	0.311	0.000746	0.331
5	0.000897	1.4	0.000234	1.26
6	0.000182	3.87	0.00008	3.21

yield is about 0.03 delayed neutrons per fission. Low yield in combination with relatively low energy of the emitted neutrons limits the practical applicability of this technique. Also, a challenge with delayed neutrons is their characteristic decrease in intensity with an increasing neutron energy [6].

Despite the very low delayed neutron yield, detecting this signature is possible, especially with high-efficiency neutron detectors such as ^{3}He tubes. The magnitude and temporal behavior of the delayed neutron signature is dependent on the fissioning isotope, allowing for characterization of mixtures of isotopes. The half-life for delayed neutrons is between 0.19 and 55 s, depending on the isotope [11]. The more neutron-rich the fission event is, the faster the β decay occurs. Depending on the decay constant of a delayed neutron precursor, delayed neutrons can be grouped instead of considering each precursor separately. Six neutron groups are frequently used, although other groupings are also possible. Table 2.3 lists data for delayed neutrons groups into six categories for thermal fissions of ^{235}U and ^{239}Pu. Note that the decay constants for ^{235}U and ^{239}Pu are similar, but yields are different, allowing for differentiation between fissile or fissionable nuclides. The yield is relatively independent of incident neutron energy up to about 4 MeV, but the yield drops by a factor of 2 for 14-MeV interrogating neutrons as compared to thermal neutrons.

The delayed neutron energy spectrum is dependent on the fissioning nuclide and its corresponding delayed family of products. It is important to note that the neutron and γ emissions occur almost instantaneously following β decay; the emission delay is due to the β decay process. The delayed neutron energy spectrum shown in Fig. 2.4 is from thermal neutron fission of ^{238}U. The spectrum associated with fast neutron fission is quite similar. One can see in the lower region the pronounced structure of the neutron energies.

The ratio of prompt to delayed neutron emission can provide a strong indication of the occurrence of fission [2]. The spectra in Fig. 2.5 show the ratio between prompt and delayed neutron emission rates, indicating why these are unique signatures for AI. Since the energy of delayed neutrons can be measured after the pulse, one can use the timing to further discriminate between the delayed neutrons and the probing radiation.

Fig. 2.4 Delayed neutron spectrum from thermal neutron fission for ^{235}U (top) and ^{239}Pu (bottom). Reproduced from Ref. [12]

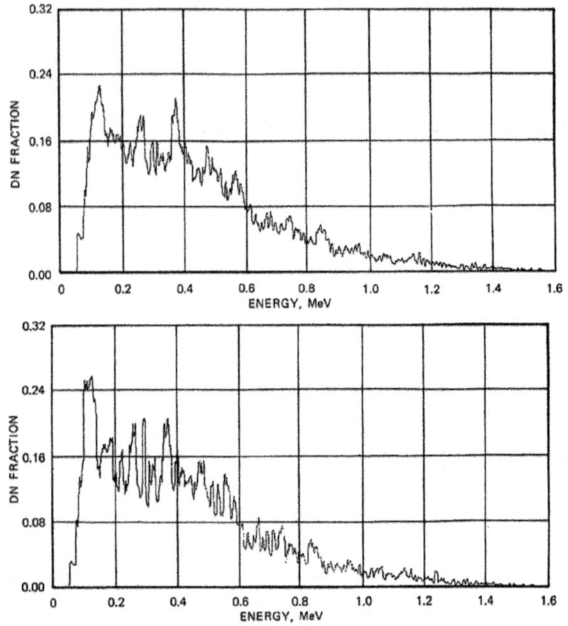

2.3.2 Delayed Gamma Rays

The delayed γ rays are emitted in a much greater abundance than delayed neutrons. Measurements of delayed γ rays can commence immediately after the interrogation beam is turned off. During this time, one expects the main source of background to be natural background and background from delayed activated nuclides, mostly from the cargo container that is being interrogated. Delayed γ rays have similar energy characteristics as for the prompt emissions: about eight γ rays are emitted with an average energy of 1 MeV. In general, delayed γ rays are more promising than the delayed neutrons because of their high intensities and longer half-lives. It is noteworthy to mention that the delayed γ rays are emitted continuously after the fission allowing for relatively long measurements. Figure 2.6 compares delayed γ-ray yield (energies above 0.51 MeV) from induced photofission of ^{235}U and ^{238}U. In AI applications, only delayed γ rays above 3 MeV are measured because the associated background is fairly high (Table 2.4).

2.4 Considerations of Background Radiation

During the SNM search, the natural background source are dominated by cosmic rays and radioactive decay of naturally present actinides like ^{238}U and ^{232}Th.

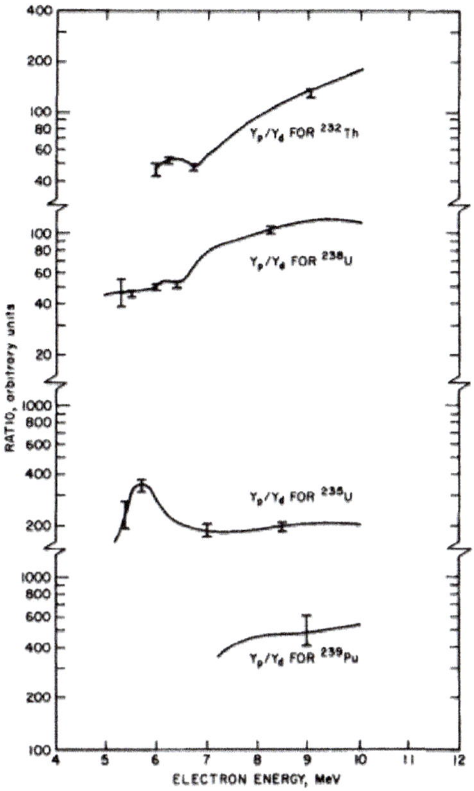

Fig. 2.5 Ratio of prompt to delayed neutron yield resulting from nuclear material interrogation with a linear accelerator. Reproduced from Ref. [1]

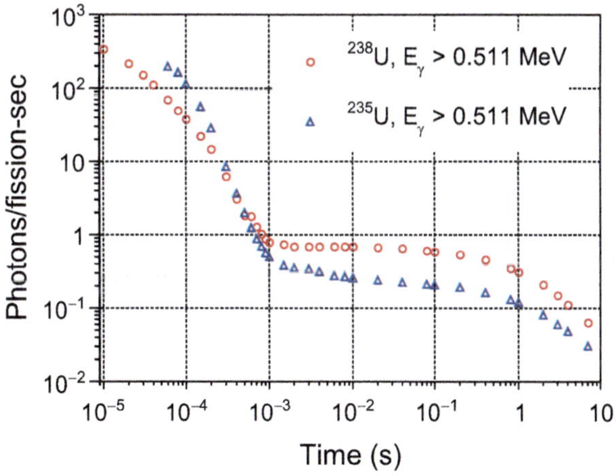

Fig. 2.6 Delayed γ rays from photofission of ^{235}U and ^{238}U as a function of time after fission. Adapted from Ref. [13]

Table 2.4 Comparison of prompt and delayed yield signatures (per thermal neutron-induced fission)

Isotope	Prompt γ rays	Delayed γ rays	Prompt neutrons	Delayed neutrons
^{235}U	0.2	0.13	2.4	0.015
^{238}U	0.2	0.11	0	0
^{239}Pu	0.3	0.07	2.9	0.0061

Gamma-ray emissions are for energies above 3 MeV. Adapted from Ref. [2]

Muons are highly penetrating and result in a significant number of fast neutrons and other types of secondary particles, including less-penetrating electrons, positrons, and protons. The fast neutrons are capable of generating photons as well as other neutrons, especially in high-Z materials. Nuclear reactions with oxygen in the environment produce $^{17}N/^{16}N$, which decays and emits neutrons and γ rays that can mask fission signals. An extensive discussion of natural background radiation, which impacts the measurements conducted passively to a much greater extend than AI techniques, is provided in Chap. 3.

For any detection system that employs AI, being able to distinguish between real signals and background, whether natural or induced by the interrogating system, is paramount. Data collected in AI without accounting for the background is of very limited use in most systems. AI systems use ionizing radiation to produce unique signatures from fissile materials. These unique signatures must be distinct (in type or intensity) from the radiation background caused by man-made materials such as medical isotopes and the interrogation system. Background signatures must be well understood for a system in order to reduce false alarms and gain a better signal to noise ratio. Accurate knowledge of the background is crucial when looking for signatures of the presence of SNM. Understanding the background can be challenging. Background measurements can be conducted prior to any introduction of SNM in the environment. Where applicable this allows the researcher an opportunity to subtract the background from measured signal when SNM is present. AI produces a significant background that may interfere with or complicate the interpretation of measured data. In situations like these, one needs to fully understand the AI system and provide techniques that are truly unique to SNM and can be distinguished even in the presence of high backgrounds. However, one can reduce the chance of more interfering background signals by carefully choosing the probing ionizing particle and supporting mechanisms such as shielding and collimation.

AI is tasked with overcoming the challenge of small signal-to-noise ratio when SNM is shielded and time of measurements is limited. Generation of signatures requires building sufficient statistics, resulting in a tradeoff between the quality of the signal read by the detection system and the dose to cargo. In particular, neutron-based AI may provide the highest quality confirmation of SNM presence, but may result in activation of cargo contents as well as masking on neutron emissions with the interrogating neutrons. Photofission may reduce the uncertainties associated with the primary neutron source. Delayed signatures are reliable in their definitive confirmation of fissile material presence. Even though γ-ray background is much higher than that for delayed neutrons, the relative intensity of γ rays with energies above 3 MeV (necessary to distinguish them from the background) is an order of magnitude higher [14] than for delayed neutrons. In addition, they are more likely to penetrate through cargo contents to be detected especially in a presence of hydrogenous materials. The combination of intensity and penetration capabilities in hydrogenous cargo make delayed γ rays 10^3–10^4 more sensitive as fission indicators than delayed neutrons [14].

References

1. T. Gozani, Active nondestructive assay of nuclear materials: principles and applications (1981). https://doi.org/10.2172/6215952. http://www.osti.gov/scitech/servlets/purl/6215952
2. T. Gozani, IEEE Trans. Nucl. Sci. **56**(3), 736 (2009). https://doi.org/10.1109/TNS.2009.2015309
3. V.V. Verbinski, H. Weber, R.E. Sund, Phys. Rev. C **7**, 1173 (1973)
4. T.E. Valentine, Evaluation of prompt fission gamma rays for use in simulating nuclear safeguard measurements. Ann. Nucl. Eng. **28**, 191 (2001)
5. V. Polhorsky, J. Kliman, J. Kristiak et al., Czech. J. Phys. **43**(8), 783–788 (1993)
6. R.C. Runkle, D.L. Chichester, S.J. Thompson, Nucl. Instrum. Methods Phys. Res. Sect. A **663**(1), 75 (2012)
7. N.E. Holden, M.S. Zucker, Nucl. Sci. Eng. **98**(2), 174 (1988)
8. A. Trkov, R. Capote, V. Pronyaev, Nucl. Data Sheets **123**(Supplement C), 8 (2015). https://doi.org/10.1016/j.nds.2014.12.003. http://www.sciencedirect.com/science/article/pii/S0090375214006826. Special Issue on International Workshop on Nuclear Data Covariances April 28–May 1, 2014, Santa Fe, NM. http://t2.lanl.gov/cw2014
9. M.S. Zucker, N.E. Holden, Energy dependence of neutron multiplicity p? in fast-neutron-induced fission for u-235, u-238, and pu-239. Technical Report BNL–38491, Brookhaven National Laboratory (BNL), Brookhaven, NY, USA (1986)
10. R.J. Tuttle, Nucl. Sci. Eng. **56**(1), 37 (1975). https://doi.org/10.13182/NSE75-A26620
11. E. Padovani, M. Monville, S.A. Pozzi, Nucl. Sci. Eng. **155**(1), 131 (2007). https://doi.org/10.13182/NSE07-A2651
12. D. Saphier, D. Ilberg, S. Shalev, S. Yiftah, Nucl. Sci. Eng. **62**(4), 660 (1977). https://doi.org/10.13182/NSE77-A15209
13. R.B. Walton, R.E. Sund, E. Haddad, J.C. Young, C.W. Cook, Phys. Rev. **134**, B824 (1964). https://link.aps.org/doi/10.1103/PhysRev.134.B824
14. E.B. Norman et al., Nucl. Instrum. Methods Phys. Res. Sect. A **521**(2), 608 (2004)

Chapter 3
Features and Limitations of Passive Measurements

Igor Jovanovic

Abstract Passive measurement of spontaneously emitted radiation is one of the simplest methods which can reveal nuclear threats such as the presence of special nuclear material. However, passive measurements are typically hampered by the relatively small magnitude of signal embedded in a large and variable background, which can be naturally occurring or originate from human activities. The magnitude of various passive signatures from special nuclear materials is reviewed and compared to typical backgrounds. Next, the concepts of gross radiation counting, energy resolution, and imaging are introduced and their value in the context of passive measurements is discussed. The difficulty of passive detection of even a sizable amount of special nuclear material in real environments is illustrated by examples, and the relative value of detection efficiency, energy resolution, and background suppression is discussed.

3.1 Magnitude of Passive Signatures and Backgrounds

Every radiation measurement is subject to the usual signal-to-background ratio constraints, which is why it is important to identify all relevant characteristic signatures and backgrounds. While the physical origins of these radiation sources were introduced in Chap. 2, the focus here is to provide the information on their detailed characteristics and typical magnitudes.

I. Jovanovic (✉)
Department of Nuclear Engineering and Radiological Sciences, University of Michigan, Ann Arbor, MI, USA
e-mail: ijov@umich.edu

I. Jovanovic, A. S. Erickson (eds.), *Active Interrogation in Nuclear Security*, Advanced Sciences and Technologies for Security Applications,
https://doi.org/10.1007/978-3-319-74467-4_3

3.1.1 Characteristic Emissions from SNM

Since the detection of special nuclear material (SNM) is of primary interest in nuclear security problems, the magnitude of the signatures from SNM must be carefully considered. In passive detection methods it is usually impossible to easily distinguish the prompt from the delayed SNM signatures. The characteristic emissions from SNM that are the most promising for its detection are X rays, γ rays, and neutrons; their intensities are listed in Tables 3.1 and 3.2. These characteristic emission strengths can serve as a basis for calculating the magnitude of detectable signatures.

One special characteristic signature of the fission process is the *neutron multiplicity*, which represents the number of neutrons emitted in either spontaneous or induced fission. For neutron-induced fission, multiplicity increases slowly and linearly with energy of the incoming neutron. Multiplicity (ν) can vary greatly from one to another fission event depending on the distribution of the excitation energy among the fission fragments, but its range is typically 0–6. Multiplicity distributions

Table 3.1 Major K X rays from uranium and plutonium [1]

X Ray	Levels (Final−initial)	Energy (keV) Uranium	Energy (keV) Plutonium	Relative intensity Uranium	Relative intensity Plutonium
$K_{\alpha 2}$	$K-L_2$	94.67	99.55	61.9	62.5
$K_{\alpha 1}$	$K-L_3$	98.44	103.76	100	100
$K_{\beta 1}$	$K-M_3$	111.30	117.26	22.0	22.2
$K_{\beta 2}$	$K-N_{2-5}$	114.5	120.6	12.3	12.5
$K_{\beta 3}$	$K-M_2$	110.41	116.27	11.6	11.7

Table 3.2 Natural radiation from uranium and plutonium isotopes [2]

Nuclide	Decay mode	Radiation, type	Intensity ($s^{-1}\,g^{-1}$)
^{235}U	α	185.7 keV, γ ray	4.3×10^4
^{238}U	α particle	1001.1 keV, γ ray	1.0×10^2
		766.4 keV, γ ray	3.9×10^1
	Fission	Fission neutrons	1.7×10^{-2}
^{238}Pu	α particle	766.4 keV, γ ray	1.5×10^5
		152.7 keV, γ ray	6.5×10^6
	Fission	Fission neutrons	2.6×10^3
^{239}Pu	α particle	413.7 keV, γ ray	3.4×10^4
		129.3 keV, γ ray	1.4×10^5
^{240}Pu	Fission	Fission neutrons	1.0×10^3
^{241}Pu	β particle	208.0 keV, γ ray	2.0×10^7
		164.6 keV, γ ray	1.8×10^6
		148.6 keV, γ ray	7.5×10^6
^{242}Pu	Fission	Fission neutrons	1.7×10^3

Table 3.3 Measured prompt fission multiplicity distributions

	^{235}U	^{238}U	^{239}Pu	^{240}Pu	^{242}Pu	^{252}Cf
Probability distribution	IF	SF	IF	SF	SF	SF
$P(0)$	0.033	0.054	0.011	0.066	0.068	0.002
$P(1)$	0.174	0.205	0.101	0.232	0.230	0.026
$P(2)$	0.335	0.380	0.275	0.329	0.334	0.127
$P(3)$	0.303	0.225	0.324	0.251	0.247	0.273
$P(4)$	0.123	0.108	0.199	0.102	0.099	0.304
$P(5)$	0.028	0.028	0.083	0.018	0.018	0.185
$P(6)$	0.003		0.008	0.002	0.003	0.066
$P(7)$						0.015
$P(8)$						0.015
$\overline{\nu}$	2.406	2.21	2.879	2.156	2.145	3.757
$\overline{\nu(\nu-1)}$	4.626	3.957	6.773	3.825	3.794	11.962
$\overline{\nu(\nu-1)(\nu-2)}$	6.862	5.596	12.630	5.336	5.317	31.812

IF induced fission, *SF* spontaneous fission [1]

$P(\nu)$ for both the spontaneous fission and induced fission can be approximated by a Gaussian distribution centered at the mean multiplicity $\overline{\nu}$:

$$P(\nu) = \frac{1}{\sqrt{2\pi\sigma^2}} \exp\left[-(\nu - \overline{\nu})^2/2\sigma^2\right]. \tag{3.1}$$

A typical value of the parameter σ that can be used is for most nuclides that undergo fission is 1.08, except in ^{252}Cf, where $\sigma = 1.21$ should be used. The measured prompt fission multiplicity distributions for important nuclides of interest are listed in Table 3.3. Reilly et al. [1], along with the calculated first, second, and third moments of the distribution ($\overline{\nu}$, $\overline{\nu(\nu-1)}$, and $\overline{\nu(\nu-1)(\nu-2)}$). It has been experimentally determined that the mean neutron energy and multiplicity are uncorrelated.

3.1.2 Naturally Occurring Radionuclide Background

The environment is rich in naturally occurring radionuclides, including the daughters that appear in their decay chains. There are approximately seventy such radionuclides which must be taken into account, and they can be broadly classified by their origin into primordial and cosmogenic.

Primordial radionuclides have been present on Earth since its formation and have relatively long half lives, for example ^{238}U, ^{235}U, ^{232}Th, ^{87}Rb, and ^{40}K. The main primordial radionuclides, their decay modes, half-lives, and elemental abundances are provided in Table 3.4.

Table 3.4 Primordial radionuclides, their decay modes, half-lives, and elemental abundances [3]

Radionuclide	Decay modes	Half-life (years)	% Elemental abundance
$^{40}_{19}$K	β^- EC β^+	1.27×10^9	0.0117
$^{87}_{37}$Rb	β^-	4.88×10^{10}	27.84
$^{115}_{49}$In	β^-	4.4×10^{14}	95.71
$^{138}_{57}$La	EC β^-	1.05×10^{11}	0.090
$^{147}_{62}$Sm	α	1.06×10^{11}	15.0
$^{152}_{64}$Gd	α	1.1×10^{14}	0.20
$^{174}_{72}$Hf	α	2.0×10^{15}	0.162
$^{187}_{75}$Re	β^-	4.3×10^{10}	62.60
$^{190}_{78}$Pt	α	6.5×10^{11}	0.01
$^{50}_{23}$V	β^- EC	1.4×10^{17}	0.250
$^{113}_{48}$Cd	β^-	9×10^{15}	12.22
$^{123}_{52}$Te	EC	$> 1.3 \times 10^{13}$	0.908
$^{144}_{60}$Nd	α	2.38×10^{15}	23.80
$^{148}_{62}$Sm	α	7×10^{15}	11.30
$^{176}_{71}$Lu	β^-	3.78×10^{10}	2.59
$^{180}_{73}$Ta	EC β^+	$> 1.2 \times 10^{15}$	0.012
$^{186}_{76}$Os	α	2×10^{15}	1.58

In contrast, *cosmogenic radionuclides* are continuously produced by cosmic rays and are relatively short-lived. Their concentration is determined by the equilibrium between the cosmic ray production and radioactive decay. One example of a well-known, common cosmogenic radionuclide is the ^{14}C, which is produced by neutron bombardment of ^{14}N in the atmosphere through the ^{14}N(n,p)^{14}C reaction.

Low-Z radionuclides are present in nature in significant quantities, and the dominant contribution to natural radioactivity comes from three of them: ^{40}K and ^{87}Rb (primordial) and ^{14}C (cosmogenic). One indication of the primordial abundance of ^{40}K is that most of the atmospheric ^{40}Ar, which constitutes approximately 1% of the Earth's atmosphere, originates from the ^{40}K decay. The other common primordial radionuclide is ^{87}Rb, which decays to the stable nuclide ^{87}Sr, which is readily substituted for potassium in minerals.

Actinides (high-Z radionuclides), including their decay chains, are also relatively abundant sources of terrestrial radioactive background. The two main elements of interest are uranium and thorium and their decay chains that produce radionuclides such as ^{222}Rn, ^{219}Rn, and ^{220}Rn. The spatial distribution of uranium concentrations in the United States is shown in Fig. 3.1. Actinide decays can be classified into four distinct series: thorium ($4n$), neptunium ($4n + 1$), uranium ($4n + 2$), and actinium ($4n + 3$).

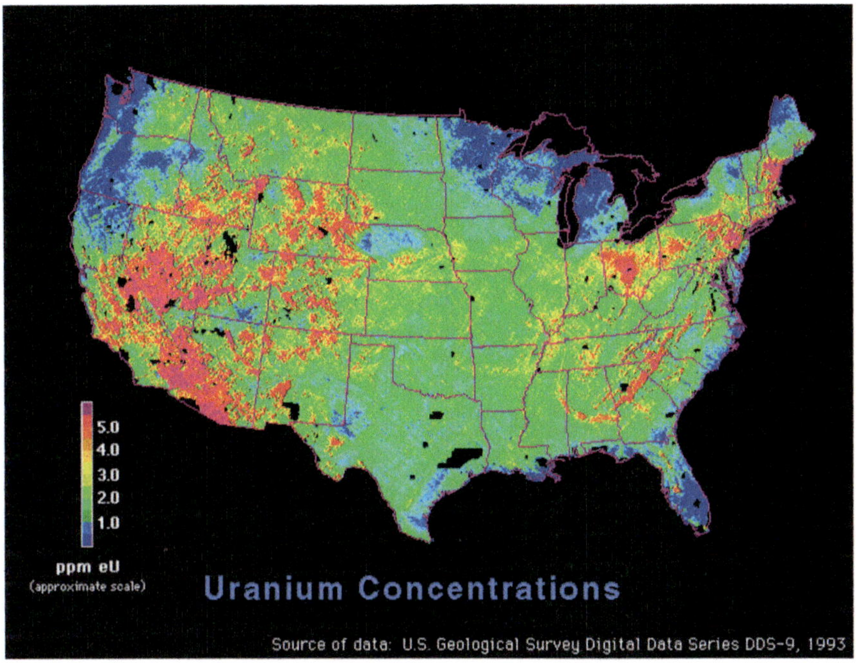

Fig. 3.1 Uranium concentration in the lower 48 United States. Reproduced from Ref. [4]

3.1.3 Cosmic Background

Much of the natural background originates from the cosmic ray interactions in the atmosphere. Cosmic rays can be either of galactic or solar origin and consist of approximately 90% protons and 10% heavier nuclei (up to iron, with traces of heavier nuclei). About 0.1% of the primary cosmic ray spectrum consists of γ rays. The primary cosmic ray spectrum is shown in Fig. 3.2.

The primary cosmic rays interact with the atmosphere to produce spallation neutrons, pions, electrons, photons, and downscattered ions (mostly protons). Pions rapidly decay to muons, many of which reach the ground, providing a flux on the order of $10^4\,\mathrm{m^{-2}\,min^{-1}}$. The mean muon energy at sea level is approximately 4 GeV, which makes the muons very penetrating (their rate of energy loss is $\sim 2\,\mathrm{MeV\,g^{-1}\,cm^2}$). Such muons deposit fairly high energy when passing through detectors; for example, a muon passing through 5 cm-long active volume of a HPGe detector deposits approximately 50 MeV.

Whereas the cosmic radiation of galactic origin is relatively constant, the solar component of the cosmic background exhibits significant fluctuations due to solar activity (including the 11-year solar cycle, yearly cycle, and the diurnal cycle). An occasional large and significant change of the solar cosmic radiation has been observed and is referred to as the *Forbush decrease*. Since the incident primary

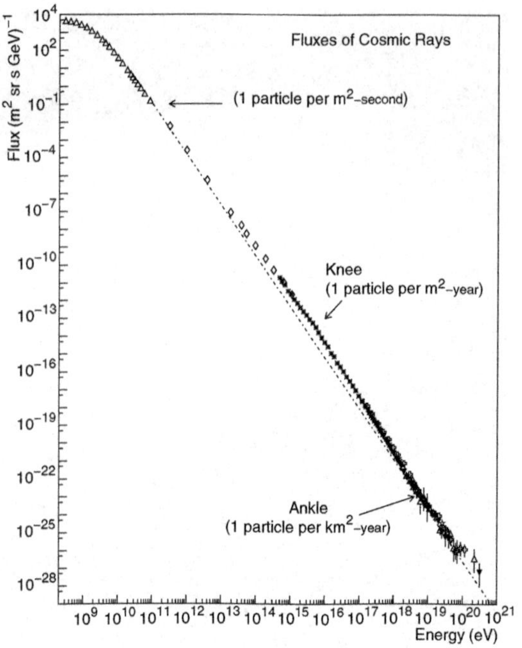

Fig. 3.2 The primary cosmic ray spectrum. Reproduced from Ref. [5]

cosmic rays are charged particles, they are subject to deflection by the Earth's magnetic field. As the particles undergo complex paths in this field, there is a significant effect of the latitude on the magnitude and spectrum of the cosmic radiation flux that reaches the Earth's atmosphere.

Another quantity of high relevance to detection is the flux of secondary particles produced by cosmic radiation, which may not be readily vetoed as in the case of muons. There has therefore been a large interest in measuring and understanding the secondary neutron flux. Recent measurements of this flux have been made [6] and, as expected, have found some variation of the magnitude of the flux depending on the measurement location. These measurements also determined that the shapes of the neutron spectra at various locations are quite similar (Fig. 3.3).

A special consideration when considering the neutron background is the effect of environment. Namely, it has been determined that large structures can increase the neutron background through cosmic ray interaction. Since this behavior is especially pronounced on or near large ships, it has been referred to as the *ship effect*. Systematic measurements of the ship effect have been made [7], finding an enhancement of the neutron background in the presence of large quantities of lead and iron (Fig. 3.4). It was also found with measurements made on ships that the water acts as a sink for neutrons and the shore acts as a γ-ray source.

While the fluxes and dose rates of secondary cosmic ray particles are known to vary significantly, it is useful to provide their approximate values at sea level (Table 3.5), which can guide the practical design of detection systems.

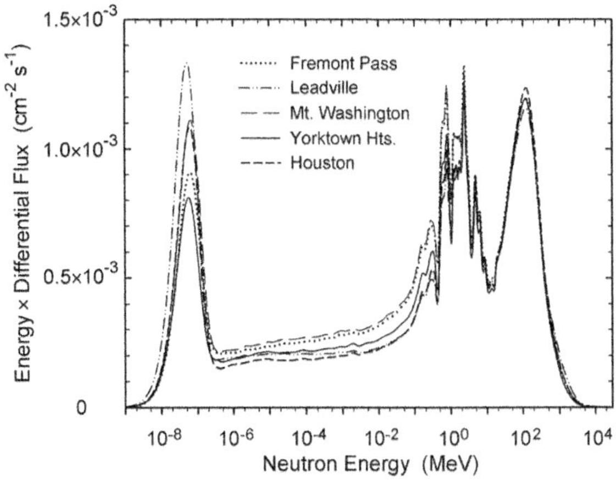

Fig. 3.3 Measured background neutron spectra at various locations scaled to sea level. © 2004 IEEE. Reprinted, with permission, from Ref. [6]

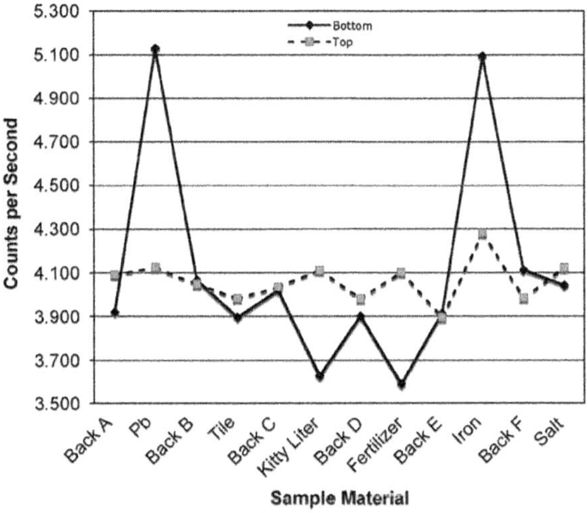

Fig. 3.4 Detected neutron background rates in the presence of large quantities of various materials. Reproduced from Ref. [7]

Table 3.5 Approximate fluxes and dose rates of secondary cosmic ray particles at sea level [8]

Particle	Flux (cm^{-2}s^{-1})	Dose rate (μrad/h)
Muons	0.019	2.5
Electrons	0.00455	0.6
Protons	0.000171	0.04
Charged pions	0.0000134	0
Neutrons	0.00646	0.04

3.1.4 Radionuclide Background Related to Human Activities

It is well-known that human activities such as construction, industrial production, and medical treatment are the significant sources of radiation background that can trigger false alarms in nuclear security systems. One example of the background

Table 3.6 Typical and maximum activity concentrations in building materials used in the European Union

Material	Typical activity concentration (Bq/kg)			Maximum activity concentration (Bq/kg)		
	^{226}Ra	^{232}Th	^{40}K	^{226}Ra	^{232}Th	^{40}K
Most common building materials (may include byproducts)						
Concrete	40	40	400	240	190	1600
Aerated and light-weight concrete	60	40	430	2600	190	1600
Clay (red) bricks	50	50	670	200	200	2000
Sand-lime bricks	10	10	330	25	30	700
Natural building stones	60	60	640	500	310	4000
Natural gypsum	10	10	80	70	100	200
Most common industrial byproducts used in building materials						
Byproduct gypsum (phosphogypsum)	390	20	60	1100	160	300
Blast furnace slag	270	70	240	2100	340	1000
Coal fly ash	180	100	650	1100	300	1500

Adapted from Ref. [9]

are the construction materials, as can be seen in Table 3.6. Since these materials are frequently shipped, passive detection systems must be properly designed and calibrated to distinguish them from SNM and other threats.

In addition to building materials, it is known that other typical commercial shipments can trigger *nuisance alarms*. They include agricultural products (bananas, Brazil nuts), glazed ceramics, camera lenses, polishing compounds and abrasives, propane tanks, kitty litter, road salt, ore, rock, vulcanized rubber, television and computer displays, medical isotopes, and smoke detectors. These materials are collectively referred to as the *naturally occurring radioactive material* (NORM). The main nuclides responsible for NORM radioactivity include ^{40}K and the uranium/thorium decay series. Another category of NORM has been defined by the U.S. Environmental Production Agency as *technologically enhanced naturally occurring radioactive material* (TENORM). TENORMs are NORMs that have been concentrated or exposed to the accessible environment as a result of human activities such as manufacturing, mineral extraction, or water processing. TENORM industries and sources include mining, energy production, water treatment, and certain consumer products such as antiques, building materials, phosphate fertilizer, and tobacco products. Even the radiation detector materials can contain naturally occurring radioisotopes in significant quantities, for example ^{207}Bi in BGO and ^{138}La in $LaBr_3$.

Another major source of radionuclide background is the use of radioisotopes in medicine. After administration of radionuclide for medical diagnostics or treatment purposes, the patient exhibits an increased radioactivity level. Radiation portal monitors may not effectively discriminate the legitimate medical radionuclides from those associated with threats. Alarms may be triggered not only by material shipments, but also by treated patients (those who are not declared or known).

Table 3.7 Decay properties of typical medical radionuclides

Isotope	Half-life	Photon energy (>30 keV) Percent Branching Ratio (>1%)							Activity (mCi)	Photons (s^{-1})
^{57}Co	217 d	keV	122.0	136.4					0.0006	3.56×10^4
		BR	85.54	10.69						
^{51}Cr	27.7 d	keV	320.0						0.075	3.34×10^5
		BR	9.83							
^{67}Ga	78.3 h	keV	91.27	93.31	184.5	208.9	300.2	393.5	10.00	5.15×10^8
		BR	2.95	37.00	20.45	2.33	16.60	4.64		
^{123}I	13.2 h	keV	30.98	31.88	158.9				0.300	1.89×10^7
		BR	12.43	2.71	83.30					
^{125}I	60.1 d	keV	30.98	35.49					N/A	N/A
		BR	20.09	4.38	6.66					
^{131}I	8.04 d	keV	80.18	284.3	364.4	636.9	722.8		41.80	1.59×10^9
		BR	2.62	6.06	81.25	7.27	1.80			
^{111}In	67.3 d	keV	22.98	26.08	26.80	171.2	245.3		3.225	2.65×10^8
		BR	24.15	11.57	2.40	90.24	94.00			
^{153}Sm	46.7 h	keV	40.90	41.54	47.00	48.50	69.67	103.1	140.0	5.47×10^9
		BR	17.96	32.36	9.42	2.85	5.32	28.30		
^{99m}Tc	6.01 h	keV	140.4						16.91	5.85×10^8
		BR	87.20							
^{201}Tl	73.1 h	keV	68.89	70.82	80.12	82.78	135.2	167.4	20.00	1.07×10^9
		BR	26.89	45.67	16.08	4.39	2.67	9.43		
^{133}Xe	5.25 d	keV	30.63	30.97	34.97	36.01	80.99		10.00	3.18×10^8
		BR	14.06	25.98	7.10	1.74	36.97			

Decay properties of typical medical radionuclides are shown in Table 3.7. The most common medical radionuclides are ^{99m}Tc, ^{131}I, ^{201}Tl, ^{67}Ga, and ^{111}In, and they are used primarily in diagnostics.

Finally, one major human activity that contributes to radiation background may be the use of radioisotope sources in industry. For example, ^{192}Ir ($E_\gamma = 320$ keV) is used for radiography of pipeline welds, boilers, and aircraft parts. In manufacturing, radionuclides have been used to dry ink on packages, cure composite materials (airline parts), as fluid level and thickness gauges, and to enhance automobile tire wear characteristics. In security, radionuclides have been considered for inactivation of potential anthrax in mail, while in food industry they are used for sterilization of vegetables, spices, and meats. There has been an extensive use of radioisotopes in oil exploration (well logging).

A non-negligible source of background is the fallout from atmospheric nuclear tests conducted in the 1950s and 1960s by the United States and Soviet Union and from several nuclear accidents, especially Chernobyl (1986) and Fukushima (2011). Nuclear weapons testing nearly doubled the ^{14}C concentration in the atmosphere at its peak in 1960s, as can be seen in Fig. 3.5.

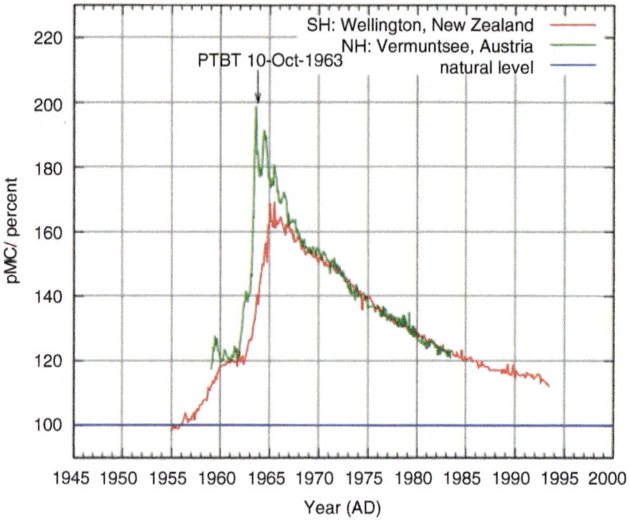

Fig. 3.5 Variation of atmospheric ^{14}C concentration as a result of atmospheric weapons testing. Reproduced from Ref. [10]

3.2 Principles of Passive Measurements

The basic concept of a passive measurement of radiation is illustrated in Fig. 3.6. In most situations, only the highly penetrating radiation (γ rays and neutrons) is suitable for detection, since other common emitted radiation (α and β particles) can be readily stopped by even a thinnest shield, which includes the intervening air or the material itself (the effect of *self-shielding*). The object that contains the material to be detected emits an intrinsic and potentially useful characteristic radiation signature of neutrons and γ rays. The important and potentially detectable characteristics of this signature include its magnitude, characteristic energy spectrum, multiplicity, and correlation with other signatures. The magnitude of the signature is primarily governed by the quantity and nature of material (for example, the specific isotope of uranium or plutonium or their mix). If measured with high fidelity, the energy spectrum of emitted radiation can be a powerful unique signature of the emitting nuclide, since it is fundamentally related to nuclear structure. The emitted γ rays typically have a narrow energy spread that is in principle immune to spectral interferences. In practice, however, the energy resolution of detectors that are used to detect γ rays is the limiting factor, which frequently leads to interferences with neighboring γ-ray lines. While the emitted neutron spectrum is similarly characteristic of the material, it is typically absent of sufficiently *sharp features* in its energy spectrum and is more challenging to measure with high fidelity. The multiplicity and correlation of the emitted γ rays and neutrons can be one of the most useful signatures that enables the detection of SNM through the occurrence of

Fig. 3.6 Illustration of the main elements of the passive radiation detection process

nuclear fission. In the nuclear fission process, whether spontaneous or induced, there is an abundant and simultaneous emission of prompt γ rays and neutrons (Fig. 2.1).

Not only the energies, but also the number of emitted radiation quanta of γ rays and neutrons is characteristic for the fissioning nuclide and to the method used to induce fission. The number of simultaneously emitted radiation quanta is referred to as *multiplicity*; some of the important multiplicity data are listed in Table 3.3. While the delayed γ rays and neutrons are also emitted following a fission process, they hold little value in passive measurements not only because they are relatively rare, but because it is not possible to determine the delay between their emission and the corresponding fission when there is no established time structure of fission present, as in the case in active interrogation. Finally, since prompt neutrons and γ rays are emitted simultaneously in fission, detected γ-neutron correlations can also be useful in the detection of SNM by fission.

While the emitted signatures are unique to the constituent radionuclides on the microscopic level, they can be significantly altered by propagation. The first barrier encountered by the emitted radiation is the object itself. While both the γ rays and neutrons can be considered as highly penetrating, they can still experience significant absorption and down-scattering. For example, in the case of ^{235}U, the characteristic 186 keV γ ray can be readily absorbed in the uranium object, such that only the γ rays of this energy emitted near the surface have an appreciable probability of escaping and contribute to the γ signature of the object. In the case of neutrons, they can undergo not only scattering, but also absorption. In the case of fissile nuclides such as ^{235}U and ^{239}Pu, they can induce additional fissions, which can multiply both the neutron and the γ emission.

Once the radiation is emitted, it undergoes propagation through the environment, which may include high-density intervening material, which is usually referred to as *shielding*. This propagation is arguably the most important factor affecting the detectability in passive radiation measurements. First, the spontaneously emitted neutrons and γ rays are emitted isotropically; as a result, the radiation flux received at the detector drops quickly ($\propto 1/r^2$). In many scenarios in nuclear security it is not possible to position the detector in close proximity to the source. In addition, radiation emitted by the source propagates through the environment, which can significantly alter the unique signature through several mechanisms, the two most prominent being absorption and scattering. In the case of absorption, the radiation quantum is removed from the emitted flux, and it is possible for re-emission of a different signature to occur. For example, a neutron can be captured by hydrogen in a hydrogenous material to emit a 2.2-MeV characteristic γ ray, or a high-energy γ ray can be absorbed to produce an annihilation γ ray at an energy of 511 keV. Radiation can undergo elastic or inelastic scattering during propagation, where the characteristic energy of a neutron or γ ray is reduced, but the particle may still reach the detector and be detected. In this case, the signature such as γ-ray energy may be altered and not counted in spectroscopic radiation detection systems. Finally, one interesting aspect of radiation interaction with the environment may even enhance the characteristic signature. An example of this is placement of an object that acts as a neutron scatterer near the measured object; in this case, the reflected neutrons may induce additional fissions in fissile material.

As the γ rays and neutrons ultimately reach the detector, some fraction of them interact with the active detector material and are detected. Detector cannot be considered as being separate from its environment, however; one can only consider it to be a fraction of the overall space that is instrumented in a fashion that allows the radiation-matter interactions to be converted into detectable signals that can subsequently analyzed and counted. Detectors exhibit their unique response functions that depend primarily on the detector material and size and reflect the various processes that can occur in the interaction of radiation with the detector. As a result of these interactions, a fraction or all of the incident particle energy can be ultimately converted into an electrical signal. Through a combination of various interactions, detector can therefore significantly alter the characteristic signature of SNM radiation incident upon it.

3.2.1 Gross Radiation Counting

The simplest method to detect the presence of nuclear material is the use of gross radiation counting. In this approach, the presence of radioactive material is inferred from the excess detected number of counts in a radiation detector over background, which is recorded over some period of time. In all nuclear security applications, and especially in cargo inspection and wide-area search, it is desirable to minimize the measurement time. For example, in cargo inspection it is necessary to maintain a screening rate that is compatible with the presented flow of traffic.

In gross radiation counting, as in all other more sophisticated counting methods that involve categorization of signal by energy (imaging) and location (imaging), a measurement can be characterized in the standard statistical framework, which leads to the definition of *minimum detectable activity* (MDA) or minimum detectable amount [11]. A detailed discussion of the concept of MDA and related statistical inference methods is widely available in the literature [12]. In Sect. 3.4 an example analysis is provided, which illustrates the level of difficulty of detection of shielded SNM using gross radiation counting.

3.2.2 Gamma-Ray and Neutron Spectroscopy

A common and powerful method that can significantly enhance the signal to background ratio is radiation spectroscopy. Similar to better-known optical spectroscopy, γ-ray spectroscopy involves measuring the energy of high-energy photons incident onto detector and thereby reconstructing the energy distribution of incident radiation. Despite the radiation interactions that modify the energy distribution of γ-ray radiation emitted by the source through processes such as scattering and absorption in the source itself, in the intervening space, and in the detector itself, there is a tendency for a fraction of the γ-ray radiation to experience no interactions prior to reaching detector and then deposit its full energy in the detector. In this situation, a highly usable signal is recorded, which is proportional to the characteristic energy of a nuclear transition. This feature is usually referred to as the *photopeak*. By detecting and measuring the relative intensity of the photopeaks in the γ-ray spectrum, it is possible to check the consistency of the measured signal with the known decay properties of a given nuclide. With some assumptions about the problem geometry, one can then also reconstruct the relative abundance of the radioactive nuclides present in the vicinity of the detector.

An example of the γ-ray spectra of highly enriched uranium (HEU) and natural uranium measured with a high-purity germanium (HPGe) detector is shown in Fig. 3.7. A 186-keV γ-ray line is prominent in HEU, which contains a large fraction of ^{235}U, whereas a 1001-keV line is present in natural uranium and originates from the decay of the much more abundant ^{238}U. In Fig. 3.8, the spectrum of plutonium metal is shown.

Onc aspect of γ-ray spectroscopy that has to be properly appreciated in this context is the importance of detector *energy resolution*. The energy resolution of a detector refers to its capacity to consistently record the same pulse height in different events in which the same energy is deposited in the detector. This property is usually limited by the statistics of production of information carriers (e.g., ionization electrons or scintillation photons) in the detector material, but can also be significantly affected by the carrier transport and electronic noise in the detection system [12]. In a detector that exhibits a better energy resolution, a greater photopeak to background ratio can be realized since a smaller fraction of the broad background is present in the measured energy range within which the signal is confined. Energy resolution thus represents an important consideration, which has to be carefully weighted with detector efficiency and cost to minimize the detectable activity in a passive measurement. While the characteristics of specific detector types are discussed at greater length in Chap. 5, here we provide Fig. 3.9 as an example that illustrates the value of detector energy resolution. When lower energy

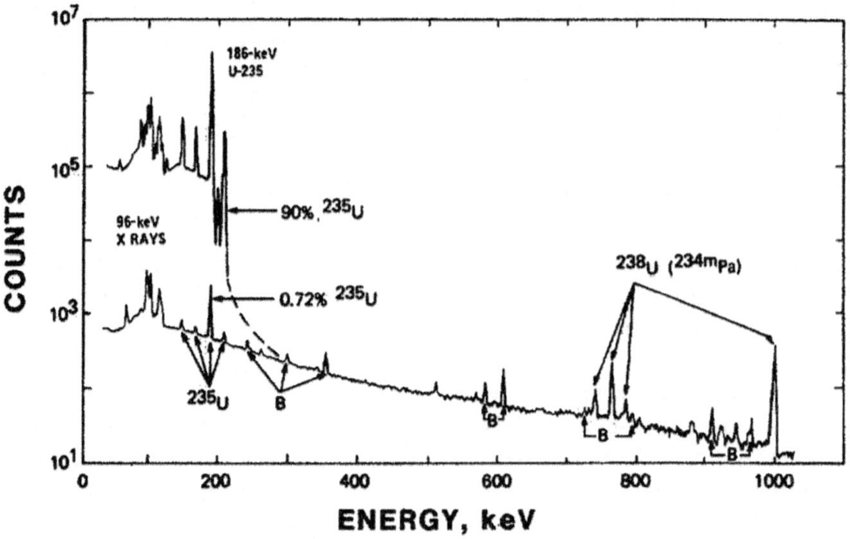

Fig. 3.7 Gamma-ray spectra from natural (0.7% ^{235}U) and 90%-enriched uranium, measured with an unshielded 1%-efficiency Ge(Li) detector. The peaks labeled ^{238}U (^{234m}Pa) are from the decay of ^{234}mPa. Background peaks are labeled B. Note the dominance in the spectrum of the 186-keV peak from ^{235}U decay. Reproduced from Ref. [1]

resolution detectors such as NaI(Tl) are used, some of the characteristic γ rays that are readily resolved with higher-resolution semiconductor detectors interfere and merge. These interferences of neighboring spectral lines can significantly reduce the number of useful lines in γ-ray spectroscopy performed with lower resolution detectors; however, lower resolution detectors are usually more practical due to low cost and scalability to large areas and volumes.

A broad continuum originating from Compton scattering is present in the measured γ-ray spectrum. This interaction type is very common and produces a full range of energy depositions in the detector up to the maximum value, referred to as the *Compton edge*. The continuum thus produced by Compton scattering of a high-energy γ ray represents a background for all γ rays whose photopeaks are located below its Compton edge. In a complex radiation environment filled with naturally occurring radioactive material, this can represent a significant problem and further increase demands on detector energy resolution. Alternative methods of detection that employ *Compton suppression* [13] have emerged, where Compton events are detected as multiple interactions in a segmented detector or a special multi-detector assembly and vetoed, reducing the magnitude of the Compton continuum. Such systems usually occupy the majority of the solid angle, which makes them poorly suited for counting in many real-world nuclear security applications (Fig. 3.10).

Neutron spectroscopy is another powerful technique that may be applied in passive measurements. However, it should be noted that SNM produces neutron

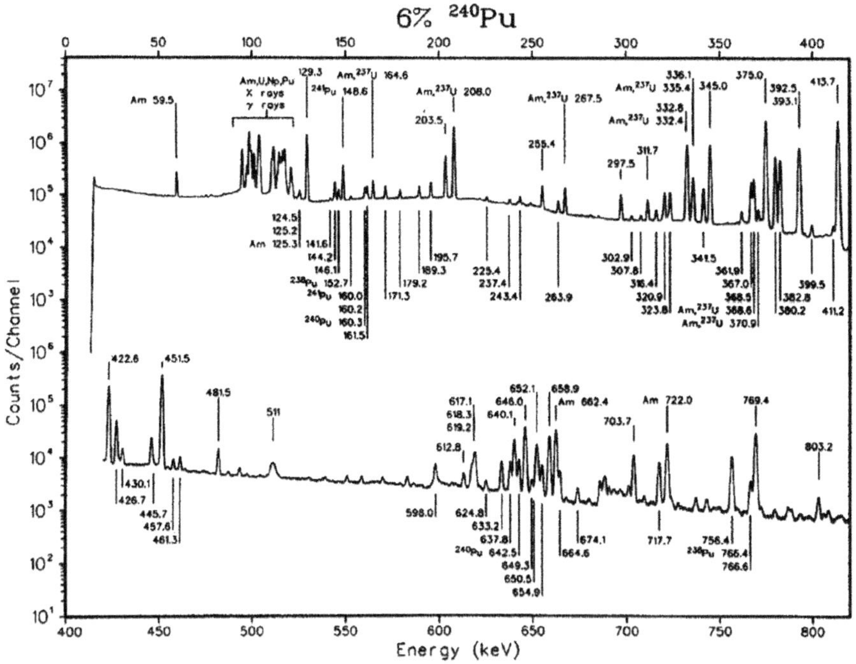

Fig. 3.8 Gamma-ray spectrum from 500 g of plutonium metal measured with a coaxial germanium detector. Peaks not labeled with a specific isotope are from ^{239}Pu. Reproduced from Ref. [1]

emissions which do not have such well-defined features in their energy spectrum as γ rays, so it may be challenging to distinguish them from background. Neutrons also readily interact in the environment and thus their spectrum can be significantly perturbed, reducing the value of neutron spectroscopy. Finally, it can be much more challenging to obtain neutron spectra than γ spectra as a result of the lack of interactions that produce the equivalent of a photopeak in γ-ray detectors. Instead, unfolding the neutron spectrum from detector response must be performed, and the result usually exhibits relatively low resolution and is sensitive to noise. Still, if an excess of relatively high-energy neutrons is observed in neutron spectroscopy, this presents a very valuable signature detection that may indicate the occurrence of a fission process.

3.2.3 Fusion of Multiple Measurement Modalities

While the passive measurements can take advantage of high-efficiency, high-energy-resolution counting and radiation imaging (discussed in more detail in Sect. 3.3.1), an opportunity that has been recognized by researchers and awaits wide implemen-

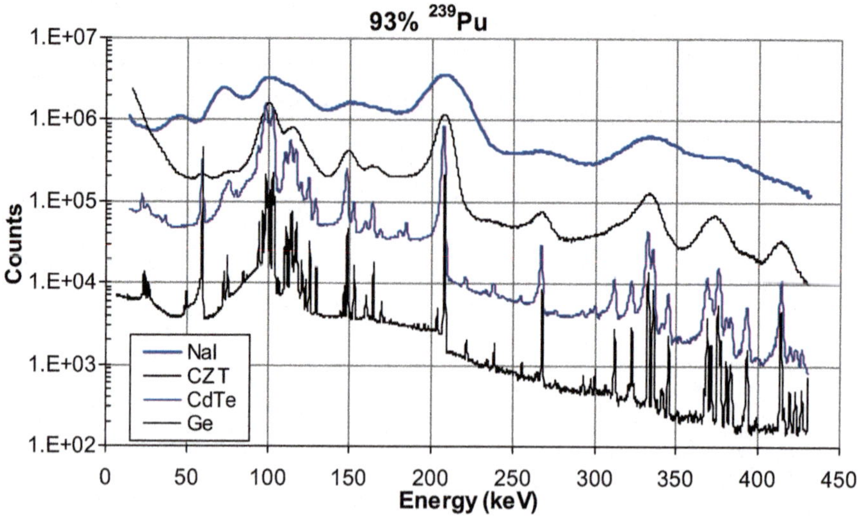

Fig. 3.9 Gamma-ray spectra of low-burnup (93% ^{239}Pu) plutonium measured with four different γ-ray detectors: NaI(Tl), coplanar grid CdZnTe, CdTe, and Ge (top to bottom). Reproduced from Ref. [1]

Fig. 3.10 Example of implementation of a Compton suppression system that employs four NaI(Tl) detectors surrounding a HPGe detector [8]

tation makes use of the combined information collected from radiation sensors and from more conventional surveillance technology, such as CCD or CMOS cameras. For example, since simple imaging in the visible or infrared part of the spectrum can provide the instantaneous location of a moving vehicle, and stereoscopic imaging or lidar could also measure its distance, this information can be combined with the time-dependent radiation measurement (gross counting, spectroscopy, or imaging) to enhance its value by use of correlation of multiple signatures. An example of this

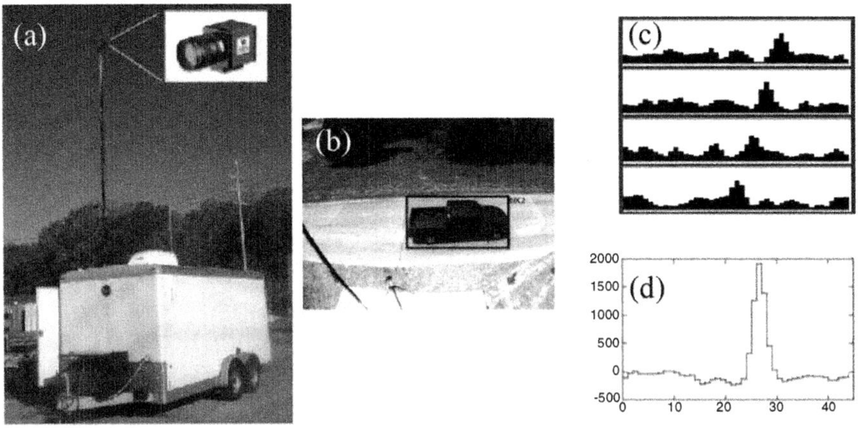

Fig. 3.11 (**a**) The Large Area Imager with the boom for deploying the video camera; (**b**) video image of a vehicle containing the 1-mCi source; (**c**) one-dimensional *images* (count histograms vs location) of a 1-mCi source transiting the imager's field of view; (**d**) Final vehicle image created by combining the data from large area imager and the video camera. © 2008 IEEE. Reprinted, with permission, from Ref. [14]

approach is shown in Fig. 3.11, where a roadside tracker has been developed that combines a large-area γ-ray imager with an independent measurement of vehicle location to harvest multiple shots from the γ-ray imager to reveal a radiation source in a moving vehicle. This would not be possible with the use of a radiation detector or imager alone because of the relatively weak emission from the source.

Approaches like these may be considered to employ the concept of *data fusion*, in which the time correlation of multiple streams of data improves the detection sensitivity.

3.3 Technology for Passive Measurements

3.3.1 Gamma-Ray and Fast Neutron Imaging

A method which can significantly enhance the performance of a passive measurement system is radiation imaging. This approach takes advantage of the typical difference in spatial distribution of a signal of interest and background. Whereas the signal of interest (for example, a relatively small sample of HEU or Pu) is very similar to an ideal point source, the source of background is usually widely distributed in space. One can draw an analogy between this condition and γ spectroscopy, where the signal of interest (a γ ray photopeak) may occupy a small fraction of the energy spectrum, while the background exhibits a broad distribution, a fraction of which is a superposition of Compton continua.

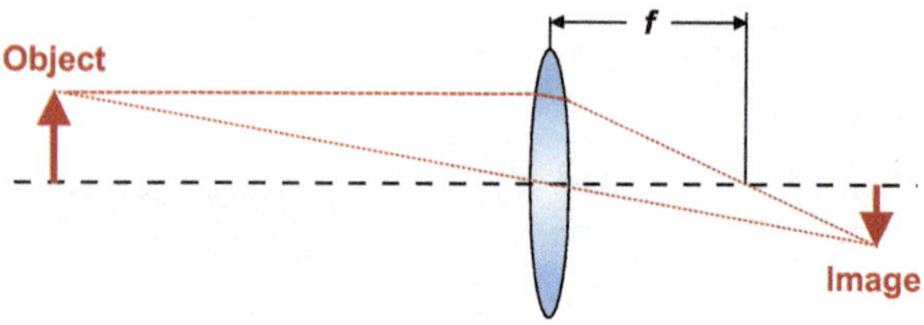

Fig. 3.12 Imaging using a refractive element (lens)

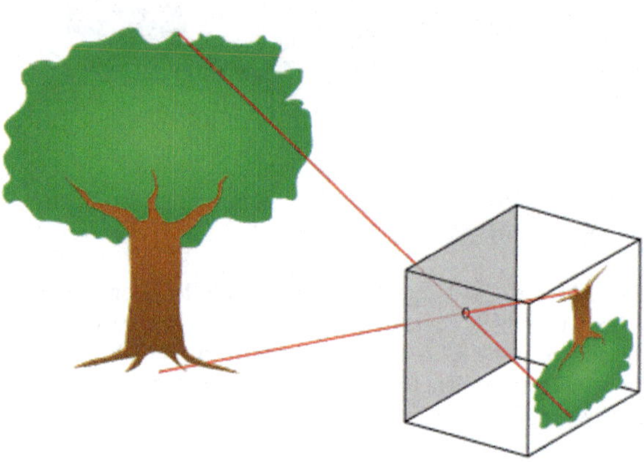

Fig. 3.13 Imaging using a pinhole

The objective of radiation imaging is to classify the events occurring in the radiation detector or a detector system based on the incident angle of radiation onto the detector aperture. If multiple such detectors are used and spatially separated, the *parallax* effect can be used to reconstruct not only the angle of incidence on the detector, but also the actual position of source that includes its distance from the detector.

In common imaging systems the goal is to map the *spatial frequency* of incident radiation (which is proportional to incidence angle) onto a position on the detector surface. The spatially dependent distribution of radiation thus incident onto the detector (or a detector array) is referred to as the *image*. In optical imaging a simple refractive element (lens) is typically used to perform this angle-to-position transformation (Fig. 3.12). This is a very efficient process, in a sense that the photons incident onto the lens aperture can be imaged with nearly 100% energy efficiency with a finite-size detector.

The use of a refractive element for imaging is usually impossible for high-energy radiation such as γ rays and neutrons. Instead, a simple\concept of *pinhole camera* can be applied. As illustrated in Fig. 3.13, a small hole in an opaque screen (shield) can restrict the directions of incident radiation in such way that there is a unique mapping of an incidence angle onto screen (detector) position. Unfortunately, this imaging method is inefficient, since only a small fraction of incident radiation can pass through the pinhole.

To address this limitation of a single pinhole imager, a related concept of *coded aperture imaging* [15] has been developed. In coded aperture imaging (Fig. 3.14), an array of opaque and transparent elements is placed in front of a position-sensitive

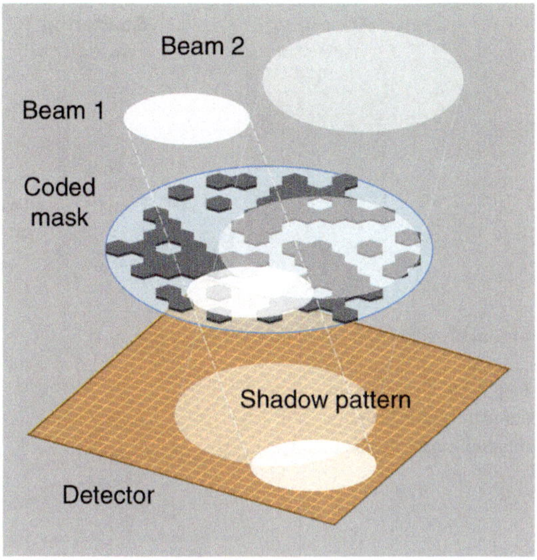

Fig. 3.14 Coded aperture imaging [16]

detector. The operation of this arrangement can be understood if one considers the fact that the illumination pattern on the detector depends on the angle of incident radiation onto the aperture (this is essentially a case of multiple pinhole cameras). The recorded pattern on the detector can be reconstructed [17] into an image while greatly improving the efficiency in comparison to a simple pinhole camera, since a much greater fraction of the coded aperture is open (transparent) to radiation. The coded aperture imagers have been developed both for γ rays [18] and fast neutrons [19]. There is little interest in imaging slow (thermal) neutrons in this context since they carry little information about their source. One variant of coded aperture imaging device that operates in the time domain has been pioneered in the field of astronomy and has been termed the *rotating modulation collimator* [20].

An entirely different concept of imaging has been successfully applied to passive radiation imaging, and is based on full or partial kinematic reconstruction of coincident interactions taking place in a detector or a detector array. Devices based upon this concept have been termed *Compton imagers* [21] (for γ ray imaging) or *neutron scatter cameras* [22] (for neutron imaging). The concept of both devices is illustrated in Sect. 3.3.1 and can be summarized as follows. Many of the interactions of γ rays and fast neutrons may result in a partial energy deposition in the detector medium. This is the case with Compton scattering of γ rays, or most fast neutron scatters, except for the special case of neutron backscatter from a hydrogen nucleus (proton). In such interactions, the incident particle emerges from the interaction at an angle that is governed by the kinematics of the scattering process (the usual simultaneous conservation of energy and momentum), as shown in Figs. 3.15 and 3.16.

In a position-sensitive and spectroscopically capable detector, the position and energy deposited by the interaction can be measured. Such detectors are frequently

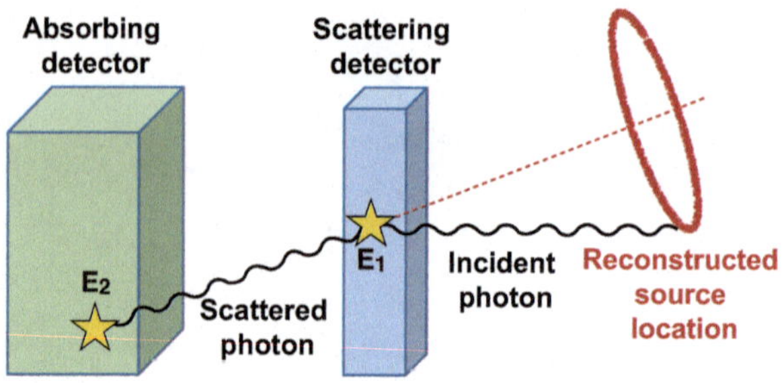

Fig. 3.15 Principle of Compton imaging

Fig. 3.16 Principle of neutron scatter camera. Reproduced from Ref. [23]

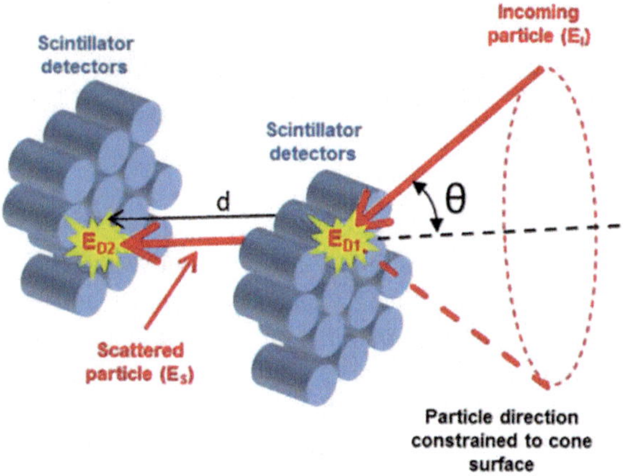

optimized to strike a compromise between efficiency of single interaction and low probability of multiple interactions. In a Compton imager or neutron scatter camera, the second interaction of the same scattered particle occur at a different position, usually in a second detector. Through a combination of interaction positions, deposited energies, and the time of flight information from two interactions (for neutrons), the direction of the incident particle can be constrained to a cone or voxel, and multiple radiation interactions may be used to improve the source localization. The reader is referred to abundant literature on Compton imagers and neutron scatter cameras, in which the details of the event analysis that permits reconstruction of directionality are discussed.

Other device types that permit γ and fast neutron imaging have been demonstrated that make use of the kinematic reconstruction, and here we mention two examples. A neutron time projection chamber (Fig. 3.17) [24] relies on either a single of multiple interaction in a hydrogen-containing gas to reconstruct the incident neutron direction. A dual-particle imager has been developed [25] by extending the concept of a neutron scatter camera to combine neutron and γ detectors into and arrangement that makes it possible to simultaneously reconstruct the incident γ and neutron direction.

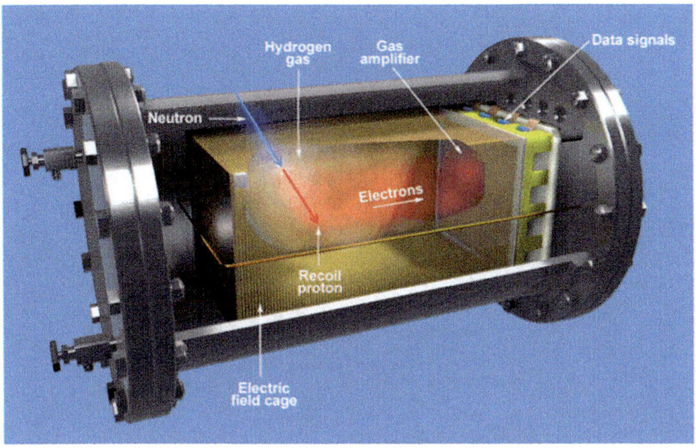

Fig. 3.17 Neutron time projection chamber. Reproduced from Ref. [24]

3.3.2 *Radiation Portal Monitors*

In recent years there has been a growing awareness of illicit trafficking in nuclear materials, which has lead to the development of a *radiation portal monitoring* (RPM) [26] program. During the early 1990s, several cases of trafficking in nuclear material that could be directly used in nuclear weapons, raised concerns that such materials could be obtained by terrorists. The attention to those concerns was greatly increased after the events of September 11, 2001. Combating nuclear smuggling requires continuous monitoring of nuclear material facilities, border locations, smuggling organizations (organized crime), and even entire countries. RPMs play a vital role in combating the illicit trafficking of radioactive material at borders, ports, and other facilities. RPMs must be designed to survive the environmental effects such as severe weather. Rapid decreases in temperature can damage (crack) NaI(Tl) crystals; polyvinyl toluene (PVT), on the other hand, is more rugged and less susceptible to environmental changes. Photomultipliers in scintillation detectors can drift (amplification can change) with a change of temperature, which can affect alarm thresholds. Working environment can be rough near RPMs, which is why they have been designed to have yellow metal or concrete columns protecting them (Fig. 3.18). This is to protect the detectors from being hit by vehicles. Even so, portals are designed to withstand some level of impact, but additional packaging around the portal monitors can affect the minimum detectable activity. RPMs are typically designed to detect both neutron and γ rays and employ packaging of multiple detectors (Fig. 3.19).

RPMs are typically placed at fixed locations, especially at traffic choke points where speeds are reduced while passing through portals. Mobile RPMs have also been developed that can be used to set up a portal at an arbitrary location and can be operated in stationary or a *mobile* mode. They are also appropriate for use at airports, rail yards, and vehicle cargo operations.

Fig. 3.18 Radiation portal monitor at a test location at Pacific Northwest National Laboratory

3.4 Limitations of Passive Measurements

3.4.1 Example of Passive Detection by Gross Radiation Counting

With many of the relevant characteristics of the signal, background, and technology for passive detection discussed, we can now turn our attention to two examples that illustrate the difficulty of passive detection of SNM in a real, shielded environment and in the presence of usual radiation background by gross radiation counting [8]. We consider the detection of 10 kg sphere of weapons grade Pu (wgPU) placed into a center of a 2-m diameter tank of water by placement of a detector at the surface of the tank (at a very optimistic distance of 1 m from the source).

In the first case, we attempt to detect an intense γ signature, which can be simplified by only considering the emission of 3.3×10^{10} γ/s at 414 keV. The emitted γ rays undergo attenuation in the source itself, which reduces the emitted γ rate by a factor of ~300. The water shield further reduces absorbs the γ rays by a factor of $\sim 3.3 \times 10^4$. Finally, the flux of the isotropically emitted γ rays is reduced at a distance of 1 m to result in the signal of ~0.02 γ cm^{-2}s^{-1}. This signal is comparable to cosmic ray muon background.

Fig. 3.19 Packaging of NaI(Tl) detectors into RPM [8]

In the second case, it is attempted to infer the presence of wgPu by detection of excess neutron signal. The spontaneously emitted neutron rate from 10 kg of wgPu is 5.2×10^5 n/s, 99.98% of it coming from the impurities such as ^{238}Pu, ^{240}Pu, and ^{242}Pu. The self absorption in the source reduces the emitted neutron rate by a factor of ~1/1.3, while the thermalization and absorption in the water along with propagation over 1 m distance result in a reduction of the neutron flux by a factor of 10^8, producing a signal of ~0.004 thermal n $\text{cm}^{-2}\text{s}^{-1}$. This is only about 1/5 of the thermal neutron background that arises from cosmic ray interactions.

It is useful to further expound upon the level of difficulty in detecting the other important form of SNM, the HEU. Considering the paucity of the spontaneous radiation signatures from ^{235}U, one can estimate in this situation that the same quantity of HEU is 20 times more difficult to detect than wgPu using γ rays, or a staggering 10^6 times more difficult to detect using neutrons.

3.4.2 Signal to Noise Ratio Scaling

We next consider the scaling of signal to noise ratio (SNR) for several models of the background. In the simplest case of no background being present, the detection is limited only by statistics:

$$SNR = n_s/\sigma_s = \sqrt{n_s}, \tag{3.2}$$

where n_s is the mean number of detected counts and σ_s is the standard deviation of the number of detected counts. Let us now examine a more complex case of a finite, constant, and known background, for which we assume the mean number of detected counts is n_b. In this situation, the SNR is determined to be

$$SNR = \frac{(n_s + n_b) - n_b}{\sqrt{\sigma^2_{n_s+n_b} + \sigma^2_{n_b}}} = \frac{n_s}{\sqrt{n_s + 2n_b}}. \tag{3.3}$$

The most realistic case is that of unknown and variable background and an SNR that is not limited by counting statistics. In this case we have

$$SNR = \frac{n_s}{n_b} \propto \frac{S\epsilon_d A_d}{B\epsilon_d A_d}, \tag{3.4}$$

where S is the signal flux, B is the background flux, ϵ_d is the detector intrinsic efficiency, and A_d is the effective area of the detector. It is apparent in this case that neither the detector size nor its intrinsic efficiency affect the SNR.

Returning to the case of constant and known background, we can consider the effect of detector size and intrinsic efficiency on the SNR when the detector is placed at a distance r from the source:

$$SNR = \frac{s(r)}{\sqrt{s(r) + b(r)}} = \frac{\frac{SA_d\epsilon_d t}{4\pi r^2}}{\sqrt{\frac{SA_d\epsilon_d t}{4\pi r^2} + A_d B\epsilon_d t}} \propto \sqrt{A_d t}, \tag{3.5}$$

where $s(r)$ and $b(r)$ are the detected signal and background counts at distance r and $B(r)$ is the background flux at the same location. As can be seen from this simple analysis, the SNR scales as square root of the detector area, indicating that a larger detector is better in this situation.

These simple arguments underline the fact that simply increasing the detector size may not lead to more reliable detection in the case of unknown, variable background.

3.4.3 Relative Importance of Detector Efficiency, Resolution, and Background

It is useful at this point to consider a more integrated framework that can provide the guidance in the selection of a detector material such that maximum detection sensitivity can be realized. Here we follow the analysis provided by Cooper [27], in which the focus was placed on the detection sensitivity of Ge(Li) γ-ray spectrometers, but with a much broader scope of applicability of the obtained result.

In this analysis, the sensitivity is defined as a measure of the ability to detect a γ-ray peak in the presence of two major backgrounds: natural radioactivity and Compton continuum from higher-energy γ rays produced by the source of interest. The minimum detectable disintegration rate, D_m, is defined as

$$D_m = \frac{N_m}{\epsilon f t}, \tag{3.6}$$

where N_m is the minimum number of the net number peak counts that can be detected (after background is subtracted), ϵ is the detector efficiency for the γ ray which is of interest, f is the fraction of disintegrations that result in emission of that γ-ray energy, and t is the counting interval. One further defines N_m as

$$N_m = A_m \sigma_{N_m}, \tag{3.7}$$

where σ_{N_m} represents the standard deviation of N_m and A_m is the reciprocal of the fractional error. Fractional error is used as a parameter in this analysis and can take a range of values (for example, $A = 100$ corresponds to a 1% fractional error, while $A = 1$ corresponds to a 100% fractional error). The final result of this analysis for a minimum disintegration rate detectable by measuring a peak at an energy E_1 can be written as

$$D_m(E_1) = \frac{A_m}{\epsilon(E_1) f_1 t}\left(\left[2bR(E_1)\left\{\overline{B}_{C_2}(E_1) + \overline{B}_N(E_1)\right\} + A_m^2/4\right]^{1/2} + A_m/2\right). \tag{3.8}$$

Here, $D_m(E_1)$ is the minimum detectable disintegration rate at energy E_1, $\epsilon(E_1)$ is the peak efficiency for detecting γ rays of energy E_1, and f_1 is the fraction of disintegrations that result in emission of the energy E_1. Other quantities in this equation include: b—factor which when multiplied by the resolution gives the number of channels included in the peak; $R(E_1)$—detector resolution at E_1; $\overline{B}_{C_2}(E_1)$—the contribution of the Compton continuum background from all higher-energy γ rays from the source at energy E_1; and $\overline{B}_N(E_1)$—natural background at energy E_1.

As can be seen from Eq. (3.8), the minimum detectable disintegration rate scales approximately as

$$D_m \propto \frac{\sqrt{\overline{B} R}}{\epsilon}, \tag{3.9}$$

where $\overline{B}$ is the average total background. This offers some guidance with respect to selection of a detector for its resolution and efficiency; specifically, the improvement in sensitivity is much more affected by efficiency than by resolution. It is also evident that the reduction of background by improved shielding, anti-coincidence veto systems [28], or Compton suppression systems [13] is less valuable than the improvement of efficiency by the same amount.

To further illustrate these scalings, let us consider the approximate ratio of the minimum detectable disintegration rates that can be achieved by a HPGe detector and a NaI(Tl) detector, where HPGe detector is assumed to have 50% of the efficiency of the NaI(Tl). It is further assumed that the energy resolution of the HPGe detector $R_{Ge} = 1.6\,\text{keV}$, while the resolution of the NaI(Tl) detector is $R_{NaI} = 80\,\text{keV}$. We further assume a uniform background γ-ray field of G γ/(s keV). We obtain

$$\frac{D_{Ge}}{D_{NaI}} = \frac{\sqrt{R_{Ge}G\epsilon}/\epsilon}{\sqrt{R_{NaI}G2\epsilon}/(2\epsilon)} = 0.2. \tag{3.10}$$

The result shows that, despite the fact that a HPGe detector has an energy resolution that is approximately 50 times better than an NaI(Tl) detector, even a modest NaI(Tl) detector with twice the efficiency of a HPGe detector is only 5 times less sensitive. These considerations must be carefully weighted when selecting detectors, shielding, and other background suppression approaches (such as shielding) to maximize the system sensitivity at a given cost.

References

1. D. Reilly, N. Ensslin, H.J. Smith, Passive nondestructive assay of nuclear materials. Technical Report, LA-UR-90-732, Los Alamos National Laboratory (1991)
2. H. Yang, Active interrogation methods for detection of special nuclear material. Ph.D. thesis, University of Michigan (2009)
3. J. Shultis, R. Faw, *Fundamentals of Nuclear Science and Engineering*, 3rd edn. (CRC, Boca Raton, 2016). https://books.google.com/books?id=KaqdDQAAQBAJ
4. J. Phillips, J. Duval, R. Ambrosiak, National geophysical data grids: gamma-ray, magnetic, and topographic data for the conterminous united states. Technical Report, DDS-9, United States Geological Survey (1993)
5. S. Swordy, Space Sci. Rev. **99**(1), 85 (2001). https://doi.org/10.1023/A:1013828611730
6. M.S. Gordon, P. Goldhagen, K.P. Rodbell, T.H. Zabel, H.H.K. Tang, J.M. Clem, P. Bailey, IEEE Trans. Nucl. Sci. **51**(6), 3427 (2004). https://doi.org/10.1109/TNS.2004.839134
7. R.T. Kouzes, J.H. Ely, A. Seifert, E.R. Siciliano, D.R. Weier, L.K. Windsor, M.L. Woodring, J. Borgardt, E. Buckley, E. Flumerfelt, A. Oliveri, M. Salvitti, Nucl. Instrum. Methods Phys. Res. Sect. A **587**(1), 89 (2008). http://dx.doi.org/10.1016/j.nima.2007.12.031. http://www.sciencedirect.com/science/article/pii/S0168900207025090
8. I. Jovanovic, K. Ünlü, Detector and source technologies in nuclear security. Nuclear Security Education Program, Pennsylvania State University (2015)
9. E.C. Directorate, Radiological protection principles concerning the natural radioactivity of building materials. Technical Report, Radiation protection 112, European Commission Directorate-General Environment, Nuclear Safety and Civil Protection (1999)
10. W. Commons, File:radiocarbon bomb spike.svg — Wikimedia commons, the free media repository (2017) Online. https://commons.wikimedia.org/w/index.php?title=File:Radiocarbon_bomb_spike.svg&oldid=251392001. Accessed 7 Aug 2017
11. L.A. Currie, Anal. Chem. **40**(3), 586 (1968). http://dx.doi.org/10.1021/ac60259a007
12. G. Knoll, *Radiation Detection and Measurement* (Wiley, New York, 2010). https://books.google.com/books?id=4vTJ7UDel5IC

13. M. Petra, G. Swift, S. Landsberger, Nucl. Instrum. Methods Phys. Res. Sect. A **299**(1), 85 (1990). http://dx.doi.org/10.1016/0168-9002(90)90752-R. http://www.sciencedirect.com/science/article/pii/016890029090752R
14. K.P. Ziock, J. Collins, M. Cunningham, L. Fabris, T. Gee, J. Goddard, F. Habte, T. Karnowski, IEEE Trans. Nucl. Sci. **55**(6), 3654 (2008). http://dx.doi.org/10.1109/TNS.2008.2007489
15. M.J. Cieślak, K.A. Gamage, R. Glover, Radiat. Meas. **92**, 59 (2016). http://dx.doi.org/10.1016/j.radmeas.2016.08.002. http://www.sciencedirect.com/science/article/pii/S1350448716301524
16. W. Commons, File:HURA hexagonal coded aperture mask principle.svg — Wikimedia Commons, the free media repository (2016) Online. https://commons.wikimedia.org/w/index.php?title=File:HURA_hexagonal_coded_aperture_mask_principle.svg&oldid=191131685. Accessed 7 Aug 2017
17. J.J.M. in't Zand, A coded-mask imager as monitor of Galactic X-ray sources. Ph.D. thesis, Space Research Organization Netherlands, Sorbonnelaan 2, 3584 CA Utrecht, The Netherlands (1992)
18. M. Gmar, M. Agelou, F. Carrel, V. Schoepff, Nucl. Instrum. Methods Phys. Res. Sect. A **652**(1), 638 (2011). http://dx.doi.org/10.1016/j.nima.2010.09.003. http://www.sciencedirect.com/science/article/pii/S0168900210019169. Symposium on Radiation Measurements and Applications (SORMA) XII 2010
19. P. Hausladen, J. Newby, J.F. Liang, M.A. Blackston, The deployable fast-neutron coded-aperture imager: demonstration of locating one or more sources in three dimension. Technical Report, Oak Ridge National Laboratory (ORNL) (2013)
20. H.W. Schnopper, R.I. Thompson, S. Watt, Space Sci. Rev. **8**, 534 (1968). http://dx.doi.org/10.1007/BF00175005
21. D. Xu, Gamma-ray imaging and polarization measurement using three-dimensional position-sensitive cadmium zinc telluride detectors. Ph.D. thesis, University of Michigan (2006)
22. N. Mascarenhas, J. Brennan, K. Krenz, P. Marleau, S. Mrowka, IEEE Trans. Nucl. Sci. **56**(3), 1269 (2009). http://dx.doi.org/10.1109/TNS.2009.2016659
23. J.E.M. Goldsmith, M.D. Gerling, J.S. Brennan, Rev. Sci. Instrum. **87**(8), 083307 (2016). http://dx.doi.org/10.1063/1.4961111
24. N. Bowden, M. Heffner, G. Carosi, D. Carter, P. O'Malley, J. Mintz, M. Foxe, I. Jovanovic, Nucl. Instrum. Methods Phys. Res. Sect. A **624**(1), 153 (2010). http://dx.doi.org/10.1016/j.nima.2010.09.010. http://www.sciencedirect.com/science/article/pii/S0168900210019728
25. K. Walter, Going deep with MEGa-rays. S&T Rev. (2011). https://str.llnl.gov/AprMay11/pdfs/4.11.3.pdf
26. R.T. Kouzes, E.R. Siciliano, J.H. Ely, P.E. Keller, R.J. McConn, Nucl. Instrum. Methods Phys. Res. Sect. A **584**(2), 383 (2008). http://dx.doi.org/10.1016/j.nima.2007.10.026. http://www.sciencedirect.com/science/article/pii/S0168900207022085
27. J.A. Cooper, Nucl. Instrum. Methods **82**, 273 (1970). http://dx.doi.org/10.1016/0029-554X(70)90361-7
28. T. Schroettner, M. Schwaiger, P. Kindl, Appl. Radiat. Isot. **61**(2), 133 (2004). http://dx.doi.org/10.1016/j.apradiso.2004.03.034. http://www.sciencedirect.com/science/article/pii/S0969804304000958. Low Level Radionuclide Measurement Techniques - ICRM

Chapter 4
Foundations of Active Interrogation

Haori Yang

Abstract In this chapter the fundamental basis of active interrogation (AI) as a detection technique is presented. The chapter starts with a brief overview of limitations of passive measurement techniques and how detectability could be improved via AI in many scenarios. The follow-up discussion on AI techniques is limited to those based on induced fission reaction and nuclear resonance fluorescence, because these have shown the most potential for practical implementation in the field to detect special nuclear materials. Availability of new cross section data has recently enabled more reliable simulation of the AI process, beneficial to system design, development and evaluation. Some recent notable achievements are summarized here. In principle, AI techniques are not limited by the weak intrinsic radioactivity of certain nuclear materials. Thus, they could provide improved detectability. However, AI techniques still face many limits, such as signature attenuation, intense active background and interference introduced by the interrogating radiation. These are discussed in detail at the end of this chapter.

4.1 Introduction to the Active Interrogation Technique

AI techniques, in a broad sense, have been used for more than 100 years since the discovery of natural radioactivity. The primary focus of these techniques has been radiographic imaging. Various AI systems have been developed during the last century. Until recently, these techniques have not been widely used except in medical and special industrial applications. The main reasons are the cost and complexity of required equipment, the success of passive detection methods to provide accurate results, and the concerns of radiation safety issues related to the deployment of radiation emitting systems outside of lab-coat environments.

H. Yang (✉)
Oregon State University, Corvallis, OR, USA
e-mail: Haori.Yang@oregonstate.edu

I. Jovanovic, A. S. Erickson (eds.), *Active Interrogation in Nuclear Security*,
Advanced Sciences and Technologies for Security Applications,
https://doi.org/10.1007/978-3-319-74467-4_4

In the last decade, especially after 9/11, homeland security has become a dominant concern. Among all the possible terrorist tactics, those involved with high-Z fissile materials seem to evoke the most concern, probably because of the small volume of material needed to make a catastrophic impact. Ninety percent of the trade goods brought into the U.S. each year enter through its 361 seaports [1]. The west coast ports of Los Angeles, Long Beach, Oakland, and Seattle are currently processing more than 11,000 containers per day, or 8 per minute on a 24/7 basis. The concern is that terrorists could use any one of these containers to smuggle special nuclear material (SNM) into the country. Currently, less than 2% of these containers are ever opened and inspected by customs service. On the other hand, based on an OECD report, a successful attack at a seaport would shut down the entire maritime shipping system for a period up to 10 days and would produce an economic loss of $58 billion in the U.S. alone [2]. Thus, the ultimate goal is for every container entering the U.S. to be screened using techniques having an extremely high probability of detecting concealed SNM. It remains a technical challenge to reliably detect intentionally concealed highly enriched uranium (HEU) with passive techniques. The potential for AI to address this longstanding concern has drawn significant resources and attention to this area of research and development.

In addition to cargo screening, AI techniques have also been developed for nuclear safeguards and forensics, such as nuclear material accountancy (NMA) and treaty verification. In these applications, the goal is not to detect the presence of SNM, but to quantify or characterize SNM samples. Such a measurement is commonly referred to as *active non-destructive assay*. Samples under assay are normally of known characteristics. Shielding might be present for radiation protection purpose, rather than source concealment. Passive techniques for quantifying plutonium based on its neutron emission have long been developed and implemented for safeguards. Measuring neutrons emitted by uranium is far more challenging due to the low spontaneous fission rate. Thus, to provide a similar capability for uranium normally requires the introduction of an interrogation source to induce fission [3]. AI is normally utilized to increase signal to a detectable level unachievable while solely relying on spontaneous emission. Contrary to this, it has also been studied for assay of spent nuclear fuel. In this case, passive detection is paralyzed by the high radiation field produced from fission products. Active techniques involving pulsed interrogation sources provide a way to improve signal to noise ratio via temporal discrimination. Examples include measurements of delayed γ rays and neutrons following induced fission and induced fission rate measured in a lead slowing down spectrometer (LSDS) [4, 5].

Recent development of AI techniques targeting SNM has focused on detecting signature signals from induced fission reactions with either fast neutron or high-energy photon sources. With cargo screening being the major driving force, the priority here is to achieve high detection sensitivity and selectivity in a reasonably short time period with consideration of a lot of unknowns, such as shielding around the target and attenuation by cargo. Techniques based on induced fission enable much better detectability by enhancing fission signatures beyond spontaneous emission. In addition to detecting the presence of SNM, isotopic analysis is achievable, for example, through measurement of fission product yields. Another AI technology worth further discussion is nuclear resonance fluorescence (NRF). It is the nuclear analogy of atomic fluorescence. Being isotope-specific, this method potentially provides higher fidelity in detection and characterization. Currently, its application is hindered by the lack of bright tunable monoenergetic photon sources. However, such sources are under active development and would make NRF a very attractive technique once become available [6]. In a broader sense, AI could include any measurement involving an external source of ionizing radiation, such as X-ray or neutron radiography. However, such techniques merely rely on changes in the interrogation radiation (e.g. intensity, energy, direction) and not on signatures resulting from induced reactions inside the target. The following discussion in this chapter will thus be limited to induced fission and NRF only.

4.2 Impact of Active Measurements on Detectability

4.2.1 Passive Techniques and Their Limits

Passive nondestructive detection is based on the detection of naturally emitted γ rays and/or neutrons from SNM. Because of the absence of an interrogating radiation source, the passive techniques are much simpler to deploy. Detection systems utilizing this approach deliver no radiation dose to the target.

As a result of the spontaneous α or β decay in heavy nuclei, γ rays are emitted during the de-excitation process. These γ rays have energies that are characteristic of the source nuclei. Some of these γ rays have sufficient energy to penetrate fissionable materials and potential shielding around them. Detection of these γ rays is the basis of passive γ-ray assay. The energy information can be used to distinguish these γ rays from signals belonging to the natural background. In addition, isotope identification is possible via γ-ray spectroscopy. The net number of counts is interpreted through a predetermined calibration to estimate source activity present in the sample. Such passive methods have found applications in the assay of materials in DOE waste containers [3]. They are also utilized in radiation portal monitors developed for security applications, for example, RadSentry by Canberra, AT-900 by SAIC and Detective by ORTEC, together with passive neutron detection techniques.

A transmission density measurement is usually used as a supplement when an accurate γ-ray assay is needed. In this measurement, the sample is irradiated by an external source of photons of energies similar to the energies of interest. The measured attenuation of the photons by the object is used to correct the γ-ray assay results for the expected loss of emitted photons. This can extend the applicability of passive γ-ray assay to denser materials, through which only a fraction of the spontaneously emitted γ rays can reach the detectors. If the object under investigation is close to the detector, it is usually rotated during the scan to improve the uniformity of the sample response. Otherwise it is placed at a sufficient distance from the detector so that the source position variations can be ignored.

Various types of detectors find applications in passive γ-ray detection. These range from detectors with medium energy resolution (e.g., NaI(Tl), CdTe, BGO detectors) to ones with excellent energy resolution (e.g. HPGe detectors). Although passive γ-ray assay can be used for identification and quantification for certain isotopes, such as ^{238}U and ^{238}Pu, there are cases where its application is impractical. For example, shielded HEU is extremely hard to detect using passive γ-ray techniques. The main spectral feature in the bare HEU γ-ray spectrum is the photopeak from ^{235}U at 185 keV. When lead shielding is placed between the HEU sample and the detector, this 185 keV peak is greatly attenuated. The primary spectral feature left is due to the isotopic impurity of ^{232}U. ^{232}U decays rapidly and enters the thorium decay chain. The final nuclide in this chain is ^{208}Tl, which contributes to the photopeak at 2615 keV. These 2615-keV γ rays are very penetrating, thus easy to detect using passive detection techniques. However, its intensity is low and it is part of the natural background in the environment.

Neutrons are emitted as a result of spontaneous fission or as a result of the (α, n) reaction between the α-particles and the nuclides of low atomic number in the matrix material. These neutrons typically have high energies and can penetrate the object and shielding with a high probability. In passive neutron detection methods, these neutrons are detected by multiple detectors surrounding the container. Since neutron energies are not isotope specific, energy information is usually not measured in such techniques. In total neutron counting, only gross rate is measured to determine the gross content of neutron-emitting nuclides. The amount of a particular nuclide can still be calculated given its specific neutron emission intensity and relative abundance. This technique is limited to the detection of ^{238}U and some plutonium isotopes, where specific activity of spontaneous fission is high. The abundance of elements in the matrix materials surrounding the SNM that can be potential targets for (α, n) reaction will also affect the gross neutron counts.

In passive neutron detection, the coincidence technique is often used to distinguish fission neutrons from other sources, including (α, n) reactions in the matrix material. During a spontaneous fission event, ~2–3 neutrons are usually emitted simultaneously. In contrast, neutrons from other sources are emitted at randomly spaced intervals. With a well-designed setup, the probability of detecting multiple fission neutrons within the coincidence time is relatively high. On the other hand, the probability of detecting an (α, n) neutron or background neutron in this time interval is no greater than it would be in any other time interval of the same duration. Thus, detectors working in the coincidence mode have a preference to detect fission neutrons. The two types of detectors mainly used in passive neutron detection are gas filled detectors and plastic detectors. When plastic detectors are used, fission neutrons and γ rays are both counted with high probability but can be separated via pulse shape analysis [7]. Plastic detectors, like most organic scintillators, have several advantages over other neutron detectors, such as faster time response. However, they usually have poorer stability caused by gain drift in the PMT and variation in temperature. They are also less rugged and sensitive to γ rays. Passive neutron detection techniques are usually used together with passive γ ray detection techniques. It has found applications in waste assay and security applications [8].

Table 4.1 Natural radiation from uranium and plutonium isotopes

Nuclide	Decay mode	Radiation, type	Intensity (s^{-1} g^{-1})
^{235}U	α	185.7 keV, γ-ray	4.3×10^4
^{238}U	α particle	1001.1 keV, γ-ray	1.0×10^2
		766.4 keV, γ-ray	3.9×10^1
	Fission	Fission neutrons	1.7×10^{-2}
^{238}Pu	α particle	766.4 keV, γ-ray	1.5×10^5
		152.7 keV, γ-ray	6.5×10^6
	Fission	Fission neutrons	2.6×10^3
^{239}Pu	α particle	413.7 keV, γ-ray	3.4×10^4
		129.3 keV, γ-ray	1.4×10^5
^{240}Pu	Fission	Fission neutrons	1.0×10^3
^{241}Pu	β particle	208.0 keV, γ-ray	2.0×10^7
		164.6 keV, γ-ray	1.8×10^6
		148.6 keV, γ-ray	7.5×10^6
^{242}Pu	Fission	Fission neutrons	1.7×10^3

Although most SNMs naturally emit neutrons and/or γ rays, the intensity of the spontaneous radiation is low, and the energies of the γ rays are fairly low in most cases, as shown in Table 4.1. In security applications, it is prudent to assume that the SNM is well shielded to circumvent passive detection. Detecting well shielded SNM using passive techniques, with high detection sensitivity and selectivity in a reasonably short time period remains a major technical challenge.

4.2.2 Improved Detectability by Active Interrogation

Both neutrons and γ rays can induce fission in SNM. On average, two to three energetic prompt neutrons and about eight prompt γ rays are produced in each fission reaction. These radiations are emitted within 10^{-15} s of the time of fission. The fission products continue to emit another six to seven γ rays and approximately 0.01 to 0.02 neutrons per fission. This radiation from the decay of the fission products is called delayed fission radiation. Both prompt and delayed fission radiation can be detected as unique signatures indicating the presence of SNMs.

A variety of active nondestructive assay techniques can be formulated by combing different incident radiations and observed radiations. Many are being explored and seem promising for security and safeguards applications. AI methods, as compared to passive methods, offer the advantages of adjustable energy and intensity of the interrogating radiation to provide optimal response; a higher intensity of the radiation from induced reactions; and the techniques are applicable to all fissile and fissionable materials. Most AI techniques rely on the detection and measurement of the delayed radiation, although prompt radiation is much more abundant. Techniques that depend on prompt radiation detection have a much higher sensitivity. However, if prompt radiation is to be detected, the fission neutrons and γ rays have to be distinguished from the interrogation radiation. While the interrogation radiation emitted by the neutron or γ-ray source is usually collimated, it can be scattered into the detector by shielding material, the surrounding environment, or the SNM itself. Discrimination is usually done based on the different energies of the incident and fission radiation, or the different response of the detectors to different particles. It is also possible to distinguish induced fission radiation using coincidence counting, similar to that in passive interrogation techniques. The detectors have to be well shielded from the source so that accidental coincidences do not create a large background.

4.3 Technology for Active Measurements

4.3.1 Induced Fission

Nuclear fission induced by γ rays or neutrons is the most important multiplicative process involved in nondestructive AI. The process produces the most easily recognizable signature unique to SNM and actinides. Neutron-induced fission cross sections are on the order of barns within the energy range commonly utilized in AI, i.e. between D-D and D-T neutron energies. The cross sections remain the same magnitude down to thermal energies for fissile materials, but drop quickly below 1 MeV for fissionable isotopes. In addition to neutrons, high-energy γ rays can also excite a nucleus and cause fission, in a process known as photofission. Photofission cross sections are normally an order of magnitude smaller than those the (n,f) reaction. A wide resonance exists between 10 and 15 MeV as the result of giant dipole resonance in actinides. Below roughly 5 MeV, the cross section values become too small for practical applications. In spite of the smaller cross section, photon interrogation is as effective as neutron activation if not more. The total induced fission reaction rate in the sample is the convolution of energy-dependent cross section and beam flux. Electron linear accelerators can usually produce much higher photon fluxes than neutron generators. In addition, high-energy photons can also liberate neutrons from high-Z samples via (γ,n) reactions, which further increases induced fission rate.

4.3.1.1 Fission Signatures

During the induced fission process, the excited nucleus splits into two large fragments following the absorption of a neutron or a high-energy photon. At the same time, two to three prompt neutrons and several γ rays are emitted. Prompt radiation results from particles liberated during the scission, while delayed radiation originates from subsequent decay of fission products. Fission product yield as a function of atomic number is isotopic specific. Thus, energies and decay constants of delayed signature signals could theoretically serve the purpose of isotope identification and quantification.

Compared to the energy ($\sim$200 MeV) released during the fission process, the energy of the interrogation radiation is very small. Thus, neutrons and γ rays emitted from neutron- or photon-induced fission reactions are almost identical to those from spontaneous fission. The signatures emitted following neutron-induced fission are relatively better studied and more experimental data is available. For photofission, currently the majority of the existing data is based on nuclear models and there have been limited experimental measurements performed to date. A common means for approximating emissions following photofission reactions is to utilize the reaction-channel analogs. After formation of the excited nucleus, the remaining energy above the fission barrier can be randomly distributed between the different states without restriction. As a result, the fissioning system has no memory on the configurations before the barrier as long as the general conservation laws are fulfilled. Thus, for calculations and modeling purposes, the photon-induced fission of a nucleus with atomic number A can be approximated with the neutron-induced fission of a nucleus of the same element but with atomic number $A - 1$.

Prompt neutrons could be used as the signature signal for SNM detection, when it is possible to distinguish them from the probing irradiation and the neutrons produced in surrounding matrix materials. The distinction can be made on the basis of the energy difference between fission neutrons and source neutrons or based on the high multiplicity of the fission neutrons. The average number of prompt neutron emitted per induced fission is higher than that for spontaneous fission and increases with the energy of the incident neutron. The energy spectrum of prompt neutrons from induced fission is largely similar to that from spontaneous fission.

The yields and energy spectra of the prompt fission neutrons from photofission are not as well-known as those from its neutron-induced counterpart. The prompt neutron yield observed in photon interrogation is typically the sum of the fission neutrons and the photoneutrons from (γ, n) reactions. At low incident photon energies, large difference in prompt neutron yields were reported among various isotopes [9–11]. These observations could be useful for isotope identification.

For several reasons, delayed neutrons may be better suited as signatures than prompt neutrons. The advantage of the delayed neutrons over the prompt neutrons is that they can be detected sometime after fission reactions. Hence, they can be distinguished from source neutrons and photoneutrons using time discrimination. Additionally, the value and the temporal behavior of the delayed neutron yield vary significantly among different isotopes, which make material discrimination possible. Conventionally, delayed neutrons are registered into six groups based on their decay constants. The individual group yields present significant variance among different isotopes. The absolute delayed neutrons yields are very small compared to those of prompt neutrons, ranging from 0.0065 per fission in ^{239}Pu to about 0.045 per fission in ^{238}U.

In addition to prompt and delayed neutrons, γ rays emitted during the fission process can also be utilized for SNM detection. When a photon interrogation source is used, prompt γ rays are buried by the much more intense probing radiation. In neutron interrogation, prompt fission γ rays can also be largely interfered by γ rays following inelastic neutron scattering and neutron absorption. Alternatively, delayed γ rays are more practical for detection purpose. Delayed γ rays are emitted seconds or even minutes after the irradiation. Thus, they are not likely affected by the probing radiation. The total energy carried out by delayed γ rays per fission is about 6–8 MeV. Approximately 6–8 delayed γ rays are emitted following each fission process, with an average energy of 1 MeV. This yield is over 100 times bigger than that of delayed neutrons. As a result, delayed γ rays are more attractive than delayed neutrons for the purpose of detection due to their higher intensities and longer half-lives. However, it is worth mentioning that the ambient background is significant higher in delayed γ-ray measurements. Due to the difference in attenuation mechanism, delayed γ rays may or may not have a better chance to reach the detectors.

It is important to note that unique γ-ray intensity distributions exist for each fissionable isotope [12, 13]. These can be used to identify nuclear materials. The delayed γ-ray energy spectra are rich and complex, but it is possible to resolve individual lines with high resolution γ-ray spectrometers. The relative amplitudes of certain lines vary significantly among different isotopes. Discrimination ratios on the order of 3 were observed between ^{235}U, ^{239}Pu, and ^{238}U [13, 14]. Some examples are shown in Table 4.2.

4.3.1.2 Interrogation Sources

The most popular interrogation radiations are neutrons and photons. Both can be produced by either radioisotope sources or charged particle linear accelerators. Radioactive isotope γ-ray sources are widely used in transmission radiography applications. The γ-ray energies from radioactive decay are high enough to penetrate objects for imaging, but not high enough to induce fission in heavy

Table 4.2 Intensity ratios of the fission product γ rays induced by 10-MeV bremsstrahlung photons and measured between 13 and 100 ms after the irradiation

Fission product	Gamma-ray energy (keV)	Intensity ratios			
		^{239}Pu	^{238}Pu	^{235}U	^{232}Th
^{138}Cs	1436.0	100.0	100.0	100.0	100.0
^{94}Sr	1428.3	136.4	235.7	186.8	171.3
^{136}I	1313.0	72.7	240.5	123.0	109.3
^{89}Rb	1248.2	13.1	85.4	92.7	228.0
^{90}Kr	1118.7	39.5	110.5	108.5	171.0
^{97}Y	1103.0	113.7	148.1	115.2	68.9
^{89}Rb	1031.9	66.5	116.4	135.9	265.1

Table 4.3 Photoneutron and photofission thresholds for materials of interest

Material	Isotope	Natural abundance (a%)	Photoneutron threshold (MeV)	Photofission threshold (MeV)
H	^{2}H	0.01	2.22	2.22
C	^{12}C	98.93	18.72	
	^{13}C	1.07	4.95	
N	^{14}N	99.63	10.55	
	^{15}N	0.37	10.83	
O	^{16}O	99.76	15.66	
	^{17}O	0.04	4.14	
	^{18}O	0.2	8.04	
Al	^{27}Al	100	13.06	
Fe	^{54}Fe	5.85	13.38	
	^{56}Fe	91.75	11.20	
	^{57}Fe	2.12	7.65	
	^{58}Fe	0.28	10.04	
Pb	^{204}Pb	1.4	N/A	
	^{206}Pb	24.1	8.09	
	^{207}Pb	22.1	6.74	
	^{208}Pb	52.4	7.37	
U	^{234}U	0.0055	6.84	5.06
	^{235}U	0.72	5.3	5.31
	^{238}U	99.27	6.15	5.08
Pu	^{239}Pu	0	5.65	5.31
Th	^{232}Th	100	6.44	5.40

isotopes. Linear accelerators (linac) with bremsstrahlung targets are commonly used as interrogation photon sources. They extend the probing photon energy well beyond photofission threshold and into the giant dipole resonance region. Bremsstrahlung photons have a continuous spectrum. The maximum photon energy is the energy of the electrons leaving the linac. For a reasonable portion of the generated photons to induce fission, the endpoint energy has to be well above the fission threshold. However, photoneutrons will inevitably be produced from (γ,n) reactions in benign materials under this condition. Table 4.3 shows threshold energies of photofission reaction in heavy metals and threshold energies of (γ,n) reactions in materials of interest [15]. As a result, fission is generated inside SNM samples both directly via photofission and indirectly by photoneutrons. In the United States, regulations and standards exist providing guidelines regarding the maximum allowed photon energy in certain cases, such as 21 CFR 179.21 and ANSI/HPS N43.14. Linacs usually work in pulse mode with typical pulse widths of a few μs. Normally, delayed radiation is measured to avoid the interference from probing photons and photoneutrons produced directly by the bremsstrahlung photons. Certain nuclear reactions can produce better-defined photon spectra with sufficient energy to induce fission. For example, the ~6–7 MeV γ rays generated through the $^{19}F(p,\alpha\gamma)^{16}O$ reaction have been studied as the probing source for photofission [16].

Neutron sources commonly used in AI include (1) spontaneous fission sources, e.g. ^{252}Cf, (2) (α,n) sources, e.g. Am-Li, Am-Be sources, (3) photoneutron (γ,n) sources, utilizing either radioactive isotopes such as an Sb-Be source, or an electron accelerator with a neutron converter, and (4) reactions from accelerated charged particles, for example, (d,d), (d,t), (d,n), and (p,n) reactions. Among all these options, D-D and D-T neutron generators have probably seen the most use in the field, because of their compact size, low cost and low power consumption. The energy of the probing radiation is very important in the AI techniques, since both the fission cross section and the attenuation of the radiation are energy dependent. For example, neutrons with higher energies are more penetrating and give a more uniform response throughout the target. However, the fission cross section decreases with energy between 1 and 5 MeV in fissile materials, which could then lead to a lower reaction rate. Photoneutron sources produce neutrons whose energies can be adjusted by varying the energy of the incident photons. Similarly, the energy of neutrons from (p,n) or (d,n) reactions can be adjusted by using different targets. Neutron energy can also be modified by adding materials between the interrogation source and the sample, to either cutoff low energy neutrons, or slow down high energy neutrons.

4.3.1.3 Techniques Involving Detection of Delayed Radiation

A unique signature of SNM is the delayed neutron emission following a fission reaction. A fission reaction can be induced by interrogating γ rays or neutrons. A few systems utilizing accelerators have extended the interrogation γ-ray energies above the photofission threshold ($E_\gamma = 6$–7 MeV) to deliberately induce fission in SNM. Some work in this area has been done by researchers at Idaho National Laboratory (INL), using a tabletop electron linac with endpoint energy up to 11 MeV [17]. A similar study was carried out by researchers at Los Alamos National Laboratory (LANL) using a linac at 6, 8, and 11 MeV endpoint energies [18].

In addition to delayed neutrons, delayed γ rays are also observed following a fission reaction. These delayed γ rays are emitted during the beta decay of many short-lived fission products. Some of these delayed γ rays have relatively high energies, thus are easy to distinguish from the ambient background. The delayed γ-ray signature has some advantages over the delayed neutron signature. Firstly, the intensity of the high energy delayed γ rays ($E_\gamma > 3$ MeV) is about ten times larger than that of delayed neutrons. Secondly, the high energy γ rays are highly penetrating in hydrogenous cargo. They typically encounter 100–100 times less attenuation than delayed neutrons [19] Thirdly, intense high energy delayed γ rays are a unique signature for SNM. They are hardly observed after the irradiation of benign materials or in natural background, nor are they produced in significant amount by neutron activation of the cargo and surrounding materials.

Delayed fission γ rays following neutron-induced fission was investigated by E. Norman and S. Prussin as a unique signature for SNM [20]. Researchers from LLNL developed a *nuclear car wash* system to detect concealed SNM in cargo containers based on this signature [21]. In this system, a D-D neutron source is used as the interrogating radiation source to induce fission in SNM. High-energy delayed γ rays between 2.5 and 6 MeV emitted by fission products were collected by a large array of scintillation detectors. Their results showed that the interrogating neutron energy should be kept below 10 MeV. Otherwise, the interference caused by the activation of ^{16}O to produce ^{16}N becomes important. The 7-second β-decay of ^{16}N produces a 6-MeV γ ray which could possibly cause false alarms in a system based on delayed γ-ray detection. Their work also showed that unwanted collateral effects of the interrogation such as neutron activation are trivial, even in the case when 14-MeV neutrons were used and not moderated.

4.3.1.4 Techniques Involving Detection of Prompt Radiation

Whether induced by photons or neutrons, prompt energetic γ rays and neutrons are emitted within 10^{-15} s after the fission. For thermal neutron induced fission in ^{235}U, ~2.5 prompt neutrons and ~7 prompt photons are emitted on average. Prompt neutron and γ-ray yields from photofission reactions are of similar magnitudes. When a pulsed mode interrogation source is used, prompt radiation can also be produced between irradiation pulses from fission induced by delayed neutrons. However, since only 0.0158 delayed neutrons are emitted following a thermal fission in ^{235}U, this type of prompt radiation emission is rare [19]. In photofission reactions, the ratio of prompt to delayed neutrons increases with incident photon energy.

A technique utilizing prompt neutron detection can dramatically reduce the time and radiation dose required for a complete scan. A shorter scan time means a higher throughput of the inspection system. However, prompt radiation detection faces measurement challenges arising from the interference of the intense interrogating radiation. The magnitude of the prompt radiation is much smaller compared to that of the probing radiation. Nonetheless, with recent breakthroughs in fast data acquisition techniques, utilization of prompt radiation has become more feasible. Detection based on prompt radiation has been successfully performed by researchers at INL and Idaho Accelerator Center (IAC) [22, 23]. They developed a state-of-the-art data acquisition system to handle the extremely high event rate during interrogating pulses. The digitizer they developed has a sampling rate of 2 GS/s. Using an ultra-fast SiC detector (sub-nanosecond rising time, insensitive to photons), signals from individual prompt neutrons are recorded during each X-ray pulse. Using a fast EJ-200 (BC-408 equivalent) plastic scintillator, both prompt neutrons and prompt γ rays were measured. Integration of the total counts over time successfully differentiated depleted uranium (DU) from bismuth or when nothing was present in the beam.

In addition to the work described above, prompt neutrons have also been measured as the signature signal for detection of SNM by researchers at Argonne National Laboratory (ANL). They investigated a system called FIGARO (fissile interrogation using γ rays from oxygen) [16, 24, 25]. In this technique, ~6–7 MeV γ rays produced by the ^{19}F(p,$\alpha\gamma$)^{16}O reaction are used as the interrogating radiation. Since the photoneutron emission threshold is less than 6 MeV in nuclear materials but greater than 7.5 MeV in common materials, the (γ,n) background is low. This allows counting of prompt neutrons during irradiation. The smaller photofission cross section at ~6–7 MeV, compared to higher energies, is compensated by the much bigger intensity of prompt neutrons compared to delayed neutrons and the very low (γ,n) background. Their research has shown that this technique is a robust method for detection of SNM with a low false positive rate. Effort continues to minimize the neutron background, thus increase the system sensitivity.

4.3.2 Nuclear Resonance Fluorescence

It is desirable for detection systems to be able to not only detect the presence of SNM but also identify the specific materials. This can be done with induced fission techniques based on differences in fission product yields [13, 14, 26, 27], or possibly on differences in the temporal behavior of delayed neutrons and γ rays [9, 28–30]. However, these approaches either require complicated spectrum analysis, or dual or multiple beam energies, and the accuracy can easily be affected by attenuation due to surrounding materials. Conventional atomic X-ray imaging systems provide gross sensitivity to atomic number. These systems are based on the principle that materials preferentially attenuate or scatter photons with specific energies determined by atomic composition. The binding energy of the most tightly-bound K-shell electrons is on the order of 100 keV. The energies of the X rays have to be small enough to investigate the atomic structure. However, X rays with such low energies suffer large attenuation thus cannot penetrate objects with high density. This limits the X-ray fluorescence technique to surface surveys [31]. Increasing the energy of the interrogation X rays will make them more penetrating, but the attenuation coefficients are simply a function of electron density at these energies, and the gross sensitivity to atomic number is lost.

NRF can be considered as the nuclear analog of the atomic X-ray fluorescence process [32]. It is the process of resonance excitation of certain nuclear levels by absorption of photons and subsequent decay of these levels by re-emission of equivalent radiation. Because the energy level structure is unique for each isotope, the resonant energies can be used as fingerprints for isotope identification. Thus, the energy spectrum of the resonantly scattered photons can be used to identify the detailed isotope content of materials of interest. Since every nuclear isotope with an atomic number bigger than two has energy levels that can be excited by photon absorption, detection systems based on NRF interaction are not limited to SNM, finding wide applications in material characterization, waste management, and other areas. For homeland security, it may be useful to identify threats such as conventional explosives, SNM, toxic materials, and chemical weapons. Compared with atomic fluorescence, NRF photons have much higher energies. These multi-MeV photons are highly penetrating and easy to detect.

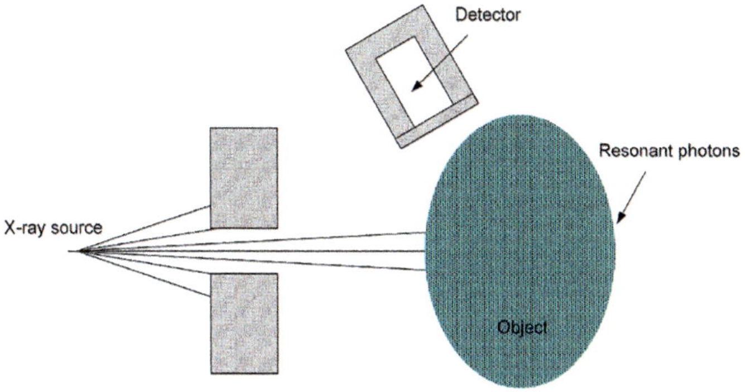

Fig. 4.1 Scattering detection method using NRF [33]

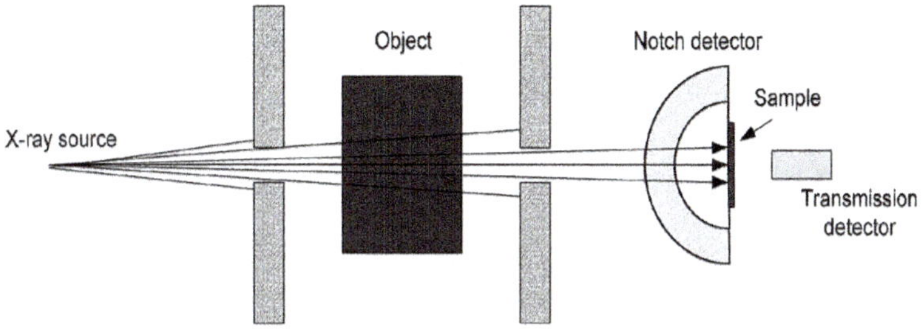

Fig. 4.2 Self-absorption detection method using NRF [33]

In a detection system based on NRF, interrogation photons are used to induce resonant excitations in nuclei, and the de-excitation photons are measured to generate an energy spectrum. The two most commonly used detection schemes are the scattering method and the self-absorption method. In the scattering method, materials of interest are placed in an X-ray beam, and detectors are located out of the beam using a backscattering geometry as shown in Fig. 4.1. All but the resonantly scattered photons at this angle are expected to have low energies, owing to the nature of Compton scattering. Thus, resonant photons can be distinguished on the basis of their energies. In the self-absorption method, the X-ray beam goes through the object, after which it strikes a sample containing isotopes of interest. A transmission detector is located downstream in the beam line to monitor the flux of the off-resonance photons. A radiation detector, called the *notch detector* is located out of the beam and is pointed at a sample of the isotope of interest in a backscattering geometry, measuring the resonant photons scattered by this sample. The system setup is shown in Fig. 4.2. The count rate under the resonant peaks registered in the notch detector is proportional to the flux of the resonant photons hitting the sample. Whenever a preferential attenuation of resonant photons is observed, the conclusion can be drawn that isotopes of interest are present in the object. This is measured as a disparity between the flux of off-resonant photons and that of resonant photons registered in the transmission detector and the notch detector separately.

One feature of NRF that makes it a powerful tool for isotope identification is the sharp profile of the NRF cross section. The width of the cross section is typically on the order of 100 meV, while the peak value is 100's of barns for strongly resonant levels. The resonance will spread over an energy range of several eV owing to Doppler shifts caused by thermal motion of the nuclei. Thus, the peak value of the cross section will be reduced to a few barns. This is usually comparable to or larger than the cross section for photo-atomic reactions. Such a distinctive resonance makes it easy to detect even a small quantity of material of interest. On the other hand, this feature also puts constraints on the implementation of the NRF technique. In most nuclear transitions the γ-ray lines are very narrow (meV), and the recoil energy loss effectively separates the absorption and emission lines. Thus, the resonance condition, i.e. the required overlap between the emission and absorption lines, is hard to fulfill when the isotope used as the source of the radiation is the same as that in the scattering or absorbing materials. Since the failure to observe resonance in the first NRF experiment by Kuhn in 1929 [34], many unsuccessful attempts to find wide γ-ray lines occurred during a time period of 20 years [35–38]. Another approach, fulfilling the resonance condition by creation of a special source (compensating recoil energy loss using a centrifuge method) was taken by Moon. He successfully observed the resonance level at 411 keV in ^{198}Hg using a ^{198}Ag source [39].

Bremsstrahlung X rays have become a widely used excitation photon source since their discovery. Schiff first proposed using electron bremsstrahlung to detect NRF in 1946, 5 years before Moon's NRF experiment [40]. The continuous energy distribution of bremsstrahlung sources overcomes all problems caused by recoil energy loss. However, the resonance spread (on order of eV) is about one millionth of the energy spread of a bremsstrahlung source. This means that less than one-millionth of the total photon flux contributes during the excitation process when a bremsstrahlung source is used. The rest generates unwanted background and noise. The source has to be strong enough to provide significant flux in a very narrow energy range. Detectors used to measure the resonant photons have to have high energy resolution. The resonant photons are almost monoenergetic, with an energy spread (eV) thousands of times smaller than the best energy resolution that modern radiation detectors can provide (keV). These resonant photons sit on a continuous background. Thus, only excellent detector energy resolution can ensure good signal-to-noise ratio. HPGe detectors provide the best energy resolution among commonly available radiation detectors. HPGe detectors with high detection efficiency are currently the best detectors for NRF experiments. In contrast, recently developed lanthanum halide scintillation detectors offer good energy resolution and a fast decay constant. The fast primary decay allows the application of these detectors at an event rate much higher than what HPGe detectors can handle. And the energy resolution is sufficient to detect strong resonances from certain isotopes.

NRF has become a popular technique to investigate composite nuclei, because both the excitation and de-excitation processes in this method only involve the electromagnetic interaction, which is the best understood interaction in nuclear physics. This method is capable of extracting information about excited states in nuclei, including their energies, lifetimes, and also angular momenta. Hence, the technique has found favor among physicists studying nuclear excited states.

Excited nuclear energy levels are unique to each isotope. Energies of γ rays emitted from these levels are often used for isotope identification. NRF seems to be an ideal tool for this with the ability to distinguish the isotope content of materials. However, despite the obvious advantage, no one has applied it in homeland security applications until recently, because of the difficulty of fulfilling the resonance condition and achieving reasonable signal-to-noise ratio in the field. Bertozzi proposed a novel approach for applying NRF to the detection of concealed or illicit materials [41]. His method involves exposing materials to a bremsstrahlung X-ray beam and detecting the resonantly-scattered photons, which have an energy spectrum unique to each isotope. The linac used in this method can produce bremsstrahlung X rays with energies ranging from 2 to 8 MeV. These interrogating photons are highly penetrating and have the potential to investigate the whole body of a cargo container. In the proposed design, scattered photons are measured using a segmented and collimated HPGe detector array. The intersection of the field of view of each detector with the beam forms one voxel of the total image. A complete scan could produce a map of isotope content in the entire cargo. Following this, Warren at PNNL measured NRF from a ^{235}U sample [42]. Researchers at LLNL studied the performance of detection systems based on NRF to detect specific isotopes. They presented false positive/negative error rates, dose delivered to the cargo, and also the detection sensitivity of these systems. The results showed a strong relationship between system performance and the properties of the interrogation photon sources. Photon sources with high intensity and a narrow energy spread showed a clear advantage over traditional bremsstrahlung sources. Thomson-upscattering (inverse Compton scattering) between energetic electrons and laser photons were proved to be the most promising approach to tunable monoenergetic photon sources. This class of source is available, but usually on a scale too large for practical deployment at ports [43]. An LLNL group is devoted to the miniaturization of such a source for homeland security purpose. Detection systems using this source based on NRF technique are also under development at LLNL [44, 45].

4.3.2.1 NRF Cross Section

In this section, general considerations related to NRF cross sections are discussed. The formulas derived here help to understand the nature of the interaction and can be used in the estimation of interaction rates. The resonance cross section is related to the width of the excited energy levels, the spin of the initial and final levels, and the excitation energy. A very narrow cross section will be broadened owing to the Doppler effect caused by thermal motion. However, the integrated cross section remains intact. Angular distribution of the resonant photons is determined by the spin of the initial and final states.

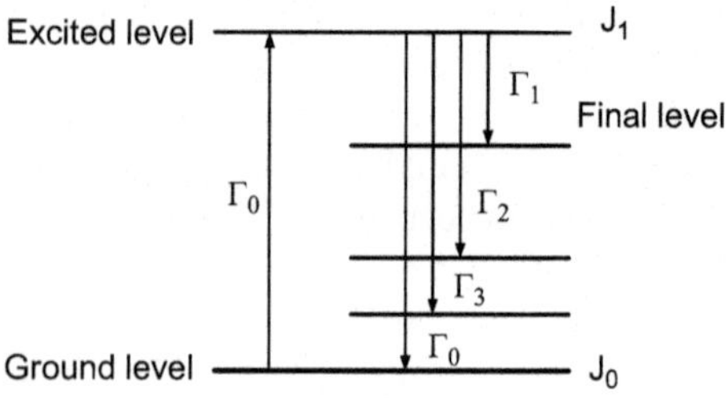

Fig. 4.3 Gamma transitions inside a nucleus [33]

Breit-Wigner Formula

The shape of the NRF cross section can be best described using the Breit-Wigner formula [46]. For a photon of energy E, the NRF cross section, i.e. the probability to excite a nucleus from a ground state with total angular momentum J_0 to an excited state with energy E_r and total angular momentum J_1, and then de-excite to a final state with total angular momentum J_1 can be written as

$$\sigma_1^0(E) = \pi \lambda\!\!\!^{-2} \frac{2J_1 + 1}{2(2J_0 + 1)} \frac{\Gamma_0 \Gamma_1}{(E - E_r)^2 + \Gamma^2/4}, \tag{4.1}$$

where Γ is the total width of the excited state, i.e. the sum of partial widths for all possible transitions to lower states, Γ_0 is the partial width for transition from the excited state directly to the ground state, Γ_1 is the partial width for transition to the final state, and $\lambda\!\!\!^{-}$ is the wavelength of the resonant photon divided by 2π. The levels are shown in Fig. 4.3.

Summing over all the possible transitions down from the excited level, we obtain the NRF absorption cross section for a photon with energy of E to excite level E_r:

$$\sigma_{abs}^0 = \sum_i \sigma_i^0(E) = \pi \lambda\!\!\!^{-2} \frac{2J_1 + 1}{2(2J_0 + 1)} \frac{\Gamma_0 \Gamma}{(E - E_r)^2 + \Gamma^2/4}. \tag{4.2}$$

In cases of pure resonance, where the only possible transition for the excited level is de-excitation to the ground level, $\Gamma_0 = \Gamma$, and Eq. (4.2) can be re-written as

$$\sigma_{abs}^0 = \pi \lambda\!\!\!^{-2} \frac{2J_1 + 1}{2(2J_0 + 1)} \frac{\Gamma^2}{(E - E_r)^2 + \Gamma^2/4}. \tag{4.3}$$

Angular Distribution

The angular distribution of the resonant photons depends on the spin of the initial and final levels of the transition. The distribution can be described using the angular correlation function $W(\Theta)$ of the scattered photons with respect to the incident photon beam.

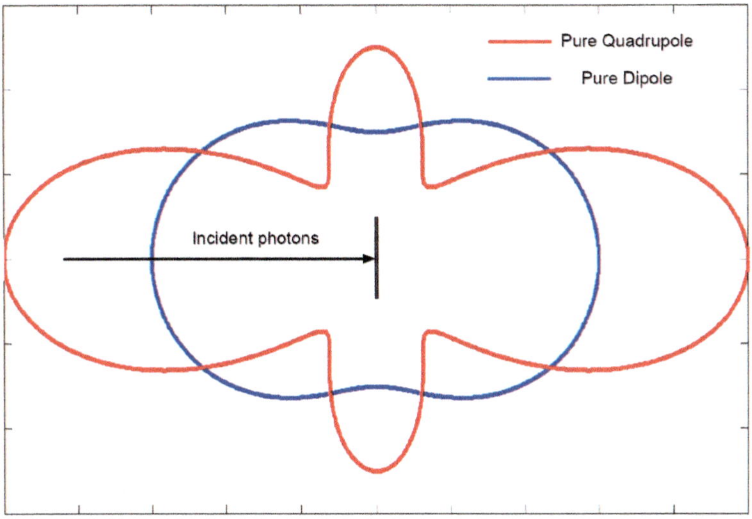

Fig. 4.4 Angular distributions of resonant photons from purely dipolar and purely quadrupolar transitions [33]

The angular distribution of photons scattered off an even-even nucleus through pure dipole transitions ($0 \rightarrow 1 \rightarrow 0$) is given as [47, 48]

$$W(\Theta) = \frac{3}{4}\left(1 + \cos^2\Theta\right). \tag{4.4}$$

For a purely quadrupole transition ($0 \rightarrow 2 \rightarrow 0$), the angular distribution can be written as

$$W(\Theta) = \frac{5}{4}\left(1 - 3\cos^2\Theta + 4\cos^4\Theta\right). \tag{4.5}$$

These distributions are shown in Fig. 4.4. The optimal angles to measure resonantly scattered photons are different for purely dipolar (90° relative to the beam axis) and quadrupolar transitions (127° relative to the beam axis). Although the distribution reaches maximum on the beam axis, the resonance photons would be buried by the much stronger interrogating beam.

For nuclei with odd atomic mass numbers, the resonant scattering is almost isotropic. In homeland security applications, which is primarily concerned with detecting odd-A nuclei, the angular distribution of resonant photons will not play a role as important as in physics experiments. But if it comes to an accurate assay, of, say, ^{238}U/^{235}U ratio, this distribution will have to be taken into account.

Doppler Effect[1]

The energy of an incident photon interacting with a nucleus varies according to the relative movement of the two. If the energy of the photon is E for a nucleus at rest, the apparent energy to a nucleus moving towards the photon source with a velocity v is larger than E and can be written as

$$E' = E(1 + v/c)/\sqrt{1 - (v/c)^2} \approx E(1 + v/c), \tag{4.6}$$

where c is the speed of light. If the velocities of the nuclei in the material follow the Maxwell's distribution function, the probability for a nucleus to have a velocity component v in the direction of the source is

$$w(v)dv = (M/2\pi kT)^{1/2} \exp\left(-Mv^2/2kT\right) dv, \tag{4.7}$$

where M is the mass of the nucleus, k is Boltzmann's constant, and T is the absolute temperature of the material.

Since the photon energy seen by a nucleus and the velocity of the nucleus have a one-to-one relationship, we can put

$$w(v)dv = w(E')dE' \tag{4.8}$$

and

$$dv = \frac{c}{E} dE. \tag{4.9}$$

Combining these, the distribution of energies seen by the nuclei in a material for a photon with energy of E is

$$\begin{gathered} w(E')dE' = w(v)dv = \\ (M/2\pi kT)^{1/2} \exp\left[-M(E' - E)^2 c^2/(2kTE^2)\right] \frac{c}{E} dE'. \end{gathered} \tag{4.10}$$

The Doppler width is defined as

$$\Delta = (E/c)(2kT/M)^{1/2}. \tag{4.11}$$

Equation (4.10) can then be re-written as

$$w(E')dE' = w(v)dv = (1/\Delta\pi^{1/2}) \exp\left[-(E' - E)^2/\Delta^2\right] dE'. \tag{4.12}$$

[1]The derivation in this section follows the work presented in [39] and [48].

The pure resonance NRF absorption cross section for a photon with energy E was derived as Eq. (4.2). The maximum of this cross section is reached when $E = E'$, giving a value

$$\sigma^0_{max} = \pi \lambdabar^2 \frac{2(2J_1 + 1)}{2J_0 + 1}. \tag{4.13}$$

Thus, the absorption cross section can be rewritten as

$$\sigma^0_{abs}(E) = \frac{\sigma^0_{max}}{[2(E - E_r)/\Gamma]^2 + 1}. \tag{4.14}$$

When the Doppler effect is taken into account, this cross section has to be averaged over all effective energies that the nuclei see with an incident photon of energy E. We define three variables: x, y, and t as follows:

$$x = 2(E - E_r)/\Gamma;\ y = 2(E' - E_r)/\Gamma;\ t = (\Delta/\Gamma)^2. \tag{4.15}$$

Then the average of the NRF absorption cross section is

$$\sigma_D(E) = \int \sigma^0_{abs}(E')w(E')dE' = \frac{\sigma^0_{max}}{2(\pi t)^{1/2}} \int_{-\infty}^{\infty} \frac{\exp\left[-(x-y)^2/4t\right]}{1+y^2} dy. \tag{4.16}$$

For large values of t, Eq. (4.16) can be approximated by

$$\sigma_D(E) = \int \sigma^0_{abs}(E')w(E')dE' = \sigma^0_{max}\left(\frac{\pi}{2t}\right)^{1/2} \exp(-x^2/4t) = \sigma^0_{max}\left(\Gamma\pi^{1/2}/2\Delta\right) \exp\left[-\left(\frac{E - E_r}{\Delta}\right)^2\right]. \tag{4.17}$$

This is the effective NRF cross section for a photon of energy E, with the Doppler broadening taken into consideration. The assumption of t being large is usually met in reality. The typical value of Δ is on order of eV, while the typical value of Γ is on order of meV. Assuming Δ =1 eV and Γ =1 meV, we can obtain the plot of σ^0_{abs} and σ_D as a function of energy. Notice the different scales in both energy and amplitude in Fig. 4.5.

The integrated cross section is defined as the integral of the cross section over the entire energy range. It can be written as

$$\int \sigma_D(E)dE = \int\int \sigma^0(E')w(E')dE'dE = \sigma^0_{max} \int \frac{1}{2(\pi t)^{1/2}} \int_{-\infty}^{\infty} \frac{\exp\left[-(x-y)^2/4t\right]}{1+y^2} dydx \frac{dE}{dx}. \tag{4.18}$$

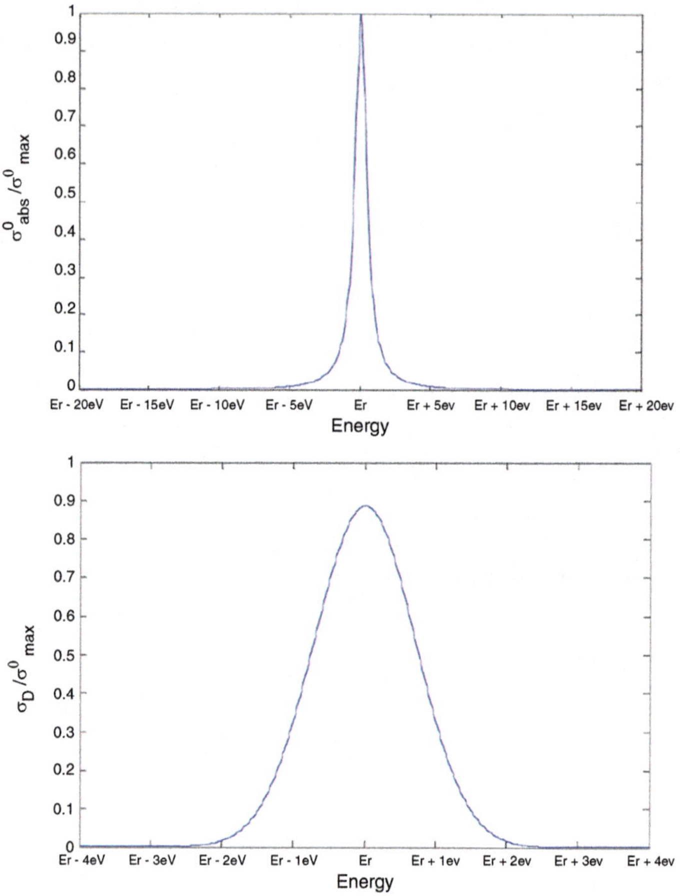

Fig. 4.5 Absorption cross section before and after Doppler broadening [33]

From the definition of the variable x,

$$x = 2(E - E_r)/\Gamma;\, dx = \frac{2}{\Gamma}dE,$$
$$\int \sigma(E)dE = \frac{\sigma^0_{max}\Gamma}{2}\int \frac{1}{2(\pi t)^{1/2}}\int_{-\infty}^{\infty}\frac{\exp\left[-(x-y)^2/4t\right]}{1+y^2}dydx. \tag{4.19}$$

The integral on the right in Eq. (4.19) has the value π regardless of the value of t:

$$\int \frac{1}{2(\pi t)^{1/2}}\int_{-\infty}^{\infty}\frac{\exp\left[-(x-y)^2/4t\right]}{1+y^2}dydx = \pi, \tag{4.20}$$

So the integrated NRF cross section is

$$\int \sigma_D(E)dE = \frac{\sigma_{max}^0 \Gamma \pi}{2}. \tag{4.21}$$

As can be seen here, the integrated NRF cross section is not affected by Doppler broadening. This result is very useful in the estimation of interaction rates, because the width of the NRF effective cross section is so narrow that in most cases the flux of the incident photon can be approximated by a constant independent of the energy in the calculation. For a thin scatterer, the NRF interaction rate can be derived as

$$r = \int N\sigma_D(E)\phi(E)dE = N\phi(E_r)\int \sigma_D(E)dE = N\frac{\sigma_{max}^0 \Gamma \pi}{2}\phi(E_r), \tag{4.22}$$

where N is the number of nuclei exposed to the photon beam.

In the derivation of the effective Doppler-broadened NRF cross section, it was assumed that the velocity of the nuclei in the absorption material follows the Maxwell distribution. This is a good assumption in gaseous materials. However, usually materials involved in NRF experiments are solids. In his study of neutron resonance, Lamb pointed out that the effective cross section in a solid has the same form as that in a gaseous material at a higher temperature, as long as the solid may be treated as a Debye continuum and that lattice binding is weak [49]. The ratio of the effective temperature to the actual temperature is

$$T_{eff}/T = 3(T/\theta)^3 \int_0^{\theta/T} t^3 \left(\frac{1}{\exp(t)-1} + \frac{1}{2}\right) dt, \tag{4.23}$$

where θ is the Debye temperature of the solid.

4.3.2.2 Sources in Nuclear Resonance Fluorescence Experiments

The natural width of an excited nuclear level is of the order of meV. Even after Doppler broadening, the equivalent width of such a level is only several eV. Thus, only photons with energies in the vicinity of the resonant energy contribute to the resonant excitation. One troublesome aspect of NRF is that the resonance emission occurs at sufficiently lower energy than required for absorption. This means that one cannot merely activate a sample of a material of interest and use its emissions as an interrogating source.

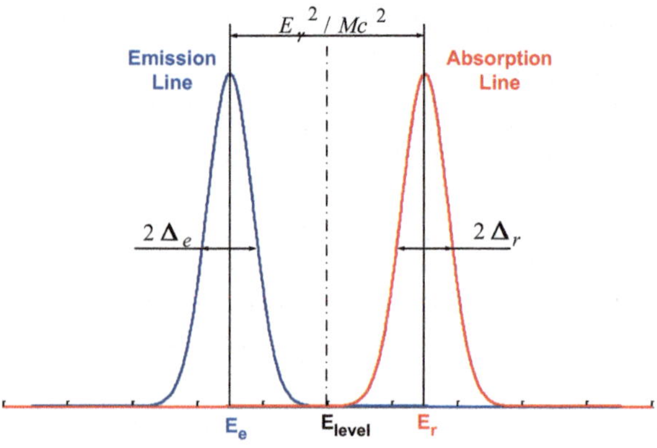

Fig. 4.6 Relationship between absorption and emission lines when nuclei are free and at rest before absorption and emission [33]

Recoil Energy Loss

When the source nuclei can be considered free, the recoil energy of a mass M after absorption or emission of a resonant photon of energy E_r is

$$\Delta E = \frac{E_r^2}{2Mc^2}. \tag{4.24}$$

Thus the absorption line and the emission line are split by an energy of $2\Delta E = E_r^2/(Mc^2)$, shown in Fig. 4.6. In the figure, $E_r^2/(Mc^2)$ is the energy loss due to recoil, while Δ_e and Δ_r are Doppler widths for the scattering photons and the incident photons, respectively. Different symbols were used in case the source and scatterer are at different temperatures. For energy levels of ^{238}U near 2.2 MeV, this splitting is 21.7 eV, which is much larger than the natural width of the levels (~meV) and also larger than the Doppler width at room temperature (1.06 eV). Because of the recoil energy loss, the absorbing material is transparent to its own resonant photons. However, this also prevents a given isotope from being used as both excitation source and absorber.

The discussion above is based on the free recoil assumption, which only stands when the bond between nuclei is weak. The problem of free recoil in solid materials has been discussed by Lamb [49]. Experiments showed that this assumption is valid in most cases.

For convenience, we separate radioactive isotope sources into two kinds. In the first kind, the mean lifetime of excited levels of the source nuclei is much longer than the time between consecutive collisions of nuclei with their surroundings. Thus, the source nuclei will be at thermal equilibrium with the surrounding particles before the emission of γ rays. Because of energy losses due to recoil, this kind of isotope cannot serve as both source and target material in NRF experiments. The energy loss can be compensated by the centrifuge method or the thermal method. In the other kind of source, where the mean life of the excited level is not significantly longer than the time between consecutive collisions, emission of γ rays happens before the nuclei reach thermal equilibrium. Some remaining momentum from previous processes leading to excited levels for the source nuclei (for example, beta decay) is left and can be helpful to compensate the energy loss due to recoil. This led to Gordzins' successful observation of NRF using Eu_2O_3 as both source and target [50], and Burgov and Terekhov's success with ^{24}Mg [51]. Radioactive isotope sources that can excite the resonance are rare because of the recoil energy loss. Their relevance to homeland security and nuclear safeguards applications is very limited.

Nuclear Reactions as a Source

Nuclear reactions, such as neutron capture, can serve as a source of NRF photons. In solids and liquids, the average time between consecutive collisions of atoms is on the order of ps (10^{-12} s), although this time is much longer in gaseous materials. In nuclear reactions, because of the higher excitation energies of nuclear levels, the majority of γ-ray transitions have half-lives shorter than 10^{-14} s. Thus, emission of γ rays following these reactions will happen well before the recoiling nuclei reach thermal equilibrium with the surrounding particles, and the emission line will be broadened due to the Doppler effect. The line broadening following nuclear reactions is usually orders of magnitude larger than the value of energy loss due to recoil. Thus, NRF should be easily observable using these γ rays as an excitation photon source. In fact, the Doppler broadening is so large that the excitation of levels in other nuclei becomes possible, as some fall in the energy range covered by the emission spectrum. For example, Rasmussen used the Doppler-broadened 4.43 MeV γ-ray radiation from the reaction to measure the lifetimes of the 4.43 MeV level of ^{12}C and the 4.46 MeV level of ^{11}B [52].

Bremsstrahlung Sources

Nowadays, the most commonly chosen excitation sources are electron bremsstrahlung X-ray sources. Bremsstrahlung X rays have wide energy spectra, ranging from zero to the endpoint energy of the incident electrons. Thus, a single irradiation is capable of exciting multiple levels in the scatterer.

In his work, Schiff confirmed the possibility of observing NRF using bremsstrahlung sources with continuous energy distributions [40]. He also pointed out the severe background problem using this kind of source. The first successful NRF experiment using bremsstrahlung X rays was performed by Hayward and Fuller in 1957 [53]. They successfully observed the resonant photons from the 15.1-MeV level in ^{12}C using a 19-MeV electron bremsstrahlung source.

Bremsstrahlung X-ray sources are widely available these days. However, most of the electron accelerators can only work in pulse mode. A typical pulse-mode linac can produce microsecond-wide electron pulses with a repetition rate of several hundred Hz. This equals to a duty cycle of 0.001%. Since nuclei return to ground state or lower excited states with very short half-life ($\sim 10^{-14}$ s), resonance photons are emitted almost immediately after the excitation. Measurement of these photons is limited to the time when the beam is active. The response time for regular spectroscopic systems is of the order of a microsecond. Thus, the maximum count rate people can get using pulse mode linac and traditional spectroscopic systems is usually the repetition rate of the beam. Considering the huge background caused by the off-resonant photons, this counting rate is too low for time-constrained portal security applications. For this reason, bremsstrahlung sources with much higher duty cycles are desired in NRF experiments. Van de Graaff electron accelerators can work in DC mode, with a duty cycle of 100%. This helps to get a reasonable counting rate with appropriate settings. Plus, the endpoint energy of these machines can be easily adjusted. Thus, it is possible to set the endpoint energy close to the energy levels of interest. This will greatly lower the background in NRF measurements generated by the down-scattering of the high-energy photons. Recent successes in implementation of NRF techniques in homeland security applications all chose to use Van de Graaff electron accelerators working in DC mode [41].

Another feature of bremsstrahlung X-ray sources worth mentioning besides duty cycle is the endpoint energy. As described earlier in this chapter, the width of the NRF cross section is of the order of eV after Doppler broadening. On the other hand, the energies of typical bremsstrahlung X rays range from zero to several MeV. Thus, only a tiny portion of the spectrum contributes to the excitation process. Off-resonant photons bring nothing but background and unnecessary dosage. When measuring resonant photons at a backscattering angle, the main sources of the background are: (1) multiple scattered photons with higher initial energies; (2) bremsstrahlung X rays generated by high energy electrons produced in the scattering material by incident photons; and (3) elastically scattered photons. Keeping the endpoint energy of the beam within 1 MeV of the energy levels of interest can greatly reduce the background in NRF measurements.

4.3.2.3 Detection Schemes

As mentioned previously, there are two detection schemes commonly used in NRF experiments: the scattering method and the self-absorption method. In this section, these two methods are discussed.

The Scattering Method

A typical experimental scattering setup is shown in Fig. 4.1: the incident irradiation is tightly collimated so that nothing else but the sample is irradiated by the beam. A detector with adequate energy resolution is placed at a backscattering angle, also narrowly collimated so that it only sees the sample. When using bremsstrahlung sources, collimation and shielding are extremely important in this kind of experiment, because scattering from surrounding objects could significantly degrade the signal to noise ratio, where the enormous counting rate of off-resonant photons can saturate the detector even at a very low resonant photon flux.

The pure resonance (ground → excited level → ground) NRF reaction rate for a thin scatterer was derived in Eq. (4.22). This reaction rate is independent of the Doppler width of the absorption line. In cases where the excited levels de-excite to an interim level instead of the ground level, the reaction rate can be re-written as

$$r = \int N\sigma_D(E)\phi(E)dE = N\phi(E_r)\int \sigma_D(E)dE = N\frac{\sigma^0_{max}\pi}{2}\frac{\Gamma_0\Gamma_f}{\Gamma}\phi(E_r), \tag{4.25}$$

where Γ_0 is the partial width for the γ-ray transition to the ground state and Γ_f is the partial width for the γ-ray transition to the interim final level.

In homeland security applications, the parameters for the γ-ray transition and the nuclear levels are usually well known. What the detector system ideally measures is the count rate of the emitted resonant photons. Thus, if the shape and intensity of the incident beam is known beforehand, the number of nuclei of interest exposed to the beam can be determined using the equations given above. However, there is no easy way to directly measure the energy distribution of the incident X rays. This limits this scattering method to just detection and isotope identification, which is sufficient for this application.

The situation gets complicated when the sample can no longer be considered as a thin scatterer. This happens when $nD\sigma_D(E_r)$ is not much smaller than unity. Here, n is the particle density in the sample, D is the thickness of the sample along the beam direction, and $\sigma_D(E_r)$ is the peak value of the NRF cross section. Under these conditions, the resonant photon flux at a certain depth in the sample will differ from that at the surface, plus the shape of the beam will vary at different thicknesses as well. In order to calculate the number of nuclei of interest exposed to the beam, the isotope content of the sample would be required. This, however, is unknown and is exactly what people are trying to measure in homeland security applications. Also, since the NRF cross section is large, resonant photons will be depleted beyond a small thickness. Material beyond this thickness makes no contribution to the resonance process, but will produce background in the measurement results.

Self-Absorption Method

A typical self-absorption experiment is shown in Fig. 4.2. A notch detector is located out of the beam and is pointed at a sample of the isotope of interest at a backscattering angle, measuring the resonant photons scattered by this sample.

In the simplest case, both the absorber and the scatterer can be considered as thin. The count rate with and without the absorber in the beam can be written as follows:

$$\begin{aligned} r(d_A) &= \int (1 - n_A d_A \sigma_{abs}(E))\,\phi(E) n_S d_S \sigma_D(E) dE \\ &= \phi(E_r)\int (1 - n_A d_A \sigma_{abs}(E))\,\phi(E) n_S d_S \sigma_D(E) dE \end{aligned} \tag{4.26}$$

with thin absorber, or

$$r(0) = \int \phi(E) n_S d_S \sigma_D(E) dE = \phi(E_r) \int n_S d_S \sigma_D(E) dE \tag{4.27}$$

without thin absorber, where n_S and n_A are the particle density in the scatterer and absorber; d_S and d_A are the thickness of the scatterer and absorber in the beam direction, $\sigma_D(E)$ is the NRF cross section with Doppler broadening, and $\sigma_{abs}(E)$ is the total absorption cross section for the absorber.

The fraction change in the count rate R is defined as

$$\begin{aligned} R = \frac{r(0) - r(d_A)}{r(0)} &= \frac{\int n_A d_A \sigma_{abs}(E) n_S d_S \sigma_D(E) dE}{\int n_S d_S \sigma_D(E) dE} \\ &= \frac{n_A d_A \int \sigma_{abs}(E) \sigma_D(E) dE}{\sigma_D(E) dE}. \end{aligned} \tag{4.28}$$

This value is independent of the resonant photon flux intensity and is proportional to the number of nuclei of interest present in the absorber.

For thick samples, the situation is much more complicated. Most systems under development for homeland security applications target mainly the detection and identification of SNM, instead of quantification [41, 54].

4.3.3 Background in Active Interrogation

Quite often, spontaneous emission of signature signals by SNM is too weak for reliable detection in the presence of ambient background through passive interrogation. Generally speaking, AI is merely a means to elevate such signals to a detectable level. Thus, AI must resolve the same ambient background as passive interrogation. However, in AI, the interrogation sources often induce additional active background and cause inference. Thus, it is paramount to manage the impact of the induced background, in order to take advantage of the increased signal level in AI. The interrogation sources and detectors are usually carefully collimated and shielded to reduce the unperturbed direct active background. The sources and detectors are chosen such that the interference from active background on the measurable is minimized. For example, in the FIGARO project, 6–7 MeV γ rays were chosen as the interrogation source to minimize photoneutron production while prompt neutrons are the targeted measurable. The temporal behavior of the source and signal is also heavily employed to minimize interference. For example, in some applications, long-lived delayed γ rays and delayed neutrons are examined when active background diminishes to a negligible level. Pulse-mode interrogation sources have also been widely deployed. Detection of signatures can be performed immediately following the interrogation pulses, where the signal is most abundant.

4.3.3.1 Ambient Background

In AI, detectors are exposed to the same ambient background as in passive interrogation. In unshielded γ-ray detectors, terrestrial and cosmic-ray radiation dominates the background. When significant shielding is present, radioactivity from the structural and shielding materials around the detector becomes much more important. Gamma rays emitted by decay daughters in the thorium and uranium series can be identified in terrestrial background spectra. Peaks at 1.46 MeV (emitted following EC decay of ^{40}K) and 2.61 MeV (from ^{208}Tl, a daughter product on the ^{232}Th decay chain) are normally the most prominent features. In addition to the naturally occurring radioisotopes, fission products as fallout from past nuclear weapons testing are also present.

Neutrons are also a component of the natural background. Any neutron sensitive detector will show response to these background neutrons at some level. The neutron background has two major sources: (a) from spallation reactions caused by high-energy cosmic protons in the upper atmosphere; and (b) secondary neutrons produced in a cascade process in the lower atmosphere or locally around the detector. Unlike the γ-ray background, significant terrestrial neutron sources do not exist. The energy of the background neutrons extends up to around 10 GeV. The energy spectra typically show peaks at several MeV from secondary neutrons and around 100 MeV from primary spallation neutrons. The spectrum and absolute flux of background neutrons vary significantly with different altitudes and latitudes. At lower altitudes, the slow neutron component is largely enhanced due to thermalization in air, earth and surrounding materials. The magnetic shielding effect causes large variations in cosmic proton flux at different geomagnetic latitudes. The neutron background level can be elevated in the presence of large amount of materials, such as the massive steel of a ship or a building structure. This increase in neutron background is attributed to the local production of secondary neutrons by fast neutrons, and to a lesser extent by cosmic-ray muons. This is the so-called *ship effect*. It has been reported that some background neutrons are produced in bursts of very short time duration. This is potentially a source of background even when coincidence counting technique is utilized.

Active Background

Besides passive ambient background, AI techniques also face the challenge of the much stronger active backgrounds resulting from the introduction of the interrogating radiation. These active backgrounds include events caused directly by the interrogating source and also those caused by secondary radiation emitted from interactions between the interrogating source particles and surrounding materials.

In photon interrogation, the intense radiation field from the source can temporarily paralyze most photon detectors. In addition, to achieve a reasonable induced fission rate, the energy of the interrogating photon source normally extends well above the photofission threshold, significantly exceeding the average energy of the fission photons. This further precludes energy discrimination as a means to separate source photons and fission photon signatures. As a result, it is mostly impossible to measure prompt fission γ rays in photon interrogation. However, if the interrogating photon energy is kept below the photoneutron threshold in benign materials, measurement of prompt fission neutrons from SNM is feasible. The direct active background from a photon source disappears almost immediately after the source is turned off, on a timescale of nanoseconds.

Source neutrons remain in the interrogation volume for a much longer period of time. The neutron die-away time is on the order of microseconds for non-moderating systems and milliseconds for moderating systems. During this time, the average energy of the neutron background decreases quickly as a result of inelastic scattering and elastic scattering in the medium. Prompt fission neutrons and delayed neutrons from the decay of fission products are emitted as fast neutrons. Thus, it is possible to exploit these neutron signature if low-energy neutrons are used as the interrogating source. Measurement of prompt photon signatures, on the other hand, is more challenging, as interference could come from γ rays following neutron capture in surrounding materials.

In addition to directly interfering with detector response, interrogation radiation may cause nuclear reactions in surroundings that produce secondary photons and neutrons. In photon interrogation, when the end-point energy of the photon source is a few MeV above the fission threshold, photoneutrons can be produced in many materials in the same energy range as fission neutrons. This leads to interference with prompt neutron measurements. Photoneutron reaction products often emit γ-rays during their decay. Since the energy of such γ rays is generally less than a few MeV, they can be distinguished from fission γ rays via energy discrimination. Delayed neutron precursors can also be produced via photonuclear reactions. Neutrons emitted from these precursors could interfere with the measurement of delayed fission neutrons. The most notable example is ^{17}N, produced via the (γ,p) reactions in oxygen present in air or water.

Neutron interrogation sources also produce active backgrounds via reactions with surrounding materials. If fast neutrons are used as probing radiation, high-energy γ-rays can be produced during the irradiation via inelastic scattering in medium. Following neutron interrogation, neutron-capture γ rays are emitted as long as thermalized source neutrons are present. Normally, such γ-ray background disappears with the neutron population and becomes insignificant after several milliseconds. However, reactions in certain isotopes could produce long-lived, high-energy γ-ray emitters, causing potential interference with the detection of delayed fission γ rays. No significant active neutron background is introduced by reactions between source neutrons and surrounding materials.

4.4 Modeling and Simulation for Active Measurements

The ability to accurately predict signatures from AI is long-overdue and highly desired for homeland security and nuclear safeguards applications. The discussion in this section is limited to active techniques based on induced fission and NRF. Neutron-induced fission process is much better understood than the photonuclear processes. Currently, much of the existing photonuclear data is based on models and there have been limited experimental measurements performed to date.

4.4.1 Simulation of the Induced Fission Process and Its Signatures

As discussed previously, delayed γ rays and delayed neutrons are often more suitable to serve as signatures of induced fission process in SNM, due to the overwhelming interrogating radiation. Delayed signals are emitted during the decay of fission products. Fission product yields (FPYs) are one of the most observable features of the fission process. They are vital to nuclear forensics and safeguards missions, such as AI technologies and post-event forensic response. The cumulative fission yields of the four nuclides, ^{95}Zr, ^{99}Mo, ^{144}Ce, and ^{147}Nd, are key parameters in estimating the effects from nuclear tests and for post-detonation nuclear forensics in general. In basic sciences, fission yields play an important role in the understanding of the fission process. For applied research, they are essential for calculation of the accumulation and inventory of fission products at different stages of nuclear fuel cycle. They are also becoming increasing important in homeland security applications, such as accelerator-driven AI systems.

In 2016, the IAEA held a technical meeting on "Fission Product Yields: current status and perspectives". It brought together an international group of experts in the field of fission yield measurements, model development, evaluation and validation. It was concluded that continued experimental efforts need to be supported and the uncertainties in the experimental measurements need to be well characterized. It was pointed out that many of the evaluated libraries are rather old and date back to the beginning of the 1990s. There is thus an urgent need to update the fission yield libraries. A separate review by a group of experts from US national laboratories was funded by DOE, which focused particularly on the cumulative fission yields from ^{239}Pu after neutron induced fission for nuclear forensics applications. The report pointed out the lack of high quality FPY data at elevated neutron energies. Particularly, it emphasized on the key importance and the difference between the cumulative yield of ^{147}Nd used by LLNL and LANL. The yields used by the two laboratories for irradiations in a *fast* neutron spectrum are about 1.97% and 2.07%, respectively, and differ by a factor of about 1.05. It also recognized the need to study energy dependence of the fission yield, especially in the energy region where $\langle E_n \rangle \sim 1.9\,\text{MeV}$. In summary, the lack of high-quality FPYs is unanimously agreed upon. Many independent reviews recommended more effort to be put into experimental and theoretical study on FPYs.

In 2006, the US nuclear data program released a new photonuclear data library as part of the ENDF/B-VII database. This new data library includes new or improved data for 24 isotopes. The improved actinide data now contains prompt and delayed-fission neutron spectra. Due to the lack of data and theoretical models for photofission, the photofission library used by MCNPX is primarily based on neutron-induced data. The model used by MCNPX assumes that target nuclei will produce fission in the same way, regardless of the type of the incident particle (e.g., neutron or photon). Reedy et al. generated delayed γ-ray spectra using fission yield from ENDF/B-VII.0 and line emission data from ENSDF. Discrepancies between measured results and simulation data were reported. In MCNPX, to enable the production of photons and neutrons from photofission reactions, the `ispn` entry of the `phys:p` card has to be changed from the default value `ispn=0` to enable photonuclear collision sampling. Also, the `fism` entry of the `phys:p` card should be set to 1 to ensure that photofission secondaries are sampled only when a photofission event occurs if one prefers analog production of delayed neutrons and γ rays. The default setting (i.e. `fism=0`) is only correct on average over a large number of interactions, thus its use is not suitable for applications where detail of secondary particles production is important, such as coincidence counting. To preserve the correlation between prompt fission neutrons and photons, the LLNL fission model must be enabled on the `phys:n` card (i.e. `fism=5`). In MCNPX, the physics module requires the ENDF/B-VII photonuclear data library `endf7u`. The `xsdir` file for MCNPX installation has to be modified to include this library for MCNPX to access this photonuclear data library. Both the `xsdir` file and the data library have to be present in the data file directory specified by the environmental variable `DATAPATH`. In addition, the `DG` entry on the `ACT` card must be set to `LINES` if individual line-amplitude of delayed γ rays is desired. This is important to the simulation of delayed γ-ray energy spectra. However, enabling this option makes the simulation significantly slower.

4.4.2 Simulation of Nuclear Resonance Fluorescence

NRF was not officially supported in MCNP until MCNPX 2.7.0. Photonuclear data, including NRF, became available for 157 specific isotopes up to 150 MeV in the `endf7u` library, but not all the data was put through extensive testing. It has been discovered that MCNPX significantly under-predicted the elastically scattered photon background. Photonuclear processes such as Rayleigh, Thompson and Delbrück scattering are not calculated by MCNPX. Furthermore, Rayleigh scattering is only considered for a limited angular range. As a result, the signal-to-noise ratio in NRF measurements at backward angles could be vastly overestimated. Ludewigt et al. enhanced the MCNPX simulation by correcting the Rayleigh scattering routine [55]. Quiter et al. determined the contribution of coherently scattered photons to the background by folding the probing bremsstrahlung X-ray spectrum with elastic scattering cross sections tabulated in the RTAB database [56].

Geant4 is one of the most widely used open-source Monte Carlo radiation transport toolkits. However, NRF process is not currently included in the public release. Jordan and Warren at PNNL have successfully implemented and tested NRF physics in GEANT4 [57, 58]. Their method was then further tested and applied by Perry et al. at Purdue University and the University of Utah [59]. Recently, Lakshmanan et al. independently developed NRF simulation capability in Geant4. Their approach was designed to be user-expandable, so users can enable NRF simulation for more isotopes without modifying the code [60]. To enable simulation of NRF process, a mechanism to provide the NRF cross sections and determine the de-excitation path is needed. In either case of the previous work, NRF cross section was calculated based on standard nuclear level information extracted from the ENSDF database. This calculation needs only to be performed once during the initialization of the Geant4 run-time class. The NRF interaction mean-free-path is then derived based on the energy of the photon and the interacting nuclei. The probability distribution of different de-excitation pathways was calculated based on the same ENSDF data used for cross section calculated. Doppler effect was also modeled to simulated the broadening of the resonance peak due to recoil energy loss. The code was then benchmarked against experimental data. In a test with ^{11}B, Lakshmanan et al. reported that the simulated integrated NRF counts agreed with experimental results to be within 20.5%.

4.5 Limitations of Active Measurements

The goal of AI technologies is to enhance signatures from SNM to a detectable level unachievable with passive methods alone. Passive methods fail in many scenarios due to significant attenuation on spontaneous emission, caused by the SNM itself, the surrounding materials, or engineered shielding designed for source concealment. In AI, although stronger signatures are expected to be produced via induced reactions by the interrogation source, one still faces the limiting factor of signature attenuation. Furthermore, the intense active background and interference introduced by the interrogating radiation pose great technical challenges in developing radiation measurement systems for signature detection.

4.5.1 Interrogation Sources

The most commonly used interrogating sources include bremsstrahlung X rays and neutrons. Guidelines and regulations exist on the maximum energy of the ionizing radiation that can be used for certain applications. NCRP also has recommended dose limit for bystanders received per inspection event during cargo screening in various commentaries based on risk benefit analysis. Both source spectrum and source intensity can be greatly altered by surrounding materials and shielding before reaching the SNM. The total attenuation is determined by two types of process: scattering and absorption. The interrogating neutron or photon is removed after an absorption interaction, while a scattering interaction changes both the energy and direction of the source particle. In the energy region of the interrogating photons ($\sim$10 to 20 MeV), the mass absorption coefficient is roughly 0.015 cm^2/g in light materials (e.g. H_2O, Al) and around 0.03 cm^2/g in high-Z materials (e.g. U and Pb). Thus, in low-density material, the mean free path of interrogating photons is around 70 cm. In denser materials, such as loosely packed UO_2 ($\rho = 2\ g/cm^3$), the mean free path decreases to less than 20 cm. Neutron interaction cross sections behave rather differently from photon absorption coefficients. While photon absorption coefficients vary smoothly with atomic number, neutron cross section varies drastically from one element to another. In addition, neutron cross sections are also highly energy dependent. The understanding of the attenuation of interrogation radiation is important not only in considering beam penetration to stimulate SNM, but also design of shielding for radiation protection.

4.5.2 Shielding and Attenuation of Signatures

Once the interrogating radiation reaches the SNM, generating signatures is usually not the most difficult problem in AI. Despite the possible reduction in flux and energy due to attenuation, one can still choose the right beam parameters so that sufficient reactions are reduced in the SNM. Photofission cross section has a broad resonance peak centered around 15 MeV. If the endpoint energy of the photon source is sufficiently high, significant photo flux remains in the resonance region because of down-scattering. In addition, photoneutrons are to be produced in many materials when the photon energy exceeds 6 MeV. These neutrons, especially when produced in vicinity of the SNM, can be more efficient than photons in inducing fission reactions due to the larger cross sections. Neutrons in various energy regions can contribute to inducing fission in SNM. Scattered source neutrons may be present in the system long after the interrogation pulse to produce fission signatures.

The survival of signatures is still the most challenging step in AI, even though the abundance is expected to be much larger than spontaneous emission in passive techniques. Low-energy photons experience large attenuation on their way to the detectors. In addition, photon background is substantial below 3 MeV, which makes detection of fission photons troublesome. Thus, detection of photon signatures has been focused above 3 MeV, where attenuation is primarily governed by Compton scattering. For NRF photons, the energy is between 2 and 3 MeV for uranium isotopes. High-resolution spectroscopy is needed to achieve sufficient signal-to-noise ratio for detection. Neutron signatures are much more effective in non-hydrogenous cargo, as their attenuation and absorption strongly depend on the surrounding materials. Using interrogating photons below 6 MeV or low-energy neutrons, prompt neutrons could be measured. High-energy prompt neutrons observe less attenuation and have better chances to reach the detection system. However, in hydrogenous cargo, a prompt neutron could fall below the detection threshold after a single scattering interaction with a hydrogen nucleus. Delayed neutrons are less energetic and undergo much greater attenuation. Fortunately, several milliseconds after the neutron interrogation, neutron background usually becomes negligible. With high-efficiency neutron detectors, detection of delayed neutrons is still a robust detection method.

In cargo screening, another complication is the variety of cargo content and possible engineered shielding for source concealment. The role of AI technique in this area remains debatable and deserves further cost-performance analysis. It is more likely that AI techniques would be deployed in the secondary or tertiary stage following passive and radiographic inspection, allowing extended assay time. Nonetheless, a successful inspection system needs to be sensitive over a wide range of cargo types and densities. For example, in hydrogenous cargo, neutron attenuation is the major concern, while in metallic cargo, photon penetration is the limiting factor. Systems combing neutron- and photon-induced fission techniques are under development to enable detection over a wide range of cargo. Furthermore, the detection sensitivity could be enhanced if both neutron and photon signatures are sought after.

4.5.3 Radiation Detection Techniques for Active Interrogation

Various radiation detectors have been deployed in AI, ranging from gas-based neutron detectors, scintillation detectors, Cherenkov detectors to high resolution γ-ray spectrometers. A unique challenge is the capability to perform measurements at very high rates. High counting rates originates from different sources depending on the chosen interrogation technique and signature measurable. For example, NRF photons need to be counted in presence of the high background from the interrogation photon source. In spent fuel assay, γ rays and neutrons signatures have to be extracted from the high radiation background emitted from the fuel.

All AI techniques discussed here rely on the detection of signature photons and neutrons, emitted by the induced reactions. Most neutron detectors are designed with selective energy response and simply perform counting. Gamma-ray detectors are also used as counting devices in some applications, but usually some level of spectroscopy measurements are involved. The most commonly used neutron detectors include gas-filled detectors (^{3}He, ^{4}He, BF_3, boron-lined tubes, etc.) and scintillators (inorganic, organic and gas scintillator). Numerous designs of moderators and packaging were developed to tailor detector response to specific missions. Bubble detectors were also used for its complete insensitivity to interrogating photons. Photon detection in AI mostly relies on scintillators and semiconductor detectors. Choices have to be made based on a few considerations, such as detection efficiency, energy resolution, response time, γ-ray sensitivity, and cost. Innovative photo detectors, such as Cherenkov glass detectors, are also being developed for AI.

Traditionally, piled-up pulses are rejected to ensure good energy resolution. To improve throughput rate, high-pass filters are normally implemented to shorten pulses. However, this reduces signal-to-noise ratio and causes degradation in energy resolution. Advanced signal processing algorithms have seen developed to handle high rates and have been implemented for AI, especially in γ-ray spectroscopy measurements. Three examples are discussed here.

Recently, pulse pile-up recovery based on template-matching has been proved to be an effective approach to achieve high throughput rate γ-ray spectroscopy [61, 62]. In an ideal model, the output signal $y(t)$ from a γ-ray detector is the convolution of the incident signal $s(t)$ and the detector response matrix M: $y(t) = s(t) * M$. Thus, if the response matrix can be accurately determined, an estimation of the incident signal can be obtained via deconvolution. The incident signal is normally modeled as a train of delta functions, with random time of arrival and amplitude. The detector response is considered to be time-invariant and can be pre-determined. This type of techniques works particular well with scintillation detectors.

A new non-linear filter algorithm, called ADONIS (Algorithmic Development FramewOrk for Nuclear Instrumentation and Spectrometry), was developed by CEA and AREVA [63]. The algorithm is based on a bimodal Kalman smoother derived from the maximum likelihood principle, describing the preamplifier output using a state space model. The system has been demonstrated with excellent resolution at count rates up to 106 cps.

In the traditional trapezoidal filtering, especially at high count rate, pile-up phenomenon is a major challenge in high-resolution high-throughput spectroscopy measurement. To obtain good energy resolution piled-up events are rejected at the expense of detection efficiency. It is highly desired to achieve high throughput rate while maintaining good energy resolution. The concept of time-variant trapezoidal filtering was tested to improve throughput rate without large sacrifice in energy resolution [64]. The principle idea of the filtering approach is that several trapezoidal filters with different shaping times are implemented in parallel and the one with largest shaping time which can resolve pile-up events is used to perform energy measurement of corresponding pulses.

References

1. C. Grassley, M. Baucus, B. Thomas, C.B. Rangel, Container security: expansion of key customs programs will require greater attention to critical success factors. Technical Report GAO-03-770, General Accounting Office, US Congress (2003)
2. M.T. Committee, Security in maritime transport: risk factors and economic impact. Technical Report, Organization for Economic Co-operation and Development; Directorate for Science, Technology and Industry (2003)
3. D. Reilly, N. Ensslin, H.J. Smith, Passive nondestructive assay of nuclear materials. Technical Report, LA-UR-90-732, Los Alamos National Laboratory (1991)

4. L.W. Campbell, L.E. Smith, A.C. Misner, IEEE Trans. Nucl. Sci. **58**(1, 2), 231 (2011). https://doi.org/10.1109/TNS.2010.2095039
5. L. Smith, N. Abdurrahman, Nucl. Technol. **140**(3), 328 (2002)
6. C.P.J. Barty, in *2011 IEEE Photonics Conference (PHO)* (IEEE, New York, 2011), pp. 638–639. IEEE Photonics Conference (PHO), Arlington, VA, OCT 09–13, 2011
7. R. Winyard, G. Mcbeth, Nucl. Instrum. Methods **98**(3), 525 (1972). https://doi.org/10.1016/0029-554X(72)90238-8
8. R.T. Kouzes, J.H. Ely, A. Seifert, E.R. Siciliano, D.R. Weier, L.K. Windsor, M.L. Woodring, J. Borgardt, E. Buckley, E. Flumerfelt, A. Oliveri, M. Salvitti, Nucl. Inst. Methods Phys. Res. A **587**(1), 89 (2008). http://dx.doi.org/10.1016/j.nima.2007.12.031. http://www.sciencedirect.com/science/article/pii/S0168900207025090
9. C. Hollas, D. Close, C. Moss, Nucl. Instrum. Methods Phys. Res. Sect. B **24-5**(1), 503 (1987). http://dx.doi.org/10.1016/0168-583X(87)90695-1
10. Application of photo induced reactions to nuclear materials safeguards problems. Technical Report GA-10106, General Atomics, San Diego, CA (1970)
11. T. Gozani, Atomkernenergie **19**(1), 63 (1972)
12. Nuclear safeguards research and development. Technical Report LA-4227-MS, Los Alamos Scientific Laboratory (1969)
13. K. Baumung, K. Boehnel, J. Klunker, M. Kuechle, J. Wolff, Investigations into non-destructive safeguards techniques. Technical Report, International Atomic Energy Agency (1970)
14. Application of photo induced reactions to nuclear materials safeguards problems. Technical Report GA-9614, General Atomics, San Diego, CA (1969)
15. I.A.E. Agency, Iaea photonuclear data library (2017). www-nds.iaea.org/photonuclear/
16. B. Micklich, D. Smith, T. Massey, C. Fink, D. Ingram, Nucl. Instrum. Methods Phys. Res. Sect. A **505**(1–2), 466 (2003). https://doi.org/10.1016/S0168-9002(03)01122-7. 10th Symposium on Radiation Measurements and Applications, Ann Arbor, MI, May 21–23 (2002)
17. J.L. Jones, W.Y. Yoon, K.J. Haskell, D.R. Norman, J.M. Hoggan, C.E. Moss, C.A. Goulding, C.L. Hollas, W.L. Myers, E. Franco, Photofission-based, nuclear material detection: technology demonstration. Technical Report, INEEL-EXT-02-0106, Idaho National Laboratory and United States Department of Energy (2002). https://books.google.com/books?id=3F-yAQAACAAJ
18. C. Moss, C. Goulding, C. Hollas, W. Myers, IEEE Trans. Nucl. Sci. **51**(4, 1), 1677 (2004). https://doi.org/10.1109/TNS.2004.832992. Nuclear Science Symposium/Medical Imaging Conference/13th International Workshop on Room-Temperature Semiconductor X-and Gamma-Ray Dectectors/Symposium on Nuclear Power Systems, Portland, OR, Oct 19–25, 2003
19. T. Gozani, Active nondestructive assay of nuclear materials: principles and applications (1981). https://doi.org/10.2172/6215952. http://www.osti.gov/scitech/servlets/purl/6215952
20. E.B. Norman, S.G. Prussin, R.M. Larimer, H. Shugart, E. Browne, A.R. Smith, R.J. McDonald, H. Nitsche, P. Gupta, M.I. Frank, T.B. Gosnell, Nucl. Instrum. Methods Phys. Res. Sect. A **521**(2), 608 (2004). http://dx.doi.org/10.1016/j.nima.2003.10.097. http://www.sciencedirect.com/science/article/pii/S0168900203028948
21. D. Slaughter, M. Accatino, A. Bernstein, P. Biltoft, J. Church, M. Descalle, J. Hall, D. Manatt, G. Mauger, T. Moore, E. Norman, D. Petersen, J. Pruet, S. Prussin, Nucl. Instrum. Methods Phys. Res. Sect. A **579**(1), 349 (2007). http://dx.doi.org/10.1016/j.nima.2007.04.058. http://www.sciencedirect.com/science/article/pii/S0168900207006456. Proceedings of the 11th Symposium on Radiation Measurements and Applications
22. B.W. Blackburn, J.L. Jones, C.E. Moss, J.T. Mihalczo, A.W. Hunt, F. Harmon, S.M. Watson, J.T. Johnson, Nucl. Instrum. Methods Phys. Res. Sect. B **261**(1–2), 341 (2007). https://doi.org/10.1016/j.nimb.2007.04.272. 19th International Conference on Application of Accelerators in Research and Industry, Ft Worth, TX, Aug 20–25, 2006

23. F. Ruddy, J. Seidel, R. Flammang, B. Petrovic, A. Dulloo, T. Congedo, in *Non-Intrusive Inspection Technologies, Proceedings of SPIE*, ed. by G. Vourvopoulos, F.P. Doty, vol. 6213. SPIE (SPIE-INT SOC Optical Engineering, 1000 20TH ST, PO BOX 10, Bellingham, WA 98227-0010 USA, 2006). Proceedings of SPIE, vol. 6213. https://doi.org/10.1117/12.664467. Conference on Non-Intrusive Inspection Technologies, Kissimmee, FL, Apr 17–18, 2006
24. C. Fink, D. Smith, B. Micklich, T. Massey, Nucl. Instrum. Methods Phys. Res. Sect. A-Accelerators Spectrometers Detectors and Associated Equipment **505**(1–2), 5 (2003). https://doi.org/10.1016/S0168-9002(03)01007-6. 10th Symposium on Radiation Measurements and Applications, Ann Arbor, Michigan, May 21–23, 2002
25. B. Micklich, D. Smith, T. Massey, C. Fink, D. Ingram, Nucl. Instrum. Methods Phys. Res. Sect. A **505**(1–2), 1 (2003). https://doi.org/10.1016/50168-9002(03)01006-4. 10th Symposium on Radiation Measurements and Applications, Ann Arbor, Michigan, May 21–23, 2002
26. D. Wehe, H. Yang, M. Jones, IEEE Trans. Nucl. Sci. **53**(3), 1430 (2006). https://doi.org/10.1109/TNS.2006.875154
27. X. Wen, H. Yang, Nucl. Instrum. Methods Phys. Res. Sect. A **821**, 34 (2016). https://doi.org/10.1016/j.nima.2016.03.051
28. N. Saurel, J. Capdevila, N. Huot, M. Gmar, Nucl. Instrum. Methods Phys. Res. Sect. A **550**(3), 691 (2005). https://doi.org/10.1016/j.nima.2005.05.058
29. M. Gmar, F. Jeanneau, F. Laine, H. Makil, B. Poumarede, F. Tola, Appl. Radiat. Isot. **63**(5–6), 613 (2005). https://doi.org/10.1016/j.apradiso.2005.05.010. 8th International Conference on Applications of Nuclear Techniques, Crete, Sept 12–18, 2004
30. L. Kull, R. Bramblett, T. Gozani, D. Rundquist, Nucl. Sci. Eng. **39**(2), 163 (1970)
31. R. Van Grieken, A. Markowicz, *Handbook of X-Ray Spectrometry*. Practical Spectroscopy, 2nd edn. (CRC Press, Boca Raton, 2001). https://books.google.com/books?id=i_iDRTp75AsC
32. U. Kneissl, H. Pitz, A. Zilges, Prog. Part. Nucl. Phys. **37**, 349 (1996). http://dx.doi.org/10.1016/0146-6410(96)00055-5. http://www.sciencedirect.com/science/article/pii/0146641096000555
33. H. Yang, Active interrogation methods for detection of special nuclear material. Ph.D. thesis, University of Michigan (2009)
34. W. Kuhn, Philos. Mag. **8**(52, 7th Series), 625 (1929)
35. L. Meitner, H. Hupfeld, Z. Phys. **67**, 147 (1931)
36. F.R. Metzger, Phys. Rev. **83**, 842 (1951). https://link.aps.org/doi/10.1103/PhysRev.83.842
37. F.R. Metzger, W.B. Todd, Phys. Rev. **91**, 1286 (1953). https://link.aps.org/doi/10.1103/PhysRev.91.1286.2
38. E. Pollard, D.E. Alburger, Phys. Rev. **74**, 926 (1948). https://link.aps.org/doi/10.1103/PhysRev.74.926
39. P. Moon, Proc. Phys. Soc. Lond. Sect. A **64**(373), 76 (1951). https://doi.org/10.1088/0370-1298/64/1/311
40. L. Schiff, Phys. Rev. **70**(9–10), 761 (1946). https://doi.org/10.1103/PhysRev.70.761
41. W. Bertozzi, R. Ledoux, Nucl. Instrum. Methods Phys. Res. Sect. B **241**(1–4), 820 (2005). https://doi.org/10.1016/j.nimb.2005.07.202. 18th International Conference on Application of Accelerators in Research and Industry (CAARI), Ft Worth, TX, Oct 10–15, 2004
42. G.A. Warren, J.A. Caggiano, W.K. Hensley, E. Lepel, S. Pratt, W. Bertozzi, S.E. Korbly, R.J. Ledoux, W.H. Park, in *2006 IEEE Nuclear Science Symposium Conference Record*, vol. 1–6 (IEEE, New York, 2006), pp. 914–917. https://doi.org/10.1109/NSSMIC.2006.355995. 15th International Workshop on Room-Temperature Semiconductor X- and Gamma-Ray Detectors/2006 IEEE Nuclear Science Symposium, San Diego, CA, Oct 29–Nov 04, 2006
43. H.R. Weller, M.W. Ahmed, Mod. Phys. Lett. A **18**(23), 1569 (2003). http://www.worldscientific.com/doi/abs/10.1142/S0217732303011216
44. J. Pruet, D. McNabb, T-rex design considerations for detection of concealed 238u. Technical Report UCRL-TR-219071, Lawrence Livermore National Laboratory (2006)
45. J. Pruet, D. Lange, Contraband detection with nuclear resonance fluorescence: feasibility and impact. Technical Report UCRL-TR-227067, Lawrence Livermore National Laboratory (2007)
46. H.A. Bethe, G. Placzek, Phys. Rev. **51**, 450 (1937). https://link.aps.org/doi/10.1103/PhysRev.51.450

47. K. Siegbahn, *Alpha-, Beta- and Gamma-Ray Spectroscopy* (North Holland, Amsterdam, 1965)
48. A. Wapstra, R. van Lieshout, G. NIJGH, *Nuclear Spectroscopy Tables*. Series in Physics (North Holland, Amsterdam, 1959). https://books.google.com/books?id=3IAAMwEACAAJ
49. W.E. Lamb, Phys. Rev. **55**, 190 (1939). https://link.aps.org/doi/10.1103/PhysRev.55.190
50. L. Grodzins, Phys. Rev. **109**, 1014 (1958). https://link.aps.org/doi/10.1103/PhysRev.109.1014
51. N. Burgov, Y. Terekhov, Sov. J. At. Energy **2**, 629 (1957)
52. V. Rasmussen, F. Metzger, C. Swann, Phys. Rev. **110**(1), 154 (1958). https://doi.org/10.1103/PhysRev.110.154
53. E. Hayward, E. Fuller, Phys. Rev. **106**(5), 991 (1957). https://doi.org/10.1103/PhysRev.106.991
54. J. Pruet, D.P. McNabb, C.A. Hagmann, F.V. Hartemann, C.P.J. Barty, J. Appl. Phys. **99**(12), 123102 (2006). http://dx.doi.org/10.1063/1.2202005
55. B.A. Ludewigt, Nuclear resonance fluorescence for nondestructive isotopic analysis. Technical Report LBNL-3890E, Lawrence Berkeley National Laboratory (2010)
56. B.J. Quiter, B.A. Ludewigt, V.V. Mozin, S.G. Prussin, IEEE Trans. Nucl. Sci. **58**(2), 400 (2011). https://doi.org/10.1109/TNS.2011.2112777
57. K.A. Jordan, J. Vujic, T. Gozani, Nucl. Instrum. Methods Phys. Res. Sect. A **579**(1), 407 (2007). http://dx.doi.org/10.1016/j.nima.2007.04.089. http://www.sciencedirect.com/science/article/pii/S0168900207006638. Proceedings of the 11th Symposium on Radiation Measurements and Applications
58. G.A. Warren, R.S. Detwiler, P.N. Peplowski, in *2010 IEEE Nuclear Science Symposium Conference Record (NSS/MIC)*. Inst Elect & Elect Engineers, Nucl & Plasma Sci Soc; IEEE (IEEE, New York, 2010), pp. 534–541. IEEE Nuclear Science Symposium (NSS)/Medical Imaging Conference (MIC)/17th International Workshop on Room-Temperature Semiconductor X-ray and Gamma-ray Detectors, Knoxville, TN, Oct 30–Nov 06, 2010
59. J. Perry, S. Xiao, T. Jevremovic, in *20TH IEEE International Conference on Tools with Artificial Intelligence, VOL 2, Proceedings*. IEEE Comp Soc; Biol & Artificial Intelligence Soc; Wright State Univ; Univ Patras; Virginia Tech (IEEE Computer Society, Los Alamitos, 2008), pp. 541–546. https://doi.org/10.1109/ICTAI.2008.94. 20th IEEE International Conference on Tools with Artificial Intelligence, Dayton, OH, Nov 03–05, 2008
60. M.N. Lakshmanan, B.P. Harrawood, G.A. Agasthya, G. Rusev, A.J. Kapadia, in *2012 IEEE Nuclear Science Symposium and Medical Imaging Conference Record (NSS/MIC)*, ed. by B. Yu. IEEE; IEEE Nucl & Plasma Sci Soc (IEEE, New York, 2012), pp. 1731–1734. IEEE Nuclear Science Symposium/Medical Imaging Conference Record (NSS/MIC)/19th Room-Temperature Semiconductor X-ray and Gamma-ray Detector Workshop, Anaheim, CA, Oct 29–Nov 03, 2012
61. P.A.B. Scoullar, R.J. Evans, in *2008 IEEE Nuclear Science Symposium Conference Record* (2008), pp. 1668–1672. https://doi.org/10.1109/NSSMIC.2008.4774724
62. X. Wen, H. Yang, Nucl. Inst. Methods Phys. Res. A **784**, 269 (2015). http://dx.doi.org/10.1016/j.nima.2014.11.008. http://www.sciencedirect.com/science/article/pii/S0168900214012728. Symposium on Radiation Measurements and Applications 2014 (SORMA XV)
63. E. Barat, T. Dautremer, L. Laribiere, J. Lefevre, T. Montagu, J.C. Trama, in *2006 IEEE Nuclear Science Symposium Conference Record*, vol. 2 (2006), pp. 955–958. https://doi.org/10.1109/NSSMIC.2006.356004
64. B.A. VanDevender, M.P. Dion, J.E. Fast, D.C. Rodriguez, M.S. Taubman, C.D. Wilen, L.S. Wood, M.E. Wright, IEEE Trans. Nucl. Sci. **61**(5), 2619 (2014). https://doi.org/10.1109/TNS.2014.2357059

Chapter 5
Active Interrogation Probe Technologies

Robert Garnett

Abstract One key component of any active interrogation (AI) system that is not present in passive measurements is the source of probing radiation. This chapter introduces the physical principles of operation of linear accelerators, cyclotrons, and novel laser-driven sources, along with the review of radioisotope sources and the use of natural radiation for AI. Some of the critical parts of these technologies are discussed in more detail. In addition, the common characteristics of various probe technologies are provided.

5.1 General Characteristics of AI Probe Technologies

Innovative AI methods to detect real and perceived threats continue to be developed to protect our borders, airports, and ports. All of these detection methods require a source of energetic particles to induce specific reactions that are key signatures for detecting conventional explosives or fissile material, or to perform imaging. Quite a broad range of accelerator technology has been applied to meeting this challenge. These accelerator systems range from relatively small, simple, and compact, to large and complex, based on the specific application requirements. They have been used or proposed as both sources of primary and secondary probe beams.

The requirements for the applicable accelerator technology can be directly derived from the AI detection methods used. These AI detection methods have been developed based on the perceived threats. The most immediate perceived threats and the materials that must be detected for each threat include:

Section 5.4 was prepared by Cameron Geddes (Lawrence Berkeley National Laboratory, 1 Cyclotron Rd, Berkeley, CA 94720, USA), e-mail: CGRGeddes@lbl.gov

R. Garnett (✉)
Los Alamos National Laboratory, Los Alamos, NM, USA
e-mail: rgarnett@lanl.gov

I. Jovanovic, A. S. Erickson (eds.), *Active Interrogation in Nuclear Security*, Advanced Sciences and Technologies for Security Applications, https://doi.org/10.1007/978-3-319-74467-4_5

1. Conventional explosives—important elements include N, O, Cl, Na, S, K, P,
2. Dirty bombs—^{137}Cs, ^{60}Co (medical rad waste), shielding materials: Pb, W, Fe, polyethylene ($(C_2H_4)_nH_2$),
3. Special nuclear materials—^{235}U (HEU), ^{239}Pu, ^{237}Np,
4. Weapons of mass destruction—^{235}U (HEU), ^{239}Pu, ^{237}Np, conventional explosives, tamper materials,
5. Chemical agents—sarin gas ($C_4H_8Cl_2S$), phosgene (CCl_2O), mustard gas ($C_4H_8Cl_2S$),
6. Contraband—people, illicit drugs and other illegal cargo.

To detect these threats requires a broad range of detection methods. These methods primarily use photon beams as the source of energetic particles to induce a particular reaction or to conduct imaging; however, some techniques also use neutrons and high-energy or exotic beams such as protons or muons, respectively. Detection methods that use photons include transmission radiography, neutron resonance absorption, neutron resonance fluorescence, and photon-induced fission. Several of the photon-based detection methods are described in detail in Bertozzi et al. [1]. Detection methods using neutrons include neutron-induced fission, neutron activation, neutron transmission radiography, and (n,γ) reactions involving capture or scattering. Long-standoff AI requires protons for radiography or muons to generate K muonic X rays in ^{233}U or ^{238}U, as described in [2]. Most integrated AI systems in use today incorporate several of these methods of detection simultaneously to broaden the range of use and to provide detection redundancy that helps minimize *false positives* by improving overall detection accuracy. Therefore, most systems incorporate both a bremsstrahlung X-ray production target and a neutron converter.

The energy thresholds for the reactions involved in each specific detection method ultimately determine the minimum probe energies (photons, neutrons or other particles) needed. Photons are generated primarily through the bremsstrahlung process by scattering electrons on a high-Z metallic target such as tantalum. The useful photon energy range for AI applications is 1–15 MeV, although present US regulations limit the energy to $\leq$10 MeV to protect humans from accidental life-threatening exposures. Neutron beams can be generated through several processes; however, the most common uses a D-D or D-T neutron generator, generating 2.5-MeV and 14.1-MeV neutrons, respectively. The accelerator requirements for these systems are defined by the beam energies needed.

A review of present AI systems reveals that electron linear accelerators are being most widely applied because electrons are easy to accelerate efficiently to high energies in a relatively compact system. Typical electron beam parameters are: energies up to 10 MeV, average beam currents up to 1–2 mA, and high beam duty cycle (50–100%). Proton beams are typically being generated using linear accelerators or compact cyclotrons. Nominal required beam energies are generally less than 10 MeV (very specific energies and small beam energy spread are required for specific resonance reactions such as for the detection of nitrogen in explosives) but can be as high as 500 MeV or greater for long-standoff AI, requiring large systems with average beam currents also in the 1–2 mA range, as for electrons.

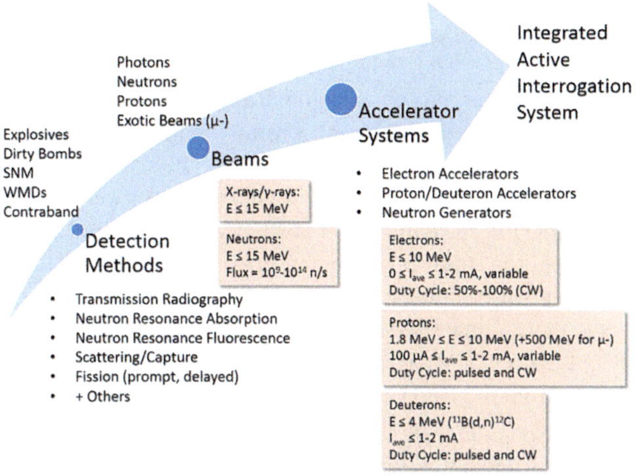

Fig. 5.1 The detection methods used define the beam and accelerator requirements for use in an integrated AI system

Pulsed and continuous-wave (CW) neutron beams can be generated with energies up to 15 MeV using a 4-MeV deuteron linear accelerator through the $^{11}B(d,n)^{12}C$ reaction. Generally, high average beam currents are desirable to improve detection statistics to speed up the AI process.

While specific accelerator requirements are derived from the detection method being applied, general accelerator requirements for a practical AI system can also be specified that need to be met in order for such a system to be fielded. These general requirements include large beam momentum and/or beam energy acceptance which allows acceleration to a wide range of final beam energies, high beam transmission which limits beam losses, reduces accelerator activation and allows for hands-on maintenance, a small or compact footprint, portability, dynamically-variable beam energy and/or beam current, variable duty factor and output beam current, high reliability and availability, and finally if possible, low cost. It is difficult to meet all of these general system requirements in any given system. Figure 5.1 summarizes the detection methods and corresponding probe beam and accelerator requirements for each of the detection methods.

5.1.1 Maturity of Applicable Accelerator Technology

Electron linear accelerator (linac) technology is readily available and is perhaps the most mature technology due to its use in medical X-ray generation and non-destructive testing in industry. Both normal-conducting and superconducting standing-wave and traveling-wave RF electron linacs are available from multiple vendors including Varian (www.varian.com, 2017), RadiaBeam Technologies (www.radiabeam.com), and Niowave, Inc. (www.niowaveinc.com, 2017) Other commercially available accelerators applicable to AI include DC accelerators such as tandem Van de Graaffs and Pelletrons, compact cyclotrons, compact neutron generators, and low-power proton and ion Radio-Frequency Quadrupole (RFQ) linacs. Advanced technology that shows promise but is still mostly relegated to the R&D environment includes various laser-acceleration schemes, induction linacs including the dielectric wall accelerator, high-average-power RFQs, Fixed-Field Alternating Gradient (FFAG) induction and RF accelerators, and Inverse-Compton-Scattering (ICS) photon sources, to mention a few. Many of these accelerator technologies show great promise for the future to meet many of the general system requirements such as compact footprint, dynamically-variable beam energy and/or beam current and lower cost. The challenge for the future is to move many of these technologies from the R&D environment to commercialization. Both new and innovative applications of present technology, as well as continuing development of new technologies is expected to impact the application of accelerators to AI in the near future.

5.1.2 Emerging Accelerator Technologies

New accelerator-based technology will be needed for next-generation AI applications. There are several emerging accelerator technologies that have the potential to meet this need in the near future. Advanced machining techniques such as additive manufacturing (3-D printing) may also enable miniaturization of accelerators beyond what is capable today or improve the accuracy of fabrication of existing compact systems [3]. One day we may see accelerators making accelerators. Several of these emerging technologies will be discussed briefly in the sections below.

5.1.2.1 Direct Laser Acceleration

During the past decade or so several mechanisms of direct laser generation and acceleration of particle beams have been demonstrated. These include target normal sheath acceleration (TNSA), radiation pressure (pondermotive) acceleration (RPA), and laser break-out afterburner acceleration (BOA). These methods have been used to generate and accelerate beams of electrons, protons, deuterons, and various ions. Intense neutron beams have been generated using deuterons and low-Z converters [4]. These TW laser-driven accelerators are being enabled by improving optical drive laser performance (higher average power and higher repetition rates), shrinking footprint, and lower cost. Present limitations that impact some application of these systems include difficulty in tailoring the final beam energy distribution—these beams generally have a large energy spread compared to beams from conventional accelerators but ongoing efforts continue to improve this, including combining laser-generated beams with conventional accelerator technology to further tailor the beam characteristics [5]. More detailed discussions of laser acceleration can be found in Refs. [6, 7]

5.1.2.2 Inverse Compton Scattering X-ray Sources

Compact high-energy X-ray sources based on inverse Compton scattering (ICS) are now being applied to AI due to the recent maturity of new technologies including high-gradient X-band (8–12 GHz) electron accelerators and performance improvements in commercial optical laser systems. These sources require a bright (small emittance, high peak charge) relativistic electron beam (typically from an electron linac) and high brightness (small waist, high energy/pulse) photons from a laser. X-band accelerating structures have now demonstrated nearly 100-MV/m accelerating gradients [8]. The repetition rate for high-power laser systems has always been a limitation for use in an AI system, but 10-Hz repetition rates are now common. ICS sources can be extremely bright and narrow bandwidth, enabling their use both in radiography and for nuclear applications requiring excitation of very-narrow, unique nuclear resonances of various isotopes. Photon energies into the MeV energy range can also be generated [9].

These sources rely on the tremendous $4\gamma^2$ upshift in photon energy that occurs as the incident laser photons are scattered from a relativistic electron beam where γ is the Lorentz factor. With $10^2 \leq \gamma \leq 10^3$, optical wavelengths can be shifted into the X-ray or γ-ray range. Figure 5.2 shows a propagating electron bunch interacting with incident photons from a laser. The forward-directed angular cone (θ) of X rays produced is also shown. The upshifted energy of the Compton scattered laser photons can be calculated from

$$E_{scat} = 2\gamma^2 \frac{1 - \cos\alpha}{1 + \gamma^2\theta^2} E_{laser\ photon}, \tag{5.1}$$

where $E_{laser\ photon} = h\nu = hc/\lambda$, h = Planck's constant, c = velocity of light, and λ is the incident laser wavelength. The maximum upshift occurs when $\alpha = 180°$ and $\theta = 0°$:

$$E_{scat} = 4\gamma^2 E_{laser\ photon} \tag{5.2}$$

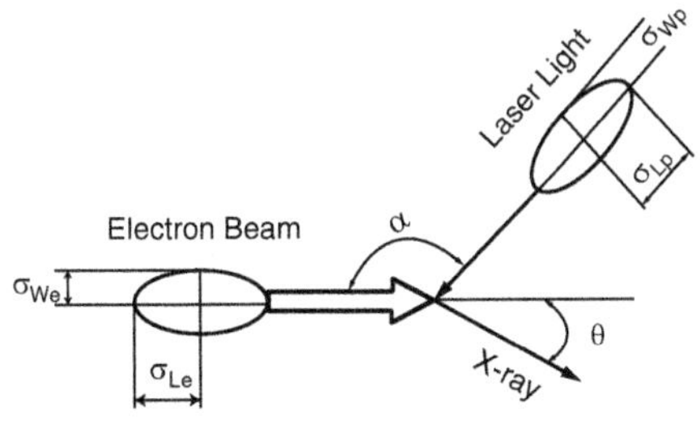

Fig. 5.2 Electron beam and laser configuration for inverse Compton scattering. Reproduced from Ref. [10]

The relativistic factor γ can be expressed in terms of the electron kinetic and rest-mass energies by

$$\gamma = (1 - \beta^2)^{1/2} = 1 + W/m_0, \tag{5.3}$$

where $\beta = v/c$, W and m_0 are the electron beam kinetic energy and electron rest mass, respectively, in MeV energy units ($c = 1$). As an example, for a beam of 100-MeV electrons, $\gamma \approx 200$. This leads to a maximum upshift in the scattered laser photon energies at $\theta = 0°$ of a factor of 80,000—a huge upshift in photon energy. It is easy to see how photons in the X-ray and even γ-ray energy range can be generated using easily achievable electron beam energies and this method. The number of upshifted photons produced depends on the quality of both the laser beam and the electron beam. These are controlled by maintaining a small laser beam waist (spot size/cross section) and a small dense electron beam in the laser-electron interaction region.

By combining a laser undulator with ICS, electron beam energies <100 MeV can be used to produce hard X-ray photons. In this case, the drive laser which supplies the initial source of photons can also provide the undulator magnetic field. A system based on this concept is operating at Lyncean Technologies, Inc. in Palo Alto, California and has demonstrated generation of 12-keV X rays using a 25-MeV electron beam. This system combines a laser undulator with a compact electron ring and ICS interaction region.

5.2 Linear Accelerators and Bremsstrahlung Sources

5.2.1 *Components of a Linear Accelerator*

Figure 5.3 shows a schematic layout of a linear accelerator (linac) and its basic components. All linacs consist of a source of particles to generate the beam (electron gun or ion source) followed by a low-energy injector (pre-accelerator) section that conditions (matches) the beam for injection into the main accelerating section, the main accelerating section, and a final beam transport section to condition the output beam to match the target or application requirements (beam spot size, divergence, etc.). All linacs also include focusing magnets of some type to perform beam conditioning and to maintain a focused beam so that it can be transported through the linac while being accelerated. Compact electron linacs typically use solenoidal focusing magnets (coils) that are placed around the accelerating structures. Proton or ion linacs typically use solenoids in the low-energy beam transport matching section and quadrupole focusing magnets in the main accelerating sections. Various focusing magnet configurations can be used [11] and optimized as needed to meet performance requirements. All linacs also require support systems such as vacuum, cooling, controls and diagnostics, as well as an RF system to drive the accelerating cavities.

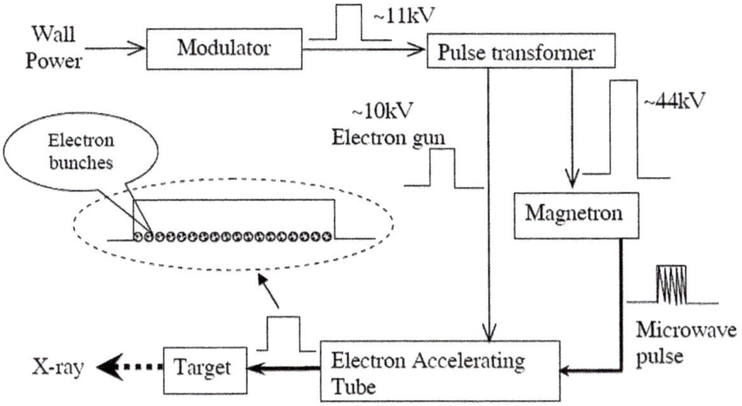

Fig. 5.3 Schematic layout of the basic components of an electron linac. Reproduced from Ref. [12]

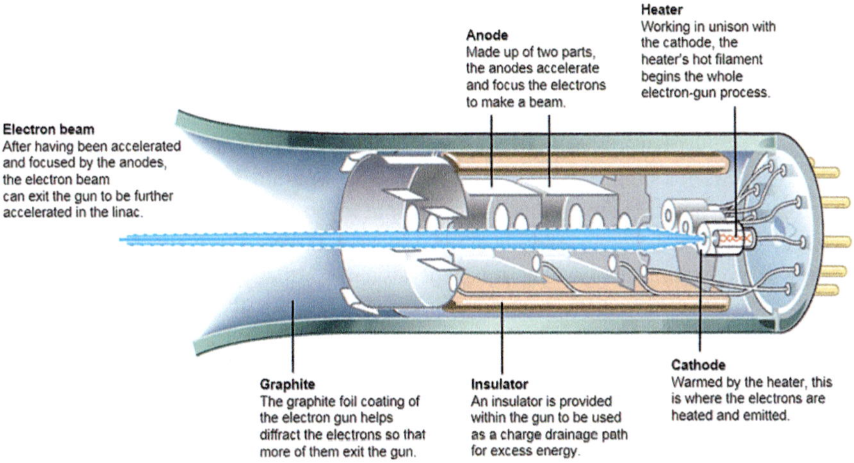

Fig. 5.4 Basic components of a thermionic electron source. Reproduced from www.howitworksdaily.com (2014)

5.2.2 Electron and Ion Sources

The simplest source of electrons for injection into a linac is the electron gun. The main process of generating electrons in an electron gun is known as thermionic emission. Figure 5.4 shows a schematic diagram of a thermionic gun. Inside the gun a small filament is used to heat the metal cathode and to release free electrons. The electrons are then accelerated rapidly by the electric field of the anode which is positively charged. There are small apertures in the anode plates which allow some

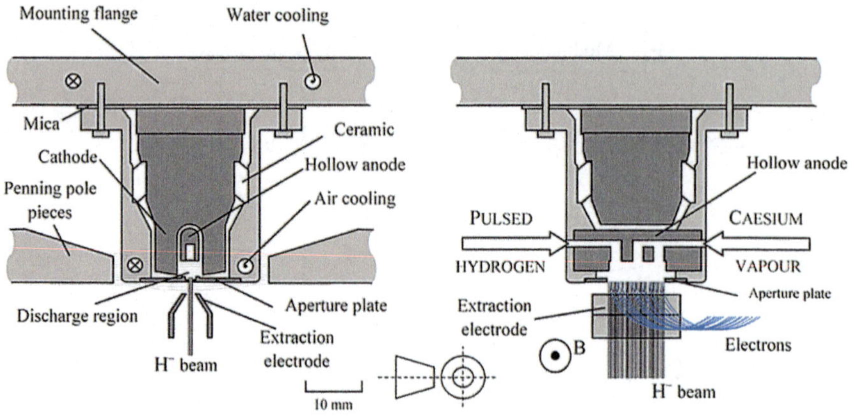

Fig. 5.5 Basic components of a Penning ion source. Reproduced from Ref. [13]

of the electrons to pass through and be concentrated into a beam. For accelerator sources, typical beam energies exiting the electron gun are in the range of 10–20 keV, but can be higher.

Another type of electron gun is based on field emission. Field emission is the discharge of electrons from the surface of a material subjected to a strong electric field. Thermionic guns are more common and operate at higher temperatures. Electron guns based on field emission operate at lower temperature and higher anode voltage, and therefore generate brighter electron beams since the initial thermal energy of the beam is lower when it is generated in comparison to the thermionic gun and the forces acting on the beam are reduced. The overall quality of the electron beam produced is therefore dependent on the physical process used, the electric field of the anode (the amount of initial acceleration), and the geometry of the beam extraction system. The highest quality beams are produced from a photocathode where a pulsed laser beam is used to cause photoemission of electrons from a metal surface. Photocathode sources are complex relative to the simple thermionic cathode and are generally used in applications where the highest beam brightness and complex time structure of the generated beam is required, and not for linac systems used for AI.

The process of generating a proton or ion beam relies on extracting heavier particles from the nuclei of atoms. There are a variety of ion sources based on several physical processes and methods. The most common ion sources in use today for generating negative hydrogen ions are Electron Cyclotron Resonance (ECR), radio-frequency (RF), or Penning sources. For high-current applications, The ECR source is the only available choice. The simplest of these negative-hydrogen-ion sources is the Penning source which is frequently also used in neutron generators to generate deuteron beams. Figure 5.5 shows a schematic view of a Penning source.

The Penning source consists of a tube-shaped anode with a magnetic field along the axis of the tube with two cathodes at either end of the tube as shown in the figure. One anode also has a small aperture to extract the beam. To generate negative hydrogen ions, the volume inside the tube is filled with hydrogen gas. An electrical discharge then generates a low-pressure plasma from which the desired ions can be extracted. The electrical discharge is created by applying an electric potential between the cathode and anode. The resulting electric field accelerates free electrons which can then ionize the hydrogen gas molecules through a process called electron impact ionization. To enhance the production of negative ions, cesium is introduced to reduce the work function of the metal surfaces inside the source volume to allow more free electrons to be generated. Penning negative ion sources using cesium are referred to as surface plasma sources and are one of the brightest of all negative ion sources [13]. These sources typically operate with voltages in the 10-kV range.

Negative ion sources were first developed to allow electrostatic accelerators to effectively double their output beam energy. For example, in a tandem generator H^- ions are first accelerated from ground to the terminal voltage, V. They are then stripped of their two electrons by passing through a thin foil. The resulting protons which already have the terminal voltage, V, are then accelerated again from the terminal back to ground, therefore doubling their voltage to 2 V.

Most cyclotrons use negative ions and stripper foils to extract the beam from the cyclotron. To extract the beam, the stripper foil is positioned near the outer perimeter of the cyclotron magnet poles. As the negative ion beam is accelerated, it circulates on larger and larger radii until it passes through the stripper foil, converting the beam from being negatively charged to positive. As a result, the Lorenz force acting on the beam due to the pole magnets is reversed and instead of the force pointing towards the center of the cyclotron, it points outwards and the beam is extracted.

In high-power proton accelerators, negatively-charged hydrogen ions are used to allow charge accumulation via multi-turn injection into a circular storage ring or synchrotron, for example. Here negatively-charged hydrogen ions from a linear accelerator pass through a stripper foil into the circular ring, leaving protons circulating in the ring. The negatively-charged hydrogen ions from the linear accelerator continue to enter the ring while the circulating beam repeatedly passes through the stripper foil unaffected. The incoming negatively-charged beam bends one way through a dipole in the injection region where the stripper foil resides and then bends out of the dipole in the opposite direction as a proton beam on top of the circulating beam. This physical process allows a local violation of Liouville's theorem [14] to be used to accumulate more and more charge in the ring. Without this negative-ion stripping, only a single pass through the stripper foil (one turn on the same orbit) could be accumulated in the ring.

5.2.3 Low-Energy Injectors

Following the electron or ion source, both electron and ion linacs require a low-energy injector section. The injector section forms the continuous beam pulse generated by the source into short beam bunches at the required accelerating RF frequency, and to transport and/or further accelerate the beam from the source to the main accelerator. Because the main purpose of the injector section is to match the beam into the main accelerator, it generally contains magnets to focus the beam and RF structures to perform bunching and acceleration. Beam bunching is achieved using dedicated buncher cavities (RF gaps) operated at the longitudinal bunching phase of $\phi_s = -90°$ followed by an appropriate drift distance. The bunchers can be operated at the fundamental RF frequency of the accelerating structure or at higher harmonics to take advantage of the non-linear fields to better tailor the beam distribution from the ion source.

In an electron injector, focusing is usually provided by an array of solenoids placed around the beam line. For ion beams, stronger focusing is generally needed where strong quadrupole focusing magnets are used. Acceleration can also easily be provided by the bunching gaps by ramping the phase to the synchronous accelerating phase. For low-energy ion beams typically used for AI, an alternative is to use the Radio Frequency Quadrupole (RFQ) which performs both the bunching, focusing, and acceleration in a single structure. The details of the RFQ will be discussed in a later section.

5.2.4 Electron Linac Technology

Figure 5.6 shows a representative electron linac-based system for AI [15]. This is a combined X ray, photoneutron source built by Rapiscan Laboratories, Inc. using a 9-MeV Varian Model L2000 S-Band electron linac. The major components of such a system are shown including the electron source (electron gun), the RF linac, the RF

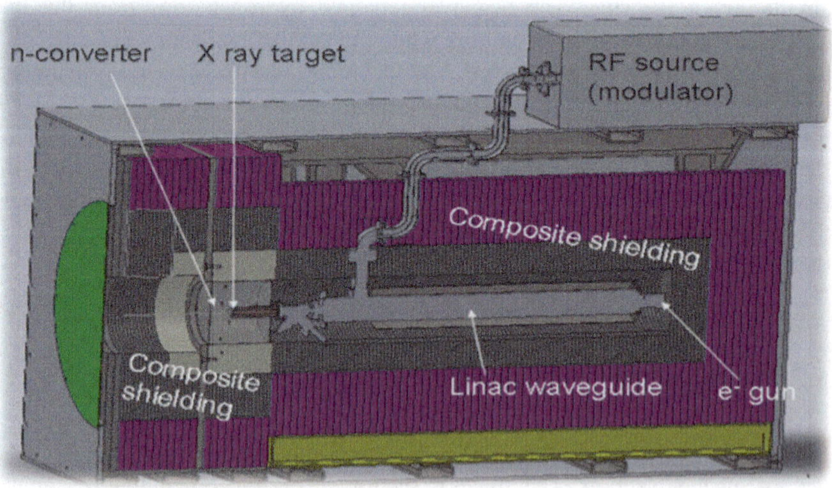

Fig. 5.6 9-MeV combined X ray, photoneutron source assembly. Reproduced from Ref. [15]

Table 5.1 Summary of typical RF microwave frequency bands used in accelerator systems

Frequency range (GHz)	Microwave bands	Typical American frequency (GHz)	Typical European frequency (GHz)
1–2	L-Band	1.5	1.17
2–4	S-Band	2.856	2.998
4–8	C-Band	5.712	5.996
8–12	X-Band	11.424	11.992

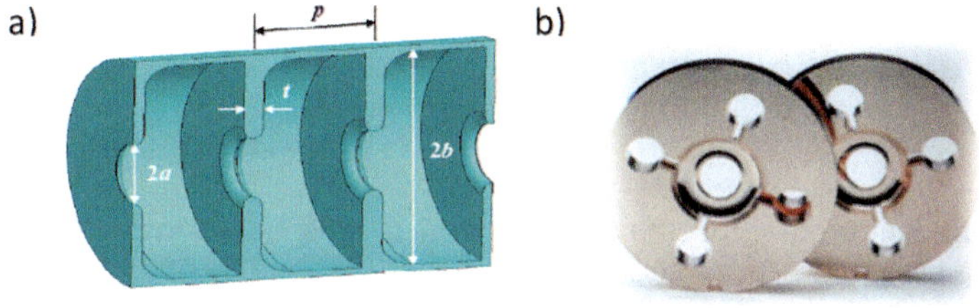

Fig. 5.7 (**a**) Typical iris-loaded linac structure used to accelerate electrons. Reproduced from Ref. [17]. (**b**) SLAC X-Band structure. Reproduced from Ref. [16]

source (gridded tube, magnetron, klystron or inductive output tube), and the X-ray target (usually tantalum) plus neutron converter (in this case, a heavy-water, D_2O converter). The dimensions of this system are approximately 1.5 m × 1.5 m × 2.5 m.

Room-temperature electron linac technology is very mature. Typical accelerating structures operate at RF frequencies of 2–4 GHz (S-band) with 2.856 GHz being the common US frequency (2.998 GHz in Europe). Table 5.1 summarizes the IEEE RF frequency bands (IEEE Standard 521-2002) for RF sources typically used to power electron linacs. The availability of RF sources has historically been determined by the communications industry or military radar system requirements, with S-band being the most commonly used for commercial applications. As a result, the accelerating structure operating frequency is selected to match RF tube availability. Room-temperature RF electron linacs have also been built in other frequency ranges, but are less common. Both C-band (5.712 GHz) and X-band (11.424 GHz) linacs have been built and RF sources are available. The higher frequencies such as X-band offer higher accelerating gradients, as mentioned earlier (up to approximately 100 MV/m), and a factor-of-four reduction in transverse structure size since the accelerating structure diameter scales proportionally to the RF wavelength.

Common features of these electron linac structures include: a thermionic cathode/injector as the source of electrons, use of an iris-loaded/disk-loaded structure which is most common due to its simplicity (see Fig. 5.7a), traveling-wave mode (RF dumped to a load), $2\pi/3$ operating mode (but other modes are also ok), energy gain up to 60 MeV/m, and use of a single RF source (typically a magnetron or klystron). Design methods for these accelerating structures are very mature with a well-established set of relationships that define the structure dimensions parameterized by RF wavelength and structure quality factor (Q). Figure 5.7b shows an X-band structure developed at SLAC [16].

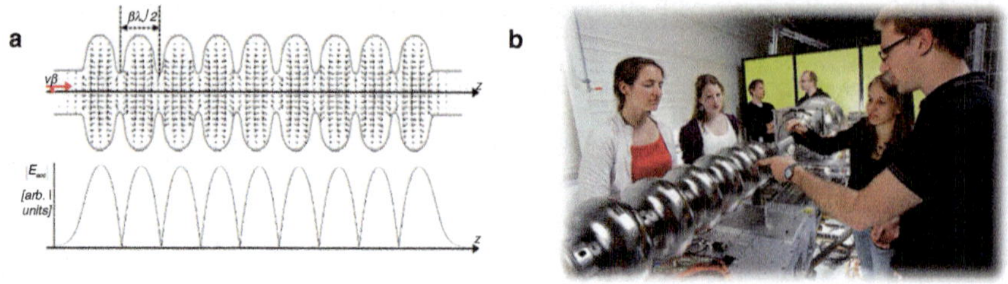

Fig. 5.8 (**a**) Typical multi-cell superconducting elliptical cavity used to accelerate $\beta = 1$ electrons. Reproduced from Ref. [18]. (**b**) Prototype 1.3-GHz ILC cavity. Reproduced from www-bd.fnal.gov

Conventional room-temperature electron linac systems are available from many commercial sources. These systems are efficient and relatively compact. Present limitations include low-average beam currents (typically less than 100 μA) and low duty factor (0.1–1.0%). High average currents and high duty factor require more attention to detail of structure cooling that generally increases structure cost and complexity.

The use of superconducting (SC) niobium accelerating structures for AI is increasing now due to the availability of these systems through commercial vendors such as Niowave, Inc (www.niowaveinc.com, 2017). These structures have been used worldwide to efficiently accelerate both electrons and ions. For electrons, simple structures can be used such as the multi-cell elliptical cavity. Figure 5.8 shows a representative example of such a cavity. A common frequency for these cavities is 1.3 GHz due to their development for the International Linear Collider (ILC). The number of cells for these structures usually ranges from 5 to 9. They can be operated at both 4 K and 2 K, but typically are operated at 4 K due to the lower cryosystem cost. These are π-mode structures (TM_{010}) with accelerating-gap spacing of $\beta\lambda/2$ where $\beta = v/c$ and λ is the RF wavelength. Other SC structure types such as the spoke resonator, half-wave resonator, or quarter-wave resonator are needed to efficiently accelerate low-velocity ions. A particular advantage of SC structures is their large apertures and the capability to operate CW (100% duty factor) due to the low ohmic losses of the structure compared to normal-conducting structures. Systems are currently available that can provide average beam currents up to 1 mA. A significant breakthrough in accelerating gradient is needed for these systems to have significantly smaller footprint. Progress in increasing superconducting gradients has stalled over the last decade but some approaches such as hydro-forming multi-cell cavities to reduce the number of surface defects that contribute to RF breakdown sites and surface coatings are beginning to show promise.

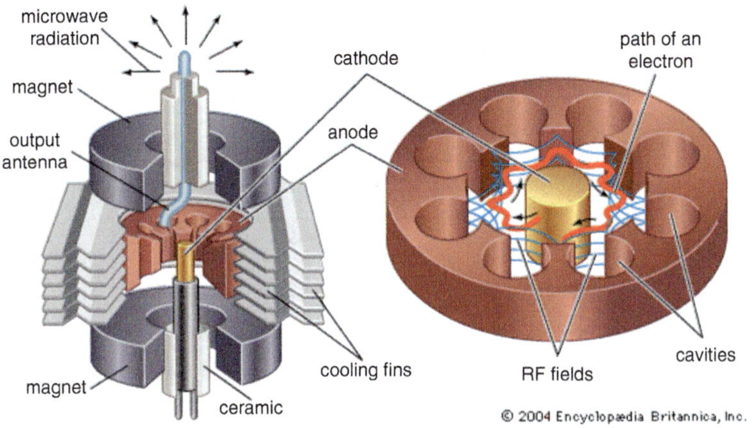

Fig. 5.9 Basic components of a magnetron RF source. Reproduced from www.britannica.com/technology/magnetron (2016)

5.2.5 Radiofrequency Systems

5.2.5.1 The Magnetron

A magnetron (see Fig. 5.9) consists of a heated cathode placed at the center of a lobed circular chamber in a high negative potential (continuous or pulsed) created by a high-voltage, DC power supply. A magnetic field parallel to the filament is generated by a permanent magnet surrounding the outside of the chamber. The magnetic field causes the electrons, attracted to the positive outer part of the chamber, to spiral outward in a circular path. Spaced around the rim of the chamber are additional cylindrical cavities. Slots are cut along the length of the cavities that open into the central, common cavity space. As electrons sweep past these slots, they induce high-frequency RF fields in each resonant cavity, which in turn causes the electrons to bunch into groups. As a result of the electron bunching, the generation of RF is enhanced.

The RF energy is extracted by a short antenna that is connected to a waveguide and the waveguide directs the extracted RF energy to the accelerating cavities. The size of the cavities determines the resonant frequency and therefore also the frequency of the emitted RF waves. A shortcoming of the magnetron is that due to the process of generating the RF, the frequency is not precisely controllable. The operating frequency varies with changes in load impedance, with changes in the supply current, and with the temperature of the tube. This is not a problem for some applications such as heating in a microwave oven or in some applications of radar where the receiver can be synchronized with the magnetron frequency. In applications where precise frequency control is needed, such as in a linac, the klystron is typically used.

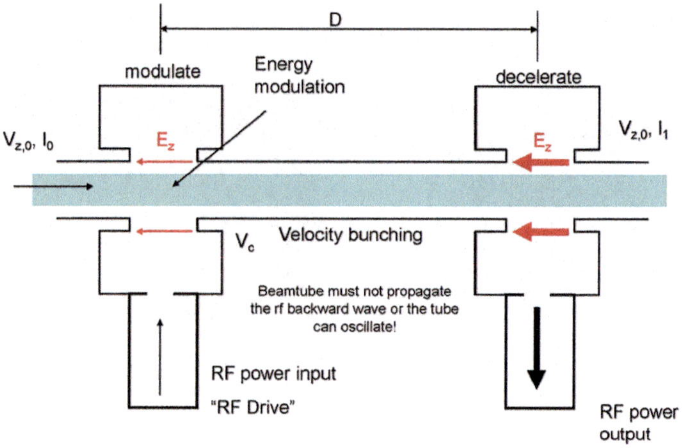

Fig. 5.10 Schematic layout of a klystron. Reproduced from Ref. [19]

The magnetron is a self-oscillating device that requires no additional external elements other than a power supply. A well-defined threshold anode voltage must be applied and synchronized with the oscillator output. The applied voltage is a function of the dimensions of the resonant cavity, and the applied magnetic field. Magnetrons have been developed that can operate over a broad spectrum of RF frequencies (S-band, L-band, C-band) and the very-high peak powers (1 kW–3 MW) in both pulsed and continuous-wave (CW) mode.

5.2.5.2 The Klystron

The klystron is the most efficient of linear-beam RF tubes. Figure 5.10 shows a schematic layout of a klystron. In a klystron, the electron beam generated by a DC thermionic cathode and accelerated by high-voltage electrodes (typically tens of kilovolts) interacts with RF fields as it passes through resonant cavities in the klystron structure separated by a cylindrical metal drift tube. The resonant cavity dimensions have been selected by satisfying Maxwell's equations for the appropriate boundary conditions to generate RF at the desired frequency. The drift tube diameter has likewise been selected to propagate the generated RF wave to the output cavity. The electron beam first passes through the input cavity to which an input RF signal close to or at the resonant RF frequency has been applied creating standing waves, which produce an oscillating RF field which acts on the electron beam. The electric field causes the electrons to bunch. To further enhance the beam bunching, the klystron structure may contain additional buncher cavities.

The energy of the bunched electron beam amplifies the RF signal, and the amplified signal is extracted from a cavity at the other end of the drift tube through a coaxial cable or waveguide. The output signal can be coupled back into the input cavity to form an RF oscillator to further amplify the generated radio waves. The gain of klystrons can be high, 60 dB (a factor of one million) or more, and with output powers up to tens of megawatts. An advantage of the klystron that makes it particularly attractive for application to linacs is the narrow output frequency bandwidth, usually a few percent. Also, klystron efficiency increases as the cathode electron beam current increases. Klystrons are available that operate over the full spectrum of RF frequencies of interest for linacs including S-band, L-band, C-band, and X-band.

5.2.6 Special Relativity, Particle Dynamics, and Accelerating Beams

In the previous section on Inverse Compton Scattering we discussed the effects of scattering laser photons by a relativistic electron beam. Some additional basic concepts need to be discussed in order to understand the differences between how electron beams and proton or ion beams are accelerated due to fundamental differences in particle rest mass. The rest mass of the electron is 9.1×10^{-31} kg (0.511 MeV/c^2), whereas the rest mass of the proton is 1.6×10^{-27} kg (938 MeV/c^2). This factor of 1836 difference in rest mass means that it is much easier to accelerate electrons using an externally applied electromagnetic field provided by an RF linear accelerator, for example, as compared to the much heavier protons. Heavy ions have even greater rest masses. Equation (5.3) above can be rewritten to express the relativistic velocity $beta = v/c$ as a function of the particle kinetic energy W in units of MeV:

$$\beta^2 = 1 - (W/m_0c^2 + 1)^{-2}. \tag{5.4}$$

Plotting Eq. (5.4) (see Fig. 5.11) as a function of particle kinetic energy for electrons ($m_0c^2 = 0.511$ MeV) and protons ($m_0c^2 = 938$ MeV) demonstrates how quickly electrons approach the velocity of light as they are accelerated to higher energies and how slowly by comparison the proton velocity increases as a function of energy. In general, electrons and other massive particles are relativistic when their kinetic energy is comparable to or greater than their rest-mass energy as given by $E = m_0c^2$.

In order to accelerate a particle, a force must be applied. The Lorentz force on a particle with charge q and velocity $\mathbf{v}$ in an electric field $\mathbf{E}$ and a magnetic field $\mathbf{B}$ is given by

$$\mathbf{F} = q\,[\mathbf{E} + (\mathbf{v} \times \mathbf{B})] = m_0\, d(\gamma v)/dt. \tag{5.5}$$

Neglecting the magnetic field, for an RF electromagnetic wave propagating along the z direction in a waveguide (or accelerating structure), the electric field on-axis can be expressed as a function of wave number, k, as

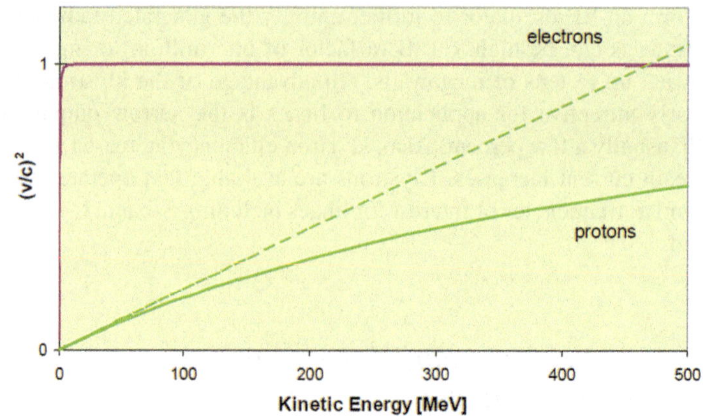

Fig. 5.11 Plot of the square of the particle velocity as a function of particle kinetic energy for both electrons and protons. Note how quickly the electrons approach the speed of light. Reproduced from Ref. [20]

$$E_z(z,t) = E(z)\cos\left[\omega t - \int_0^z dz\, k(z) + \phi\right]. \tag{5.6}$$

The wave number k can be expressed in terms of the phase velocity v_p by $k(z) = \omega/v_p(z)$. In order to efficiently accelerate a particle, the phase velocity of the electromagnetic wave must be closely matched to the particle velocity. If this condition is met then the particle arrives at a position z at time $t(z) = \int_0^z dz/v(z)$ and the electric force on the particle is given by $F_z = qE(z)\cos\phi$. This particle is called the synchronous particle and the phase ϕ is called the synchronous phase. Significant simplifications occur when accelerating electrons as compared to protons or ions. Since electrons are quickly accelerated to velocities very close to the speed of light, this allows the phase velocity of the traveling accelerating wave in the structure to be set equal to the velocity of light throughout the entire electron linac. Since protons travel more slowly, this approach of fixing the phase velocity in general does not work and requires a different approach.

Typically for protons and ions, an electromagnetic standing wave in an RF cavity is used to accelerate the beam. The accelerating structure can be fabricated as a periodic or quasi-periodic array of accelerating cells, each with a single accelerating gap where the parameters vary linearly along the structure to maintain synchronism with the beam particles as they are accelerated. A standing wave is the superposition of two waves traveling in opposite directions in the accelerating structure. For a standing wave, the electric field along the axis is given by $E(z,t) = E(z)\cos(\omega t + phi)$. In a simplified form this leads to a gain in the kinetic energy of a beam particle in a gap of length L equal to

$$\Delta W = qV_0TL\cos\phi \tag{5.7}$$

where the axial RF voltage in the gap is given by

$$V_0 \equiv \int_{-L/2}^{L/2} E(0, z)dz \tag{5.8}$$

and T is a quantity called the *transit-time factor*, which typically ranges from 0 to 1, and is essentially the normalized average fraction of the longitudinal electric field seen by a particle in the gap (see [11] for a more detailed derivation) and can also be thought of as a figure of merit of accelerating efficiency. The detailed functional form of the transit time factor depends on the RF cavity type and the resonant RF cavity mode. One of its simplest forms is for a simple TM_{010} mode pillbox cavity of length g and having small aperture radius holes on both ends:

$$T = \frac{\sin \pi g/\beta\lambda}{\pi g/\beta\lambda}, \tag{5.9}$$

where λ is the RF wavelength.

Typically, the accelerating cavity geometry is optimized and varied as a function of proton or ion beam energy to maintain high transit-time factor values above 0.5. This is not required for traveling-wave structures used for accelerating electrons where $\beta = 1$ and the cell length and gap spacing are held constant.

All accelerators operate on the principle of phase stability. In a phase-stable accelerator particles are accelerated in a series of gaps as shown in Fig. 5.12 by an alternating RF field that provides an on-axis accelerating force. In general, particles in the beam arriving with slight phase or energy errors will continue to be accelerated with small oscillations in energy and phase around the correct beam centroid energy and equilibrium phase. This automatic correction of the phases and energies of the beam distribution leads to individual beam bunches separated in time by the RF period of the accelerator (see Fig. 5.13).

In a linear accelerator, the beam bunch is accelerated in a linear path by a constant-frequency RF field. The gap separation is adjusted to the RF period such that a beam particle arriving at each gap is always at a specified equilibrium, stable phase of the accelerating field and is synchronous with the RF period. This synchronicity of the beam with the phase of the RF field is required to allow

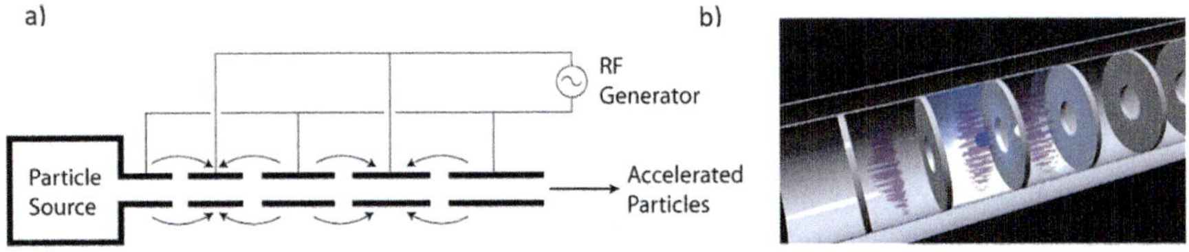

Fig. 5.12 (**a**) Schematic layout of a drift-tube linac showing the accelerating gaps and the time varying electric fields. Reproduced from Ref. [21]. (**b**) The electric-field distribution in a disk-loaded structure used to accelerate the beam. Reproduced from Ref. [22]

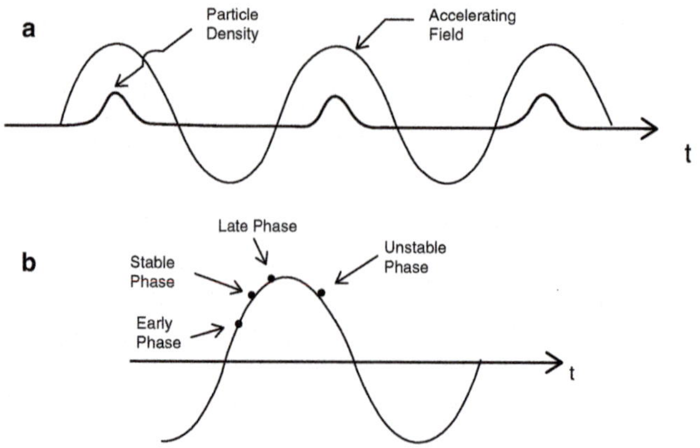

Fig. 5.13 (**a**) Individual beam bunches are formed as part of the acceleration process in a linac. Like charged particle fill RF *buckets* only on one phase of the RF. (**b**) The particles in the bunch must be at the proper phase of the RF to remain in the phase-stable region. Reproduced from Ref. [11]

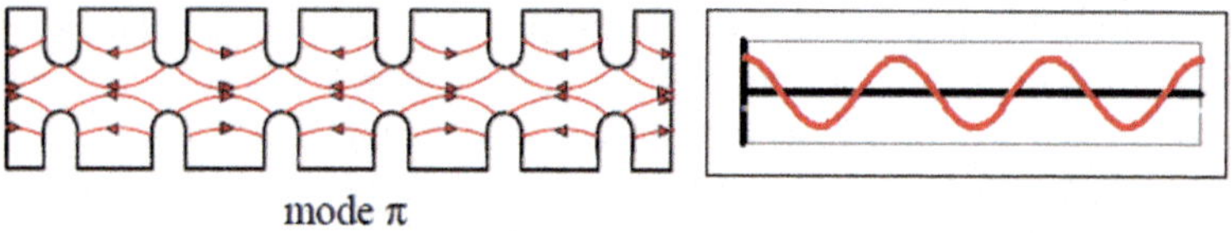

Fig. 5.14 Multi-cell RF cavity operating in the π-mode. Reproduced from Ref. [20]

maximum energy transfer to the particles and to efficiently accelerate the beam. Figure 5.14 shows a multi-cell cavity operating in the so-called π-mode where the fields in the gaps have a 180° phase difference between adjacent cells at any snapshot in time. The arrows in the figure represent the direction of the on-axis electric field resulting from the sinusoidal, time-varying RF field also shown. Therefore to maintain synchronicity of the beam position at the center of the gaps for maximum energy gain, for an RF period τ the length of a cell L must be equal to

$$L = \beta c\tau/2 = \beta\lambda/2. \tag{5.10}$$

From this example it is clear that for electrons with velocity nearly equal to that of light ($\beta = 1$), most of the linac after the initial beam injection and acceleration can efficiently accelerate the beam using a sequence of identical accelerating structures with fixed gaps separated by a distance equal to $\lambda/2$. For protons or ions where β increases more slowly by many orders of magnitude or a comparable kinetic energy, the cell lengths must increase in a well-prescribed manner as a function of beam energy to maintain synchronicity and to efficiently accelerate the beam.

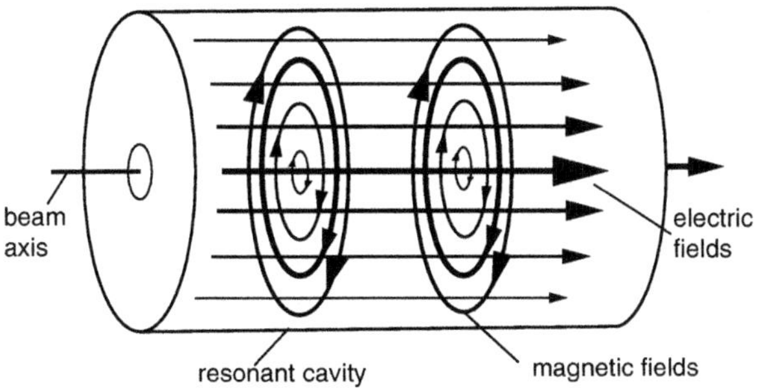

Fig. 5.15 The transverse-magnetic TM010 mode in a cylindrical waveguide. Reproduced from Ref. [24]

5.2.7 Accelerating Structures

In general, all accelerating structures are resonant RF cavities consisting of a metallic pipe that acts as a cylindrical waveguide that satisfies certain boundary conditions based on Maxwell's equations. The actual geometry of the structure is defined by the operating frequency and the RF operating modes desired. In order to accelerate particles, a longitudinal electric field must be generated on-axis as discussed earlier. From Maxwell's equations and basic RF theory [23] we know that a structure can be excited to resonate in well-defined modes that have electromagnetic field configurations of interest. The specific family of modes that have a longitudinal electric field on-axis are the transverse-magnetic (TM) modes. Figure 5.15 shows the TM_{010} mode in a cylindrical waveguide. This mode has the lowest resonant frequency. Also, the electric field of this mode points from one end of the cavity to the other and has its largest amplitude on the axis where $r = 0$. As a result, this mode is ideal for accelerating charged particles.

For a purely cylindrical waveguide without internal structure, the phase velocity is always greater than c and therefore synchronicity between the wave and a propagating beam can never be achieved. To lower the phase velocity so that synchronicity can be achieved, the structure must be modified. This leads to the disk-loaded waveguide which is a periodic structure with cell length L as shown in Fig. 5.7. Disk-loaded linacs are used to accelerate electrons and are generally designed to have a phase velocity $v = c$ for the specified operating frequency. They can be operated in either standing-wave (SW) or traveling-wave (TW) mode.

A typical traveling-wave (TW) disk-loaded linac is shown in Fig. 5.16. Electrons at velocity nearly equal to $\beta = 1$ enter the structure when the electric field on-axis is at a maximum. The beam is initially pre-accelerated and bunched prior to injection into the linac and is injected at the proper RF phase. On average the beam will travel along the linac in phase with the RF wave and as a result be accelerated. RF energy to establish the traveling wave and the on-axis electric field to accelerate the beam is provided by the input coupler also shown in Fig. 5.16. An output coupler

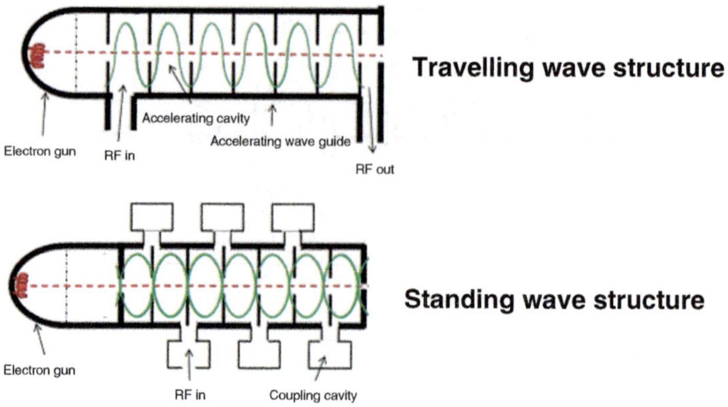

Fig. 5.16 Typical RF coupling configurations for traveling wave (TW) and standing wave (SW) linac structures. Reproduced from www.slideshare.net (2016)

is also provided to extract the excess RF power. The excess RF power (typically 25–30%) is dumped into a matched resistive load to minimize any reflected power in the RF accelerating cavity. Some RF power is also dissipated as resistive losses in the structure which is usually made of copper.

The basic design parameters for the disk-loaded linac structure have been optimized over time and are well-known [17]. Assuming the structure parameters as defined in the figure, for a given RF wavelength, $\lambda = c/f$ where f is the resonant frequency, the cell geometry of the structure can be defined:

$$b/\lambda \approx 0.383\left[1 + 20(a/\lambda)^3\right], \tag{5.11}$$

where $v_g/c \approx 2.4(a/b)^4$, $(\pi cd)/(\lambda v_g Q) = 1.26$, and $d \equiv$ distance between RF feeds.

Most disk-loaded structures used for accelerating electrons follow this prescription. The resonant frequency of the structure is given by $\omega_0 = 1/(LC)^{1/2}$, where L is the inductance and C the capacitance, respectively of a single cell in the structure. Although coupling between cells of the structure can occur along the axis through the cell apertures, most structures use stronger cell-to-cell magnetic coupling through off-axis slots cut into the cell walls (see Fig. 5.7b).

For protons and heavy ions, synchronization between the accelerating beam particles and the RF field is achieved by keeping the phase of the standing wave constant and varying the spacing of the gaps in the structure as a function of length along the accelerating cavity to maintain synchronism. The most common standing-wave accelerating modes are shown in Fig. 5.17. For the π mode the electric field is at a maximum at the center of each cell and changes sign from one cell to the next. To maintain synchronicity with the beam for this mode requires the cell length $L = \beta\lambda/2$ where $\beta < 1$. Both the 0 mode ($L = \beta\lambda$) and the $\pi/2$ mode can also be used.

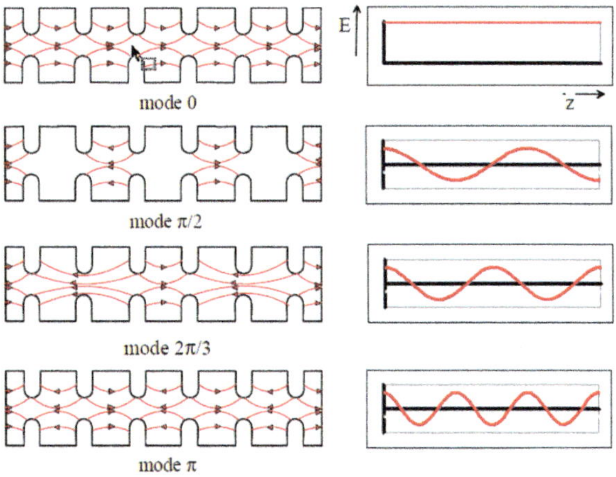

Fig. 5.17 Common SW structure accelerating modes. Reproduced from Ref. [20]

Fig. 5.18 Inside view of the Fermilab drift-tube linac. Note the vertical drift-tube supports and side-mounted post couplers that are used to stabilize the RF field in the tank (www-bd.fnal.gov)

Perhaps the most common 0-mode structure for accelerating protons at relatively low energy (below about 100 MeV) is the drift tube linac (DTL). The DTL is a standing-wave structure. Figure 5.12a shows the general configuration of the DTL. Note the increasing length of the drift tubes. The RF accelerating gaps are held at constant length but the drift tubes increase in length as a function of the velocity of the beam. As a result of Gauss's law when the beam bunch is inside a drift tube it is shielded from the electric field and therefore *drifts*. The beam only sees the on-axis accelerating field in the gaps. Typically, the drift tubes also contain quadrupole focusing magnets. In a DTL, the aperture on the beam axis has been greatly reduced in order to increase the RF efficiency. The particular configuration of the fields allows the walls between cells to be completely removed. This also maximizes the RF coupling between the drift tubes, as well as power flow inside the structure, and minimizes RF losses on the tank walls. The drift tubes are supported by *stems* from the top of the structure. Slug tuners are used to displace the volume within the resonant cavity and to actively tune the structure to keep it on resonance as its temperature fluctuates. Additionally, post-couplers provide field stabilization against mechanical errors and are set during the final structure tuning and fixed in place before the DTL is operated. Figure 5.18 shows the inside of the Fermilab 201-MHz DTL. The upper supports for the drift tubes can be seen, as can the side-mounted post couplers.

5.2.7.1 Traveling Wave vs Standing Wave Operation

Figure 5.16 shows the RF coupling schemes for both TW and SW accelerating structure for comparison. In TW operation where the RF field is introduced from one end of the accelerator, the individual cavities fill sequentially with electromagnetic energy at the group velocity of the structure. The fill time of the cavity for a single pass is approximately given by $t_f = L/v_g$, where the v_g is the group velocity and L is the length of the accelerating cavity. In SW operation, the RF field fills the whole structure simultaneously; however, multiple reflections in the cavities are required before the field reaches its full amplitude. Maximum acceleration in the TW structure occurs when the beam velocity is close to the group velocity of the structure. For the SW structure, the beam must be injected after the fields have reached maximum, usually at a time greater than the fill time for the cavity.

Before high-power RF sources were developed, the choice to operate in TW or SW mode was strongly influenced by the RF source available. Coupling together and trying to synchronize low-power magnetrons resulted in problems maintaining frequency and phase stability. These problems could be overcome by using only SW structures which allowed the backward component of the standing wave to be used to stabilize the magnetron frequency. The problems with RF phase stability were eventually solved by the development of high-power klystrons which could be driven from a single power source. Today these issues are no longer a limitation. High power magnetron and klystrons exist over a broad range of applicable RF frequencies.

Today RF sources for accelerators are designed to look as much as possible like a pure resistive load. For TW accelerators this is simple to achieve since no reflected wave appears at the input drive point to give a reactive component of the impedance. For high-Q SW systems, the RF load must always be impedance matched to avoid large amounts of reflected RF power. In general, TW linacs are less efficient and more complicated to operate. The achievable accelerating gradients in SW linacs is approximately a factor of two higher than in a TW linac, making the SW linac more compact. As a result, most AI linacs use SW structures. Table 5.2 compares typical TW and SW electron linac parameters.

Table 5.2 Comparison of TW and SW linac parameters

Parameter	TW	SW
Typical accelerating gradient	5 MeV/m	10 MeV/m
Wall plug-to-accelerating efficiency	20–50%	30–60%
Beam capture efficiency	80%	20–30%
Electron gun voltage	40 keV	5–20 kV
RF bandwidth	2 MHz	200 kHz
Low-level RF control system	Not required	Required

5.2.7.2 Phase Space and Beam Emittance

A beam bunch in a linac consists of a distribution of charged particles moving in approximately the same direction, with approximately the same positions, the same phase relative to the RF accelerating fields, and approximately the same energies. Additionally, at a given time t, every particle in the beam can be described by three position coordinates (x, y, z) and three momentum coordinates (p_x, p_y, p_z), for a total of six coordinates per particle, where each particle can be represented by a single point in the six-dimensional phase space of spatial coordinates and momenta. Since this six-dimensional space is difficult to visualize, it is easier to refer to a collection of points in two-dimensional projections of the beam phase space where the three normalized phase-space projections are $x - p_x/m_c$, $y - p_y/m_c$, $z - p_z/m_c$, where the coordinates are defined as above. Instead of measuring the transverse momenta of the particles in the beam, it is convenient to measure the divergence angles $x' = dx/ds$ and $y' = dy/ds$ which are proportional to the transverse momenta.

In longitudinal phase space $(z - p_z/m_c)$, the position and momentum relative to the synchronous particle are replaced by the phase and energy variables $\Delta\phi$ and ΔW. The projected phase-space area occupied by the beam is characterized by the emittance, ϵ, which is a measure of its phase-space extent, and emittance growth increases the minimum achievable beam size for a given focusing system. In the presence of only linear fields, the beam phase-space contours in a linac are approximately elliptical. The presence of non-linear fields can distort the particle trajectories in phase space and lead to a considerable departure from elliptical trajectories. For a distribution of particles in a beam bunch, the most useful definition of emittance in a linac is the RMS emittance. The general form of the RMS ellipse in two-dimensional phase space is defined by the Courant-Snyder (CS) parameters [25] α, β, γ and an RMS emittance ϵ_{rms}. For the un-normalized transverse emittance in $x - x'$ space, the RMS ellipse is given by the general equation for an ellipse by

$$\gamma x^2 + 2\alpha x x' + \beta x'^2 = \epsilon_{rms}, \tag{5.12}$$

where $\epsilon_{rms} = (Area)_{rms}/\pi$ and $\gamma\beta - \alpha^2 = 1$. The CS parameters can then be defined as functions of the moments of the beam distribution in the projected phase space coordinates:

$$\langle x^2 \rangle = \beta\epsilon_{rms},$$

$$\langle x'^2 \rangle = \gamma\epsilon_{rms},$$

$$\langle x x' \rangle = -\alpha\epsilon_{rms},$$

$$\epsilon_{rms} = \sqrt{\langle x^2 \rangle \langle x'^2 \rangle - \langle x x' \rangle^2}$$

Similar expressions for the CS parameters and the RMS emittance can be written for the transverse $y - y'$ and longitudinal $\Delta\phi - \Delta W$ two-dimensional phase spaces. During acceleration, the transverse normalized emittance defined by $\epsilon_n = \epsilon\beta\gamma$ is an adiabatic invariant. This is because the transverse momenta of the beam are unaffected by longitudinal acceleration while the divergences of the particles are reduced during acceleration. The Lorentz factor, γ, takes this into account. Therefore, the acceleration reduces the un-normalized emittance but does not affect the normalized emittance. In general, emittance is a conserved quantity in six-dimensional phase space (Liouville's theorem) when there are no dissipative forces, no particles are lost or created, certain collisional effects between particles can be ignored.

Emittance is a figure of merit of the quality of a beam. In general, linac designs seek to minimize emittance growth which can lead to larger beams or filamentation of the beam distribution that can lead to beam losses. This is typically more of a concern in ion linacs, but beam poor quality can also impact electron linac performance. Accelerator parameters such as the accelerating and focusing gradients used, the phase profile along the linac structure, details of the injected beam distribution including matching the input beam into the linac structure, can all be used to control or tailor the beam distribution or beam emittance.

5.2.7.3 Quality Factor and Shunt Impedance

There are several figures of merit commonly used to characterize accelerating cavities. The most common is the quality factor, or Q, of a resonant cavity. The Q of a cavity can be defined in terms of the average power loss, P, as

$$Q = (2\pi f U) / P, \tag{5.13}$$

where f is the resonant RF frequency and U is the stored energy. If a cavity has a high quality factor it is electrically efficient. Another useful figure of merit is the effective shunt impedance per unit length defined as the effectiveness per unit power loss per unit length for delivering the maximum energy gain to a particle:

$$ZT^2 = (E_0 T)^2 L/P, \tag{5.14}$$

where $E_0 = V_0/L$, V_0 is the axial RF voltage in the gap for a known gap length L. The quantity Z is called the shunt impedance. This ratio is expressed in units of meg-ohms per meter and can be approximated by:

$$Z = 10^{-6} E_0^2/P. \tag{5.15}$$

For most copper linear accelerator structures Z ranges between 25–50 MΩ/m. One of the main goals of choosing the accelerating cavity parameters is to maximize the effective shunt impedance per unit length. This is especially critical for normal-conducting cavities when RF losses can be high. Several modern accelerator and electromagnetic structure modeling codes can be used to optimize accelerating cavity parameters such as the CST Studio suite of codes [22].

5.2.8 *Beam Focusing*

In addition to the axial accelerating forces in a linac, the beam will experience the external focusing forces provided by an array of magnetic quadrupoles or magnetic solenoids. These focusing magnets are required to counteract the self-force of the beam since all particles in the beam have the same charge and from other interactions of the beam particles with the cavity RF fields that also have a net defocusing effect. In order to effectively transport and accelerate a beam in a linac without significant losses and to maintain the quality of the beam, there must be an equilibrium between the external focusing forces and the internal defocusing forces acting on the beam. In general two types of magnet focusing can be found in linacs. Solenoidal focusing is typically employed in electron linacs and whereas both types of focusing will work, quadrupole focusing lattices are typically used in ion linacs.

5.2.8.1 Space Charge Forces

The Coulomb repulsion between like-charged particles in the beam bunch is commonly referred to as the space charge force. At relativistic velocity, the space-charge repulsion in the beam begins to be compensated for by the self-generated magnetic field of the moving beam particles. The net radial force seen by a single particle in the beam bunch is given by

$$F = q\left(E_r - vB_\phi\right) = qE_r\left(1 - v^2/c^2\right) = qE_r\left(1 - \beta^2\right) = qE_r/\gamma^2, \qquad (5.16)$$

where E_r and B_ϕ are the radial and azimuthal components of the electric and magnetic fields, respectively:

$$E_r = q/r\int_0^r n(r)rdr \qquad (5.17)$$

$$B_\phi = qv/r\int_0^r n(r)rdr. \qquad (5.18)$$

As the beam velocity approaches the limit $v = c$, γ^2 approaches infinity and the repulsive forces goes to zero. However, for high-current beams at low energy the space charge force must be compensated for.

5.2.8.2 Radiofrequency Defocusing Forces

The radial impulse near the axis of an accelerating gap, assuming that for the mode of interest only the E_z, E_r, and B_θ components of the RF field are non-zero and that E_z is independent of r, can be approximated the expression [11]

$$\Delta \left(\beta \gamma r'\right) = -\frac{\pi q E_0 T \sin \phi L}{mc^2 \gamma^2 \beta^2 \lambda}, \tag{5.19}$$

where the radial component of the momentum is given by $p_r = mc\gamma\beta r'$ and $E_0 = V_0/L$, where V_0 is the voltage in the gap length, L. From Eq. (5.19) it can be seen that for typical negative operating phases of the accelerating cavities ($\phi < 0$), the net effect due to the RF field is defocusing ($\Delta > 0$). It should also be noted that as $\gamma^2\beta^2$ increases (the beam velocity approaches the limit $v = c$, γ^2 approaches infinity), the net RF defocusing force decreases. It decreases even more rapidly with beam velocity than the space-charge contribution.

5.2.8.3 Solenoidal Magnet Focusing

Figure 5.19 shows a solenoid (coil of wire) with current flowing in the wire and an electron traveling through the center of the coil. From Eq. (5.6) we see that the electrons moving down the axis of an ideal solenoid will have their velocity parallel to the axial magnetic field generated by the current in the wire. This therefore results in no force on the electrons. However, in a real solenoid, Gauss' law tells us that when the magnetic field changes along the axis, there must also be a radial component of the magnetic field that can then act on the electrons in the beam:

$$B_r = -\frac{r}{2}\frac{dB_z}{dz}. \tag{5.20}$$

Therefore, due to the fringe fields at the entrance and exit of the solenoid, in the first half of the solenoid the beam receives non-zero angular momentum. In the second half at the exit, this angular momentum is removed since the field is reversed. However, the beam distribution is finite in size and has non-zero momentum

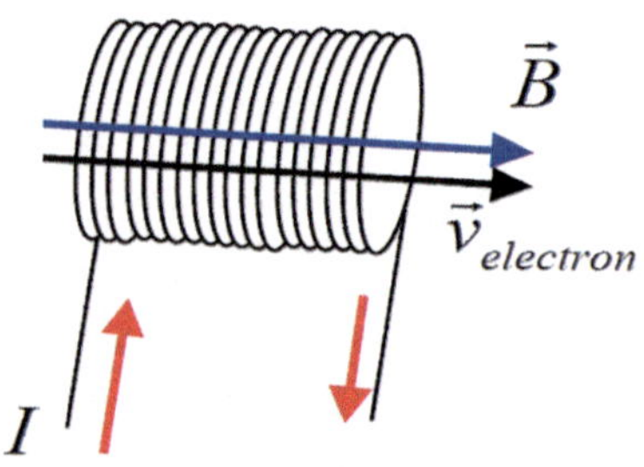

Fig. 5.19 Simple solenoid. Reproduced from Ref. [26]

components prior to entering the solenoid. Solenoidal focusing results from the interaction between the azimuthal (ϕ) component induced at the entrance fringe-field region and the longitudinal magnetic field component (B_z) in the main section of the solenoid. Therefore, the electrons experience a net focusing force from both the fringe fields and the main solenoidal field. The strength of the focusing effect is proportional to the B_r^2.

5.2.8.4 Quadrupole Focusing in a Linac

Magnetic lenses are the most common method for compensating for space-charge and RF-defocusing effects in linacs. For ion linacs the most common magnetic focusing element is the magnetic quadrupole lens. To focus the beam, the quadrupoles are placed in a periodic lattice in the linac between accelerating sections or inside drift tubes. Quadrupole focusing magnets can be electromagnets requiring a current source to drive the magnet windings or the can be a quadrupole configuration of permanent magnets. Figure 5.20 shows a cross-section of a quadrupole magnet.

An ideal quadrupole magnet produces a constant transverse gradient where

$$G = \frac{\partial B_x}{\partial y} = \frac{\partial B_y}{\partial x} \tag{5.21}$$

and the Lorentz force (Eq. (5.5)) components for a particle moving on the z axis with velocity v are given by

$$\begin{aligned} F_x &= -qvGx \\ F_y &= qvGy. \end{aligned} \tag{5.22}$$

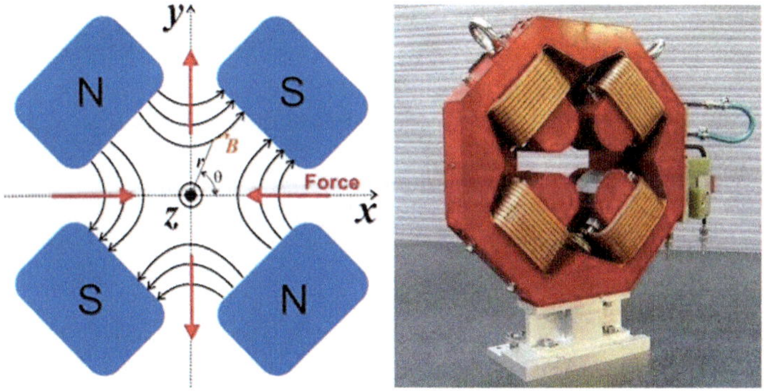

Fig. 5.20 (**a**) Cross-section view of a quadrupole focusing magnet showing the magnetic field configuration (www.quora.com, 2017). (**b**) 3-D view of a simple quadrupole (BDSIM Accelerator Beamline Simulation Tool, www.pp.rhul.ac.uk/bdsim)

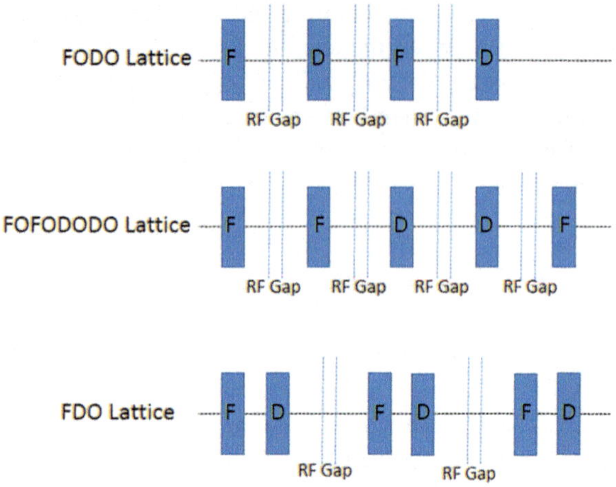

Fig. 5.21 Common magnetic quadrupole focusing lattices. An F-quad implies focusing in the x-plane and defocusing in the y-plane. The focusing is reversed for the D-quad. The net effect over a focusing period is to focus the beam transversely

Note that when a particle is on axis ($x = y = 0$) the force on the particle is zero. If qG is positive, the quadrupole lens focuses in x and defocuses in y [11]. The opposite is true if the sign of qG is reversed. Therefore, magnetic quadrupole lenses only focus in one plane. A series of alternating-sign quadrupole lenses are required to give overall focusing of a particle beam. Figure 5.21 shows several common configurations of magnetic quadrupole focusing lattices. The FODO lattice is the most common.

Assuming the linear restoring force of an ideal magnetic quadrupole lens, the motion of particles in the beam along the axial direction s can be described by Hill's equation having the form

$$x'' + K(s)x = 0, \tag{5.23}$$

where $x'' = d^2x/ds^2$ and $K(s) = qG/mc\beta\gamma$. Because Eq. (5.23) is a linear second-order equation, its solution can be written in matrix form. The matrix form of the solution to Hill's equation is called a transfer matrix. Transfer matrices exist for most magnetic beam-focusing elements and also for drifts, etc. and can be combined to quickly estimate numerically the effects of these transport elements on the beam.

5.2.9 *Induction Accelerators*

Induction accelerators use ferromagnetic (high inductive impedance) cores to provide stored energy to accelerate beams of electrons or ions. These induction cells and the beam act as a one-to-one transformer where the changing magnetic fields generate accelerating electric fields on axis in the gaps between cores. The voltage across the accelerating gap of an induction cell and the achievable pulse length are related to the magnetic flux in the core material at saturation and the core cross-sectional area, A, by Faraday's law:

$$A = VdB/dt \tag{5.24}$$

In a typical induction linac configuration, a toroidal ring of magnetic material surrounds an accelerating column, and the change in flux in this magnetic core induces an axial electric field. The stored energy (volt-seconds) available to this system for a given magnetic material depends upon the cross-sectional area of the toroidal ring. Therefore, the beam pulse length determines the available energy gain per cell. A premium is put on minimizing the cell diameter for two reasons. The first is to minimize the total weight of magnetic material and hence its cost, and the second is to minimize the power required for magnetization which is also proportional to volume. The requirement for energy uniformity during the beam pulse makes it necessary to make some provision in the pulsing system to maintain a constant dB/dt during the useful part of the pulse. This can be done by providing a current ramp in the primary pulse. Typical accelerating gradients do not exceed $\sim$1 MV/m, limited by HV breakdown initiated by stray ions or electrons striking the core insulator. Most operational data exists for electron induction linacs with very little data for ion accelerators and improving induction linac accelerating gradients is an active area of research and development.

Confining magnetic fields are generally provided by pulsed solenoids integrated into the design of the induction cells. Electrical energy to the induction cores is typically provided by pulse-forming Blumleins or compact pulse-forming networks (PFNs). Because this is a non-resonant form of acceleration, any charge-to-mass-ratio from electrons to heavy ions can be accelerated.

Historically, induction accelerator technology has been used for applications requiring kilo-ampere electron or ion beams such as for heavy-ion fusion or weapons radiography. Recent examples of this technology for large-scale application include the Neutral Drift Compression Experiment-II (NDCX-II) at Lawrence Berkeley National Laboratory and the Dual-Axis Radiographic Hydro Test (DARHT) facility at Los Alamos. However, compact induction accelerators are also being developed for AI. These include compact induction linacs [27, 28], where the required enabling technologies include use of multilayer high-gradient metal/insulating layers to provide continuous accelerating gaps and the ability to rapidly and precisely trigger stacked pulse-forming networks.

Fixed field alternating gradient (FFAG) accelerators using an induction core to accelerate the beam are also under consideration for AI applications [1, 29, 30], although not yet commercially available. Because an induction core is used, no RF system is needed as in other variations of the FFAG, therefore simplifying the design. Other advantages include a small circular accelerator footprint for beam energies $<$10 MeV, static guide magnet fields, high duty cycle (>50% possible), dynamically-variable output current, and conservative induction core requirements ($\sim$50 keV/turn). Several FFAG topologies are possible including scaling, non-scaling, radial, and spiral. A recent example of an applicable FFAG is the RadiaBeam Radiatron FFAG Betatron.

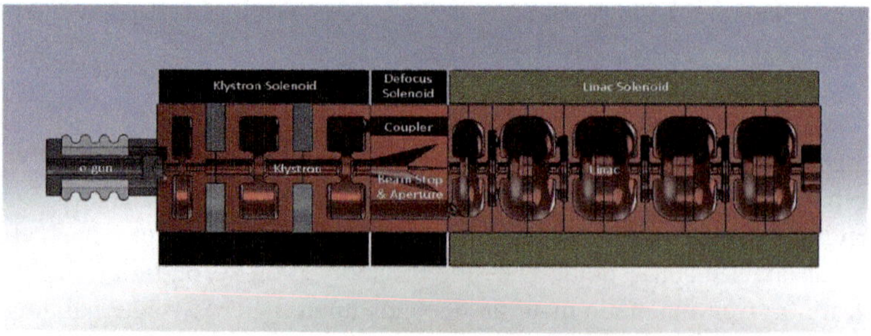

Fig. 5.22 Common magnetic quadrupole focusing lattices. An F-quad implies focusing in the x-plane and defocusing in the y-plane. The focusing is reversed for the D-quad. The net effect over a focusing period is to focus the beam transversely

5.2.10 KLYNAC

An interesting electron accelerator concept being developed at Los Alamos National Laboratory that it is worth noting is a compact linear accelerator that integrates the klystron RF source and a coupled-cavity linear accelerator into a single resonant system using a single electron beam. This device has been coined the KLYNAC [31]. The goal in developing the KLYNAC is to provide a compact, less-expensive alternative to existing, more conventional electron linacs in the beam output energy range up to 10 MeV. Figure 5.22 shows the details of the KLYNAC concept.

The klystron section of the KLYNAC is essentially a conventional klystron structure with an input cavity, some number of intermediate cavities and an output cavity. The accelerator structure is, likewise, a conventional on-axis coupled structure. What makes the KLYNAC unique is the means of coupling the klystron output cavity to the accelerator. The coupler is a resonant coupler rather than an ordinary transmission line. If the klystron and accelerator are coaxial, the need for a separate cathode for the accelerator can be eliminated by injecting some of the klystron beam into the accelerator. The key element of the KLYNAC concept is that the klystron output cavity is resonantly coupled to the input of the input cavity of the accelerator. The coupler section of the Klynac uses a resonant coupling concept developed at Los Alamos National Laboratory (LANL) in 1967 as part of the Los Alamos Meson Physics Facility (LAMPF) design [32].

An 8-cell KLYNAC is has been tested at LANL and shows promising results. The device operates at 3 GHz (S-band). The klystron section is driven by a 75-keV, 21-mA electron beam. A Pierce type electron gun is powered by a 48-stage, solid-state diode-directed Marx modulator [33] provides the electron beam. Test results have demonstrated approximately 850 kW pulsed RF extracted from the electron beam and coupled to the accelerating section. The goal is to develop a system capable of generating a 10-MeV, 40-mA, 5% duty (20-kW average beam power, 10-μs pulse length, 5-kHz repetition rate) KLYNAC. A total of 44 linac cells will be required to reach 10 MeV. Such a system would be ideal for AI applications.

5.2.11 Frontier Accelerators

Significant effort would be required to discuss in detail the various frontier accelerator concepts that could eventually have application to AI. Most of these are still primarily in the R&D phase and are not yet ready for commercialization, but offer the promise of realizing ultra-compact devices to accelerate electrons to high energies through miniaturization. These include: laser-plasma acceleration [34], dielectric-wake acceleration [35], and dielectric laser acceleration [36].

5.2.12 Generating Bremsstrahlung X Rays

Electron linacs are common bremsstrahlung photon sources used for AI imaging. Bremsstrahlung photons are generated when the negatively-charged electrons provided by linac are deflected by the positively-charged atomic nuclei as the electron beam strikes a high-atomic-number (high-Z) target, typically tungsten or tantalum. The deflection of the electrons involves a loss of kinetic energy (acceleration) of the beam. This energy loss is emitted as electromagnetic energy. Figure 5.23 shows a characteristic bremsstrahlung spectrum. It is a continuous spectrum with an energy cut-off equal to the maximum energy of the electrons in the beam. Therefore, MeV electron beams can generate MeV photons. The distribution also contains spikes at the characteristic X-ray energies of the target material. Characteristic X rays are emitted from heavy elements when their electrons make transitions between the lower atomic energy levels. These transitions can be stimulated by the electron beam from the linac.

Optimization of the photon source includes selection of the target material and thickness. Generally, increasing the target atomic number improves the efficiency of X-ray production and the energy of the characteristic and bremsstrahlung X rays. The intensity of the X rays at a given point depends on the distance of the detector from the X-ray source. This intensity decreases as $1/r^2$ where r is the distance from the source to the detector. Additionally, photons from the source are absorbed by the object to be interrogated, further reducing the intensity of X rays detected. To compensate for these effects, the output beam current of the linac is adjusted.

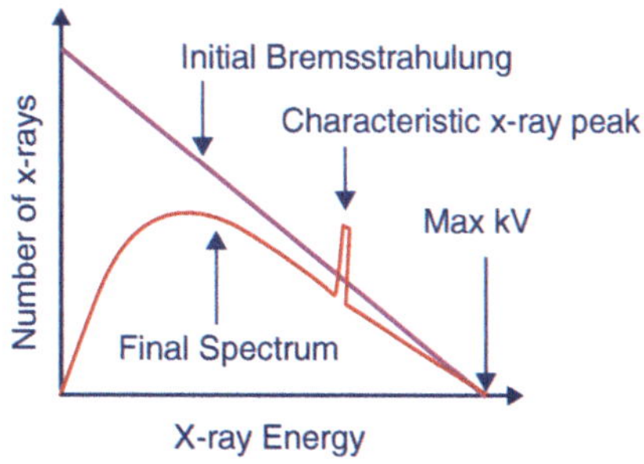

Fig. 5.23 A characteristic bremsstrahlung X-ray spectrum (www.instructables.com)

The intensity of the X rays is directly proportional to the beam current (number of electrons per unit time) striking the X-ray production target. The beam current does not change the bremsstrahlung energy distribution.

5.2.13 *Bremsstrahlung Based Photoneutron Production*

Neutron beams can be generated through a multi-stage process using the photons produced by an electron linac. More direct processes need a source of protons or heavier ions. The most common commercially-available X-ray sources are based on electron linacs with beam energies below 9 MeV. Neutrons are generated when photons interacting with a nucleus in the target have an energy above the nuclear binding energy of the target material. These photons can excite a nuclear giant dipole resonance which after decaying emits a photoneutron or the material may under go a photon-induced fission (photofission). Typical target materials are similar to those high-Z materials used to generate X rays; however, photon energies above 7 MeV are typically required to produce neutrons and heavier target materials produce more neutrons. Target materials include: W, Ta, Be, D_2O, Pb, and ^{238}U.

The most efficient neutron production is through the use of a 2-stage target such as using a tungsten target to perform the electron to photon conversion and a beryllium target for photon to neutron conversion. Figure 5.24 shows such an example target configuration. The neutron beam characteristics can be tailored somewhat by including neutron reflector and moderator materials in the target configuration and optimizing the reflector/moderator geometries [37].

Several high-performance systems are being investigated to improve the detection of nuclear materials. These systems require both higher neutron energies to penetrate sufficiently into potentially highly-shielded containers and to maximize the fission rate to improve the speed and accuracy of detection of ^{235}U in large cargo containers [15]. Using an electron linac in the energy range less than 10 MeV with an optimized Be or D_2O (γ,n) converter in combination with ^{238}U to modify the

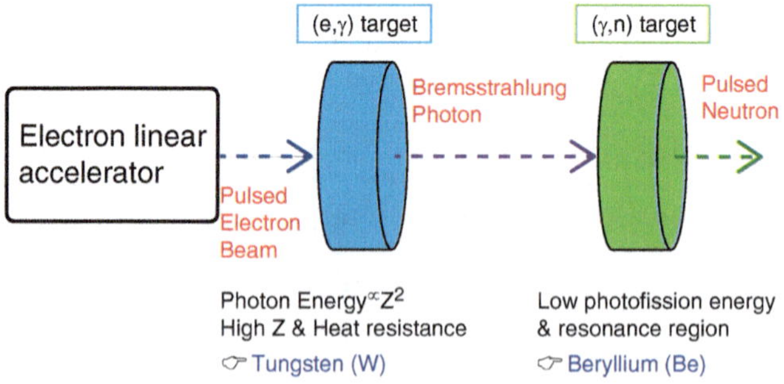

Fig. 5.24 A generic two-stage system to generate neutrons via photofission. Reproduced from Ref. [38]

neutron spectrum has been studied and shows promise to provide an intense, fast neutron source with desirable spectral features. Commercially available electron linacs capable of 100-μA average current operation are sufficient to compensate for the relatively low (γ,n) cross-sections. At 9 MeV, neutron fluxes of more than 10^{11} n/s are expected based on simulation results.

5.3 Ion Accelerators and Low-Energy Nuclear Reactions

5.3.1 Neutron Generators

Compact neutron generators are available over a broad range of neutron fluxes. This is a well-developed technology generally incorporating the use of a Penning ion source producing a supply of deuterons which are then accelerated in an electrostatic field and scattered from either a deuterium or tritium target to produce neutrons through the reactions D + D $\rightarrow$ ^{3}He + n or D + T $\rightarrow$ ^{4}He + n, generating 2.5- and 14.1-MeV neutrons, respectively (see Fig. 5.25). Alternatively, cold-cathode sources have also been used. These compact neutron generators offer steady-state neutron production, high pulse repetition rates up to $\sim$20 kHz and nominal duty factors from 5% up to 100%. Very high neutron fluxes, up to 10^{14} n/s (D-T) and 10^{12} n/s (D-D), have been demonstrated by Phoenix Nuclear Labs; however, neutron fluxes in the 10^8 n/s range are more common. Several commercial vendors exist making this technology available for several applications, including AI. Also, in recent years significant advances have been made to improve operability and integration of these devices into AI systems through the application of modern digital controls, smart diagnostics capable of monitoring most performance characteristics, and GUI controls interfaces. These advances have also led to significant improvements in safety systems needed and system interlocks needed for instrument monitoring.

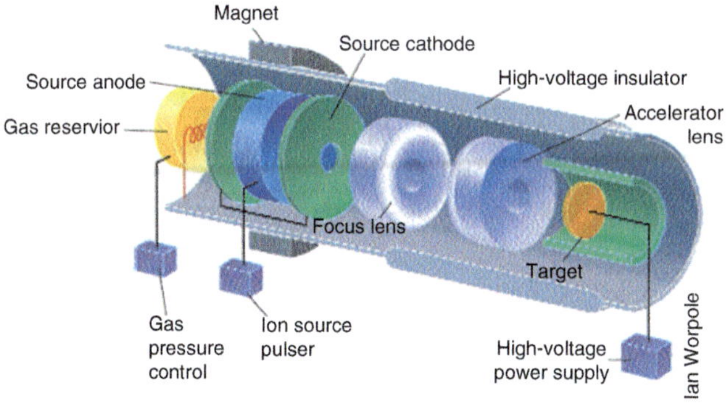

Fig. 5.25 Schematic layout of a compact neutron generator incorporating a Penning ion source to produce a supply of deuterons [39]

The most common application of neutron generators for AI involves the measurement of prompt γ-ray photons resulting from fast neutron inelastic scattering and thermal neutron absorption with the chemical elements found in explosives, narcotics, and chemical weapons. Neutron generators are also used for prompt neutron measurements following fast or thermal neutron-induced fission for identifying fissile materials. More details about compact neutron generators can be found in Ref. [39].

5.3.2 Compact Cyclotrons

In a classical cyclotron as shown in Fig. 5.26, a static and nearly-uniform dipole magnetic field B is applied perpendicular to a vacuum chamber that sits between the poles faces of the magnet. Two hollow electrodes are mounted inside the vacuum chamber. The electrodes are driven by an applied RF voltage. Ions that are injected at the center of the cyclotron inside the vacuum chamber are accelerated by the electric field across the gap due to the applied voltage and follow a circular trajectory due to the applied magnetic field. While inside the electric field free region of the hollow electrodes, the ions are only affected by the uniform magnetic field and travel in circular orbits in a plane normal to the magnetic field. Both the RF frequency and magnetic field are adjusted to satisfy a resonance condition to allow acceleration of the non-relativistic ion beam. This requires the time for an ion to complete a half-revolution in the magnetic field to be equal to the time for reversal of the RF field. Using the Lorentz force equation (Eq. (5.5)) for a charge q moving at a velocity v

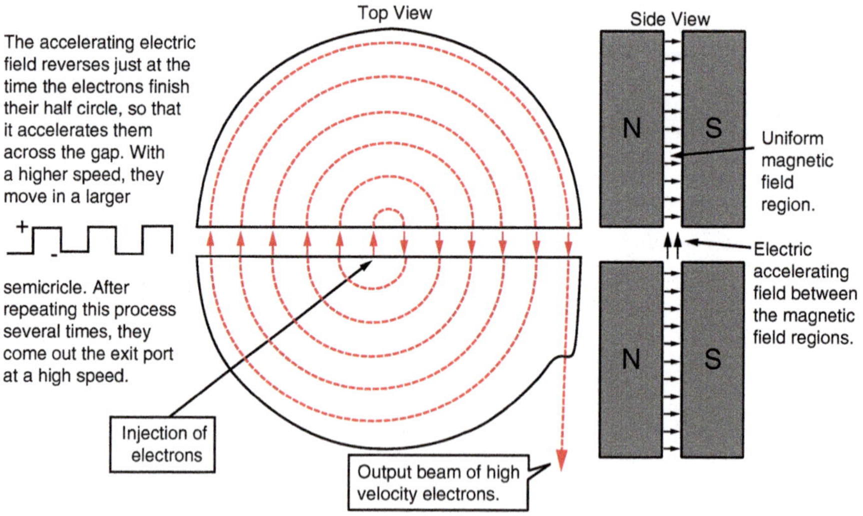

Fig. 5.26 Schematic layout of a classical cyclotron (hyperphysics.phy-astr.gsu.edu, 2016)

perpendicular to a magnetic field B, an ion will move in a circular path or radius r whose value can be obtained by balancing the centrifugal force and the magnetic force:

$$mv^2/r = qvB. \tag{5.25}$$

This leads to the cyclotron frequency

$$f = \frac{v}{2\pi r} = \frac{qB}{2\pi m}. \tag{5.26}$$

In order to assure stability of the beam in the axial direction, the magnetic field must decrease as a function of r and the accelerating voltage must be high enough for the beam particles to reach the final design energy before they are out of the accelerating phase of the RF cycle. Rarely today is a classical cyclotron used. Today generally rather than having large dipole magnets, sector magnets are used that lighten the overall weight of the cyclotron and also allow the magnetic field to be tailored in such a way that there is some net axial focusing of the beam. The final energy of particles accelerated in a cyclotron depends on the accelerating voltage across the gap, the RF frequency, the radius of the cyclotron, and the magnetic fields. The output particle beam from a cyclotron does not have a bunched time structure like beams in most other RF accelerators but is instead continuous wave (CW).

Beam extraction at the output of the cyclotron is typically accomplished through the use of an electrostatic deflector to initially change the direction of the orbiting beam and to direct it into a magnetic extraction channel. This extraction system is usually complicated. To allow variable output energy of proton cyclotrons, the concept of accelerating negative hydrogen ions (H^- ions; a proton and two electrons) was introduced in 1962 at the University of Colorado [40], where a thin foil was introduced to strip the electrons from the beam, reversing the direction of bend and quickly sending the beam out of the cyclotron. This is a very simple mechanical extraction system compared to the previous approach that required several active extraction elements. Use of a stripper foil for beam extraction is common practice today in all H^- cyclotrons. For variable output energy, the foil is made movable and can be inserted at different radii to change the output energy of the cyclotron over a broad range.

The ability to extract the beam and the maximum output current are affected by the quality of the beam in the cyclotron and how good the turn-to-turn separation of the CW beam is. The current limit (maximum current that can be accelerated) for a cyclotron can be calculated by assuming fully overlapping turns within the cyclotron (*current sheet* approximation) by the following expression:

$$I_{limit} = \epsilon_0 f_0 \frac{\Delta E}{e} \Delta\phi \Delta z v_z^2, \tag{5.27}$$

where ϵ_0 is the permeability of free space, f_0 is the orbit frequency, ΔE is the energy-gain per turn, $\Delta\phi$ is the phase spread of the beam, Δz is the vertical size of the beam, and v_z is a quantity called the axial tune which is proportional to the axial focusing. For typical cyclotron operating parameters, $I_{limit} \approx 10\,\text{mA}$; however, this limit is never achieved due to the requirement to have adequate turn separation within the cyclotron to be able to achieve efficient acceleration and extraction. Also since cyclotrons are weak focusing devices, there is significant beam loss that results in activation of the structures, making it difficult to operate and maintain. More typical cyclotron output currents are actually in the range of 10–250 μA. The turn spacing as a function of radius r in the cyclotron magnetic field is given by

$$\Delta r(r) = \frac{m\Delta E}{qB^2 r}, \tag{5.28}$$

where ΔE is the energy-gain per turn.

Clearly there are competing effects when attempting to make a cyclotron as compact as possible. For example, for a fixed radius r, as the magnetic field is increased to achieve as high an output energy as possible, eventually practical limits for beam extraction will be reached as the turn separation decreases. In addition, overlapping turns will cause internal beam scattering and space-charge effects that will enhance beam losses. Nonetheless, innovations in cyclotron design continue with the goal of increasing extracted beam currents, increasing beam energy while reducing footprint, and reducing overall cyclotron weight.

An example of an innovative design is the modern resistive coil cyclotron. These are efficient ferromagnetic structures that can operate at high magnetic fields and have been designed to fields as high as 9 T. An operating example is the IBA C230 proton cyclotron in use for proton beam radiation therapy shown in Fig. 5.27a. It operates at 3 T, is 3.5 m in diameter, and costs \$4 M to complete. Also shown in Fig. 5.27b is an alternative design, a 250-MeV superconducting isochronous cyclotron built by Varian, also for medical proton therapy. While not compact enough to be portable, they certainly meet many of the requirements for AI use such as in a port inspection facility or on a ship for long standoff AI. Extracted beam currents are generally <1 mA, but a broad range of beam energies can be delivered using the appropriate design. Attempts at ultra-compact footprints (≤1 m) for high

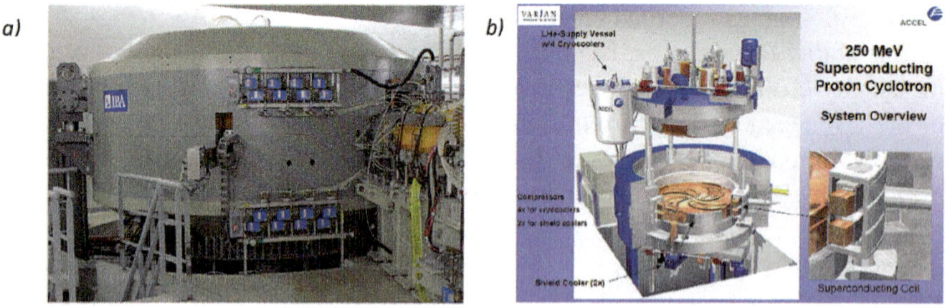

Fig. 5.27 (**a**) IBA C230 resistive-coil proton cyclotron. (**b**) Varian 250-MeV superconducting proton cyclotron

beam energies have been generally unsuccessful due to limitations in turn-to-turn separation of the particle trajectories and associated beam extraction difficulties. An example of an ultracompact, 10-MeV superconducting design developed at the Massachusetts Institute of Technology can be found in Ref. [41].

5.3.3 The Radiofrequency Quadrupole (RFQ) Accelerator

The radiofrequency quadrupole (RFQ) accelerator is a compact accelerating structure for proton and ion acceleration that captures, bunches, and simultaneously focuses and accelerates low-energy beams. It can accelerate beams over a broad mass range from protons to low-charge-state heavy ions, and over a range of energies from a few keV/nucleon up to several MeV. The RFQ accelerator concept was proposed by Kapchinskij and Teplyakov in 1970 [42] and demonstrated in 1974 at the USSR Institute for High Energy Physics at Protvino, Russia. Design of the RFQ in the US quickly followed in 1977 at Los Alamos National Laboratory with the first proof-of-principle test in 1979 by Stokes et al. [43].

Figure 5.28 shows schematically the features of an RFQ accelerator. There are several types of RFQ accelerators, the most common being the 4-vane RFQ which is typically used at higher RF frequencies (>200 MHz) for protons or deuterons and at high duty factor because its vane geometry accommodates the necessary cooling channels, and the 4-rod RFQ, typically used at lower RF frequencies (<200 MHz) to accelerate heavy-ion beams. A main advantage of the 4-rod RFQ is its simplicity of design and electromagnetic mode separation that makes it simpler to fabricate and tune. Attempts are being made to extend the design and operation of the 4-rod RFQ to higher RF frequencies (near 300 MHz) and to higher RF duty factors [44].

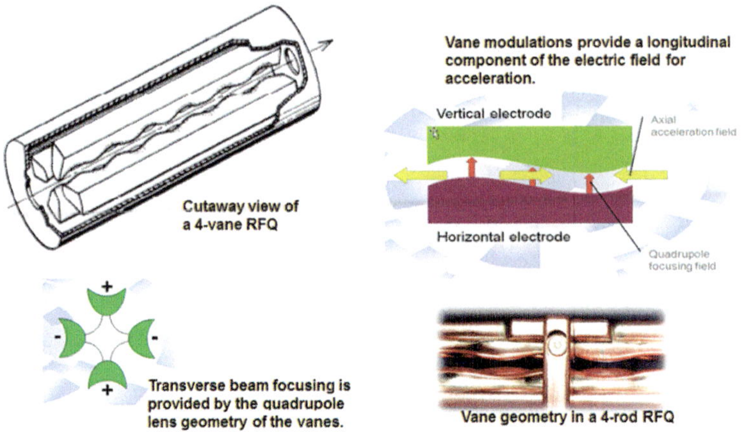

Fig. 5.28 General description of RFQ accelerator features [11]

The design of the RFQ accelerator is well-formalized, allowing designs over a broad range of performance parameters. The parameters of the RFQ can be tailored, for example, to provide very narrow output beam energy spread if needed to induce a particular nuclear resonance such as the $^{13}C(p,\gamma)^{14}N$ nuclear resonance absorption proton-capture reaction in which a well-defined 1.75-MeV proton is captured in ^{13}C to form the 9.17 MeV excited state of ^{14}N used to detect the nitrogen component in high explosives [45].

Common to all RFQ structures are the 4 vanes that are excited in an electric quadrupole mode. This quadrupole field configuration provides the transverse focusing for the beam as it traverses the RFQ structure. Modulations in the vanes (see Fig. 5.28) that vary along the length of the RFQ (in the direction of the beam) in a prescribed manner provide longitudinal components of the electric field to accelerate the beam. At the entrance of the RFQ the vane modulations are small and the beam is primarily focused as it enters the RFQ. The modulations are then gradually increased to bunch the beam while maintaining strong transverse focusing. As the beam is accelerated, the magnitude of the vane modulations is increased to more efficiently accelerate the short beam bunches that have formed. For high-intensity beams, the bunching process causes an increase in the beam density, which increases the space-charge forces within the beam, and often leads to a blow-up of the transverse beam emittance. To overcome or minimize this effect, the RFQ can be designed to bunch the input DC beam from the ion source adiabatically (gradually).

This is done in the *gentle buncher* section of the RFQ as shown in Fig. 5.29. Also shown in the figure are the other main sections typically found in the RFQ. The first section is labeled the radial matching (RM) section where the vane modulation is zero and the transverse quadrupole focusing field is small. The RFQ aperture is reduced in the RM section to gradually increase the focusing field and to transform the DC input beam to a radially time-varying beam matched to the focusing structure of the time-varying quadrupole field. The next section of the RFQ is the *shaper* section, an intermediate section between the shaper and gentle buncher, used to pre-bunch the beam by transitioning from only a focusing field to focusing and bunching. It should be noted that much of the RFQ structure is used to capture and condition the input beam relative to the length of structure used to apply the maximum accelerating gradient.

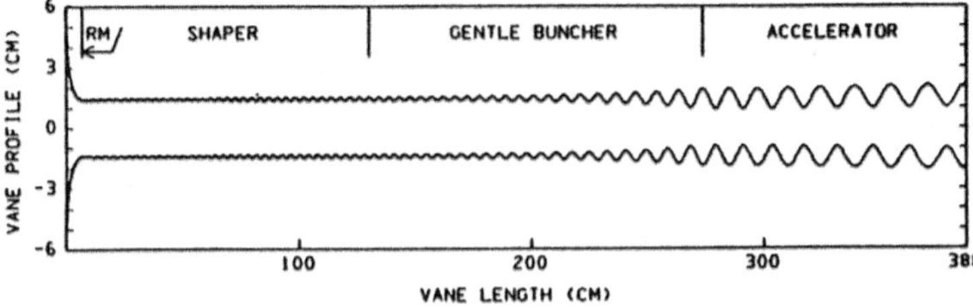

Fig. 5.29 Example of the main sections of a RFQ accelerator as a function of distance along the RFQ [11]

Design and fabrication of RFQ accelerators is a mature technology widely applied world-wide for many applications. However, the design process is complicated since many of the design parameters are coupled, requiring the use of well-benchmarked codes to optimize a design (PARMTEQM (laacg.lanl.gov), Toutatis [46], etc.). Accelerating gradients are typically in the 1–2 MV/m range. Cost of the structure is high, approximately \$1 M/m for 4-vane structures. An RFQ accelerator is generally designed for a specific particle charge-to-mass ratio, to operate efficiently at a particular design beam current and vane voltage, up to the design current limit, which is typically chosen to be double the design current. Although most commercially-available RFQ accelerators are designed for low-average output beam currents and low RF duty factor, high-intensity, high-duty-factor designs are possible. An example is the Los Alamos 700-MHz Low-Energy Demonstration Accelerator proton RFQ which operated at an output-beam-energy of 6.7 MeV and 100-mA CW (100% duty factor) [47].

A recent interesting example of pushing RFQ design parameters can be found in Vretenar et al. [20]. The linac group at CERN has designed and built a prototype 750-MHz RFQ, one of the highest in RF frequency, to be used as an injector for hadron therapy linacs. Because of the high RF frequency used, the structure is very compact with an outer diameter of only 134 mm. The RFQ reaches a proton output beam energy of 5 MeV in only 2 m. Additional segments can be coupled together to reach an output energy as high as 8 MeV. The total weight is only 220 kg. Four solid-state inductive output tubes (IOTs; [48]) are planned to be used to power the structure. The design injection energy that must be provided by the ion source is 40 keV. The RFQ has been designed to operate at up to 5% duty but will operate at lower average beam current due to the lower hadron therapy requirements. Demonstration of this RFQ at the higher duty may make it attractive for other applications such as AI. For a peak output beam current of 1 mA, this system could deliver an average beam current of 50 μA at the maximum duty [49].

5.3.4 *Measuring Low-Energy Nuclear Reaction Signatures*

Because of the ability to design an RFQ to have very specific output beam properties, it can be used for a broad range of AI detection applications. An example of this is the use of an RFQ beam to excite a very narrow nuclear absorption resonance of γ rays in nitrogen to detect explosives [50]. The major subsystem components include the RFQ and associated beam transport system, a high-power photon production target, and a tomographic detection and imaging system as shown in Fig. 5.30. One of the requirements of this system was a high proton beam current ($\sim$15 mA) to generate a resonant γ-ray flux sufficient to perform high-speed luggage interrogation at airports. Such high output beam currents are easily achievable in an RFQ.

The measurement technique is based on a narrow energy state in the nucleus of ^{14}N that results in a strong resonance in the photonuclear cross section for the

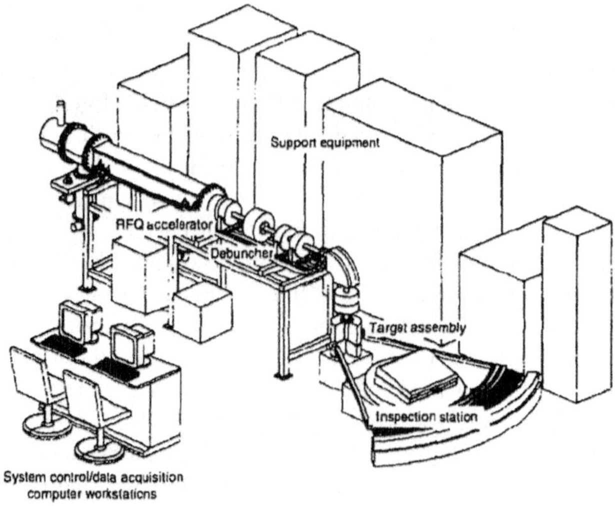

Fig. 5.30 RFQ-based system for explosives detection. Reproduced from Ref. [50]

reaction $^{14}N(\gamma, p)^{13}C$ at 9.17 MeV. Gamma rays are absorbed by the nitrogen, followed by the prompt emission of a proton and a ^{13}C nucleus. The transition rate from the ground state of ^{14}N is unusually large (a resonance) and as a result, 9.17-MeV γ rays are highly absorbed, resulting in a strong indication of the presence of nitrogen.

The essential feature of this approach is the use of the inverse reaction $^{13}C(p,\gamma)^{14}N$ to produce γ rays at the resonance energy. A 1.75-MeV proton beam from an RFQ is incident on a ^{13}C target and produces γ rays at the resonance energy within a narrow angular cone with respect to the forward proton beam direction. The nitrogen resonance is very narrow in energy and has a large integrated cross section which makes the signature easy to detect and relatively insensitive to other radiation backgrounds. In addition, the 9.17-MeV γ rays are highly penetrating, making it difficult to shield the explosives from detection. The output energy spectrum from the RFQ has been designed to be very narrow so as to generate a very narrow γ-ray spectrum, further enhancing the performance of the system; however, the option to broaden the energy spread of the beam in the downstream beam transport before the target also allows non-resonant imaging at other γ-ray energies. The system was tested in collaboration with the FAA and successfully demonstrated the ability to detect low levels of nitrogen-based explosives.

RFQ accelerators have also been used in AI systems to generate low-energy neutrons to detect highly-shielded enriched uranium [51]. In this system, a portable RFQ linac is used to generate a 1.93-MeV proton beam that strikes a ^{7}Li target to produce a forward directed 60-keV neutron beam via the $^{7}Li(p,n)$ reaction. The 60-keV neutron beam induces fission in ^{235}U which then emits a spectrum of fission neutrons. The highest energy neutrons (>500 keV) are detected with a neutron and γ-ray pulse-shape discriminating (PSD) scintillator detector. Other similar systems have been explored using proton or deuteron beams generated by an RFQ accelerator [41, 52].

5.3.5 High-Energy Systems and Large Standoff AI

Large standoff AI assumes standoff distances of 100 m or more and requires high-energy beam sources. These proposed systems would be primarily focused on the detection of nuclear threats well in advance of their reaching close proximity to US borders. A possible system configuration to sweep a beam throughout a large structure such as a sea-going vessel would entail fitting a high-gradient or compact high-energy accelerator (500–1000 MeV, 1-mA average beam current) on a large ship coupled with a helicopter-borne detection system [2].

Approaches using intermediate-energy protons (~1 GeV) to perform radiographic imaging and the delayed-neutron signature from SNM have been proposed. Laboratory measurements have been made [53] to demonstrate proof-of-principle, and to understand measurement limitations. Energetic protons are an attractive alternative to photons and neutrons for AI of nuclear threats because they have large fission cross sections, long mean free paths and high penetration even in high-density materials, require much lower doses to produce a signal, and can be manipulated with magnetic optics.

Nuclear threats can also be detected using muons to perform AI [2]. There are several distinct features of the muon interaction with matter that make their use attractive although generation, capture, and acceleration of the muons are complicated. The unique features of the muon include a low nuclear interaction cross section, allowing long-range propagation of a muon beam, and deep penetration of materials until stopped by ionization loss in a short distance. For large-standoff AI of a large object, a large beam footprint on target is needed, but at high muon flux. This is possible with muons since a focused muon beam can be propagated through the atmosphere at a range limited primarily by beam-size growth due to scattering with essentially zero attenuation. In contrast, as an example, a 12-m diameter beam of photons propagating through an average density of 0.33 g/cm^3 attenuates by a factor of $>10^6$ before reaching shielded HEU (50 cm PE, 2.5 cm Pb).

A muon-beam intensity of $>10^9$/s is required for efficient interrogation. Muons are generated through pion decay by scattering a proton or electron beam with energy above the pion-production threshold (~140 MeV) from a refractory target such as carbon (p + C $\rightarrow$ π $\rightarrow$ μ^-). Figure 5.31a shows the mission space for large-standoff muon AI parameterized by accelerator beam energy and mass-depth penetration. Figure 5.31b shows the energy spectrum of K-muonic X rays in ^{235}U and ^{238}U that are the primary signature for detecting special nuclear materials by active muon interrogation.

Figure 5.32 shows the Los Alamos National Laboratory concept for muon AI. This concept requires a 1-mA, 500-MeV proton accelerator to produce the secondary muons, 0-mode pill-box cavities (aperture closed by thin metal windows;

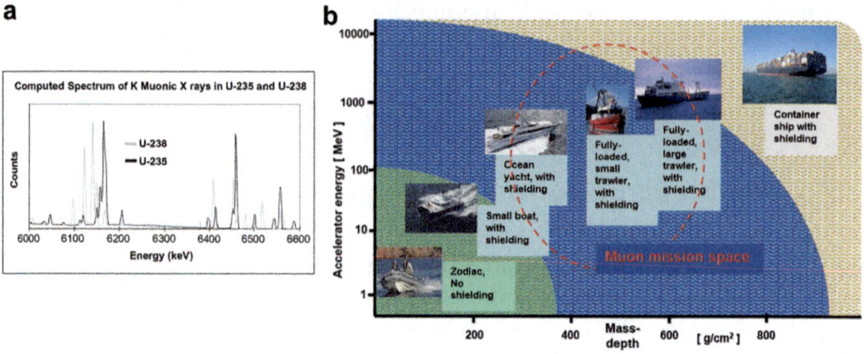

Fig. 5.31 (**a**) Mission space for muon active interrogation. (**b**) Computed energy spectrum for K-muonic X rays from ^{235}U and ^{238}U. Reproduced from Ref. [2]

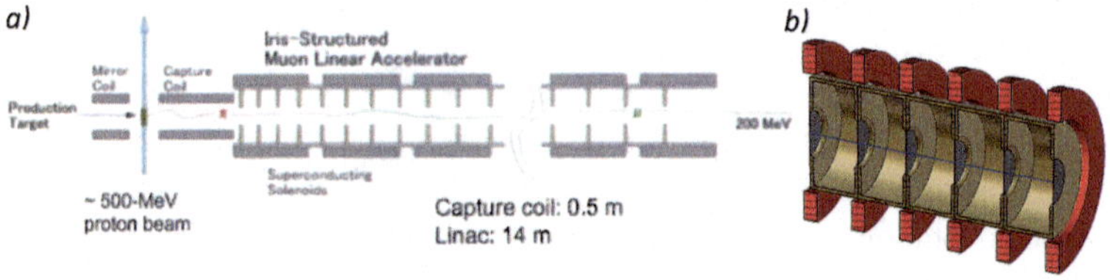

Fig. 5.32 (**a**) Los Alamos concept for muon AI [2]. (**b**) 0-mode pill-box cavities used to accelerate the captured muons [54]

also shown in Fig. 5.32) used to accelerate the captured muons to 200 MeV, and the 3–5 T capture/confinement field produced by a long segmented solenoid magnet. The muon linac is then followed by a conventional superconducting linac using elliptical cavities to accelerate the μ^- beam to high energy (500–1000 MeV) to be used for long-standoff AI.

Compact accelerator technology to produce high-energy protons, for example, does not currently exist, although emerging concepts such as the Fixed-Field Alternating Gradient (FFAG) accelerator are showing promise. Present normal-conducting accelerator technology applied at low energy coupled with superconducting accelerator technology at high energy can meet present requirements but requires a large footprint for implementation to reach beam energies near 1000 MeV.

5.4 Laser-Based Radiation Sources

5.4.1 Introduction: Laser Driven Sources for AI

Laser-based methods have potential to enable new capabilities in AI and related security applications by generating compact, directional sources of nearly monoenergetic photons as well as directional neutrons, and may also enable miniature bremsstrahlung sources. Laser scattering is used to produce monoenergetic photon beams, while laser-driven acceleration techniques enable compact systems for both photon sources (whether monoenergetic or bremsstrahlung) and neutron sources.

Near-monoenergetic photon sources (MPSs) offer improved sensitivity at greatly reduced dose for AI as detailed in Chap. 10 as well as new capabilities in related fields such as treaty verification, non-destructive assay (NDA) of spent nuclear fuel, and emergency response. The ability to select energy, energy spread, flux, and pulse structure to deliver only the photons needed can resolve many of the issues with current broadband bremsstrahlung sources, including unnecessary dose that can interfere with the signatures to be detected and/or restrict operations [55]. MPS benefits are maximized for sources that also emit beams of very narrow angular divergence, which concentrates flux and mitigates scattering issues. To improve performance for missions leveraging photofission, as well as for nuclear resonance fluorescence (NRF) and radiography, application assessments show that production of 10^{10}–10^{12} photons/s at selectable energies between 1 and 15 MeV and in narrow bandwidth is important [56]. Use of energies below 9 MeV is desirable to avoid photoneutron production, both to enable use of the strong photofission prompt neutron signal and for personnel safety. In this case the ability of an MPS to deliver all its photons near a single energy is particularly important to achieve efficient photofission excitation, as well as radiography with Z contrast.

Monoenergetic, narrow angular divergence sources can in particular be generated by scattering a laser from a relativistic electron beam, known as Thomson or Compton scattering. Applications of monoenergetic photon sources motivate development of more compact technologies for the required electron accelerator, which is of much higher energy for an MPS (e.g., 0.6 GeV electrons to produce 9 MeV photons) than is required for bremsstrahlung sources. This high electron energy requirement has motivated development to reduce size of conventional accelerators. Replacing conventional accelerators by a laser driven plasma accelerators, however, now offers the potential [57, 58] for much smaller accelerators, as well as related techniques to reduce the size of other key source elements: the scattering laser used to produce monoenergetic photons, and the electron accelerator beam dump. Applying compact laser-plasma accelerators at lower energies could also enable very small bremsstrahlung sources (which have otherwise similar spectral and angular properties to other bremsstrahlung sources). Directional neutron sources can be driven by energetic ion beams or monoenergetic photon beams. As is the case for monoenergetic photon sources, laser-plasma acceleration of ions or of electrons offers a potential path to enable such sources at scales compatible with AI and security missions. An ion beam, or a photon beam produced from an electron beam, is used to drive a secondary target to produce neutron beams (via, for example, (d,n) or (γ,n) reactions).

This section reviews the operation principles and unique properties of laser driven sources for AI, with a focus on monoenergetic photon sources and in particular those based on laser-plasma acceleration. The first subsection describes the principles of monoenergetic photon sources based on laser scattering, and their relevant properties for AI, based on conventional electron accelerators. The second subsection discusses laser plasma electron acceleration and its application to render monoenergetic photon sources compact. Application to bremsstrahlung sources is also noted. The last subsection briefly reviews laser driven directional neutron sources. For each source type, unique source properties, limitations, and development needs are discussed in relation to ongoing projects.

5.4.2 *Monoenergetic, Narrow-Divergence Photon Beams Using Laser Scattering*

Thomson scattering, or Compton scattering (sometimes also referred to as inverse Compton scattering) of a laser from a counter-propagating electron beam is a well-established source of quasi-monoenergetic photons with narrow angular divergence, attractive properties for AI [56]. The energy E_{ph} of a photon near-backscattered by an electron at $\gamma_e \approx 2E_e$ [MeV] is relativistically upshifted [59]:

$$E_{ph} = \frac{4\gamma_e^2}{1 + \gamma_e^2\theta^2 + a_0^2/2} E_L, \tag{5.29}$$

with E_L being the laser photon energy and θ the scatter angle. For an electron bunch, radiation adds incoherently due to the very short scattered wavelength, and the resulting collimated photon energy spread convolves electron angular spread and energy spread. Contributions from energy spread in E_L can be kept small because laser pulses many periods long should be used to produce high photon yield. Available high power lasers typically have E_L in the range of 1.5 eV (800 nm), requiring 0.2 GeV electrons for 1 MeV photons, 0.6 GeV electrons for 9 MeV photons, and 0.8 GeV electrons for 15 MeV photons. Frequency conversion allows modestly higher E_L up to the 3 eV (400 nm) range, doubling photon energies from the same electron energies. These electron energies are nearly a hundred times those required to generate bremsstrahlung photons with the same endpoint energy, which requires large accelerators using conventional systems. Development of compact accelerators would enable AI with the corresponding benefit that all photons can be delivered near the desired energy. Energy selection alone can reduce radiation dose required to achieve a given signal by factors ranging from threefold for radiography to more than 50-fold for photofission [56]. As noted below, narrow angular spread provides additional benefit.

Photon emission is primarily within narrow angles $\theta \leq 1/\gamma_e$, which is in the miliradian range for energies of interest. This provides beam spot sizes at the cm-scale for target distances in the few to tens of meters range typical of AI applications. The angular dependence in Eq. (5.29) results in photon emission energy that varies with angle, which means that for narrow bandwidth (BW) the source must be collimated to less than this angle (Fig. 5.33). Collimation then transmits a fraction of the total photons approximately equal to the fractional BW. Hence narrower BW requires

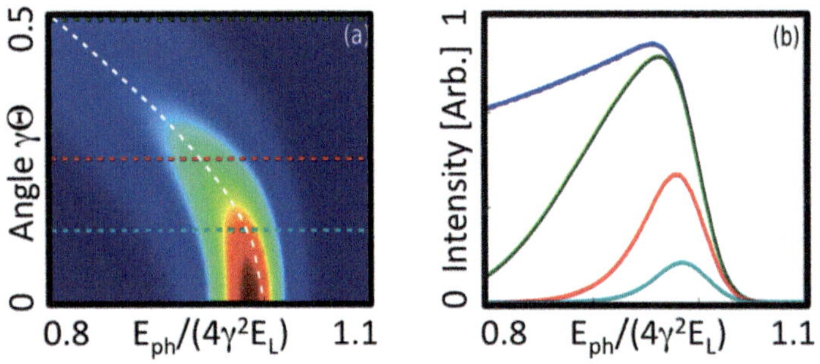

Fig. 5.33 A typical Thomson scattering photon spectrum, showing (**a**) intensity (colorscale) as a function of normalized energy and scattering angle; (**b**) spectra integrated over different collimator angles. The colors correspond to the collimations shown by the respective dashed lines in (**a**), illustrating that narrow collimation is required to achieve low bandwidth, down to the limit of the on axis bandwidth beyond which tighter collimation affects only intensity (Reprinted from Ref. [60])

more scattering events to generate the same number of photons within BW, e.g. one photon in 1% BW requires ≈100 total events. The collimation required, and the fraction of photons discarded, to generate a narrow divergence beam from such a source is much less than for a bremsstrahlung source.

Thomson scattering sources are well developed in large, fixed research facilities, and MeV-class sources include HIGS at Duke [61]. A new generation includes sources under construction at the Extreme Light Infrastructure project in Romania [62] and at the Japan Atomic Energy Agency [63], and proposed facilities at FNAL (http://asta.fnal.gov/files/asta_proposal_october_2013.pdf), SLAC (https://portal.slac.stanford.edu/sites/ard_public/facet/documents/facet-ii%20proposal%20v6.pdf), the Canadian Light Source, and BNL (http://www.bnl.gov/atf/). Thomson scattering is so predictable and well understood it is routinely used as a beam diagnostic on conventional electron accelerators such as the ALS linac at LBNL [64], HIGS at Duke and at Helmholtz Zentrum Dresden Rossendorf [65].

Photon yield is typically limited by the small scattering cross section and the need to keep $a_0 = eE/\omega m_e c^2 << 1$, where E and ω are the laser electric field and frequency respectively, to avoid broadening due to the nonlinear $(a_0^2/2)$ term in Eq. (5.29). Fixed facility MPSs run high electron current in conventional radio-frequency accelerators. Typically a relatively low energy laser pulse is recirculated to interact with multiple electron microbunches. Photon yield per electron is then low, which requires high accelerator power and shielding to achieve a given photon flux. This approach may be prohibitive in weight for transportable applications. For a transportable MPS, development of scattering methods approaching or even exceeding one photon/electron (10^7–10^8 photons/s for typical accelerator bunch charges) using realistic laser energies is important.

The photon beams have narrow (mrad) angular divergence, which will likely require rastering of the beam across an object for many applications. Achievable intensity within this narrow divergence is high. For the quoted examples, at kHz repetition rate, yields could be in the range of 10^{10}–10^{11} photons/s. Due to the narrow (mrad) emission angle, a large fraction of the flux can be delivered to the target at moderate energy spreads in the 10–50% range. Intensities in a static beam are high, with the yields cited above corresponding to a range of 10^{15}–10^{17} photons/s/sr. Rastering would typically reduce averaged intensity at the target plane to 10^{12} to 10^{16} photons/s/sr by sweeping the beam across the target. While such a beam would be very intense if held stationary on one position, rastering reduces the number of photons required to extract signal from each area of the target relative to use of a bremsstrahlung fan beam, hence reducing the required dose further. Adapting flux as such a beam is scanned across a target can allow extraction of signal using the physics-limited minimum dose, typically much less than used in a conventional bremsstrahlung fan beam. Combining effects of energy selection and narrow-angle emission, for many applications such a source can deliver higher penetrating power and signal contrast while using one to two orders of magnitude lower radiation dose [56]. To accomplish rastering while preserving low energy spread, the electron beam and the collimating aperture would be scanned together. The electron beam could be deflected using steering magnets to accomplish rastering, or using plasma-based techniques as detailed below.

Energy of the MPS beam could be adapted from shot to shot either by changing the electron accelerator energy (continuous tuning) or by frequency converting the scattering laser (typically integer multiples of energy such as 3, 6, or 9 MeV). This enables efficient photofission excitation, high contrast differentiation of materials in radiography using multiple energies, and selection of NRF lines [56]. Photon dose and polarization could be controlled continuously via the scattering laser.

Photon emission will be in short bursts (ps to fs in duration), and repetition rate must be matched to the application. For radiography, conventional detectors must be operated in current mode. For photofission applications, the desired signatures spread over longer times, allowing use of conventional detectors [56]. For NRF, typically the source yield must be controlled to limit detector rate. The short pulse length makes direct counting inefficient ($\leq$1 photon/shot/detector), and photons are spaced too closely to allow conventional pixelization. At energy spreads below the percent level, calorimetric detectors could significantly improve performance [66]. Concepts capable of reading out the energy of many MeV γ-ray hits simultaneously are also being developed. These disperse the γ-ray flux using Compton scattering in a foil, and subsequently analyze the resulting electrons by CCD tracking or a magnetic spectrometer [67, 68].

It is useful to compare Thomson MPSs to systems currently in use. Bremsstrahlung-based systems produce continuous energy spectra up to a cut-off energy given by the energy of the electron beam used to generate them. Typical energies include 3, 6, and 9 MeV. The beam is forward directed with large divergence, and is typically collimated into a *fan beam* to mitigate scattering issues, ease detector setup, and reduce detector cost. Typical angles include approximately

2° by 30°. Beam hardening filters can be used to cut off the lowest-energy photons, but also attenuate higher energy photons hence reducing usable flux. Linacs are the most commonly used accelerators, typically producing tens of μA of average current (delivered in pulses at ≈100 Hz), giving photon yields of order 10^{13} photon/s/sr on axis and hence 10^{11} photon/s into a fan collimator of 2° by 30°. Higher current (multi-mA), nearly continuous beam Rhodotrons are under development and can produce yields approximately two orders of magnitude higher than linacs but are large systems. An alternative at low energies is a ^{60}Co source, emitting lines at 1.17 and 1.33 MeV. These energies are not as penetrating, and flux is of order 3.5×10^9 photons/s/sr. All Thomson sources eliminate the converter target used in bremsstrahlung and some other sources, and produce narrow angle, tunable beams that reduce collimation requirements. These properties reduce undesired radiation production and corresponding shielding bulk.

Thomson MPSs also have important advantages relative to other source candidates. Annihilation of a relativistic positron beam [69] can produce a moderately low energy spread (typically order of 10%) and moderate divergence (10s of mrad level) photon beam but with significant low energy background and other issues [70]. Nuclear reactions can be used to produce monoenergetic photons (*Active interrogation using low-energy nuclear reactions*) [71–73] of broad divergence and at specific energies (often accompanied by neutron production) via interaction of an ion beam with a target. To meet application needs for flux on target, and given realistic source fluxes, the collimated beam will likely need to subtend a much larger angular spread than Thomson sources.

Use of Thomson MPSs is currently limited for AI applications by the need for high-energy linacs, which are large fixed facilities using conventional technology. The required electron energies for Thomson-based MeV monoenergetic sources are two orders of magnitude higher than those required for equivalent energy bremsstrahlung photons, and energy spread must also be lower. For current conventional radio-frequency accelerators, this results in large systems. The ELI-NP system under construction in Romania will for example use a linac approximately 70 m long to reach photon energies near 20 MeV [62]. Higher accelerating gradients are the path to reducing size. The T-REX/MEGA-Ray project at LLNL [74] is developing a source based on high gradient conventional radio-frequency accelerator cavities, towards a concept of a system designed to fit into a cargo container for photon energies near 2 MeV. Heavy shielding, which can dominate overall system size and weight as accelerators are made smaller, is conventionally also required due to the high electron beam energy. This is compounded by the low photon production cross section, which has hitherto required either high electron current (and thus heavier shielding) or very large scattering lasers. Energy spreads at the few percent level have been accessed on the facilities described above, and are adequate for photofission and radiography applications. For NRF applications, development of narrower energy spread, at or below the 2% level, via precise control of the electron beam is important. Energy spread at the 0.5% level is a near-term goal of the ELI-NP facility now under construction [62] and there is long term development towards the 0.1% level [66]. Current Thomson sources have (macro-pulse) repetition rates typically at or below 100 Hz for conventional linear accelerators (linacs). However, it is anticipated that future lasers will allow higher rates of 1 kHz to potentially 200 kHz over time (note that ring-based, energy recovery, or superconducting RF systems can offer much higher repetition rates, but are large for the required energies, and hence not considered here).

5.4.3 Laser-Plasma Acceleration for Compact Monoenergetic Photon Sources

Laser-plasma acceleration (LPA) produces the required 0.2–1 GeV electron energies for a Thomson MPS in plasmas less than 3 cm long, allowing a total source head length including laser focusing of a couple of meters (excluding the laser driver). The plasma can also be used to guide the scattering laser to increase the distance over which it interacts with the electron beam, reducing required laser size and/or electron current required per photon produced. LPA-based systems can further be applied to allow deceleration of the electron beam after production of the desired photon beam. Combined, these three techniques have the potential to substantially reduce system size to transportable scales. Realizing these advantages for many field applications requires compact, durable lasers at or above kHz repetition rates, which are being developed.

Accelerating fields orders of magnitude greater than conventional accelerators are achieved in LPAs using an electron density wave driven in a low density plasma by the radiation pressure of an intense laser to accelerate particles. The resonance condition for driving the wave results in predictable scaling for beam and laser energies [57, 58]. Over the past decade, quasi-monoenergetic electron beams were demonstrated at energies of 0.1 GeV [75–77] followed by 1 GeV [78], with few-percent energy spreads and charge typically in the range of 10^8 electrons/pulse. These electron energies enable photon energies spanning application needs from 1 MeV to above 20 MeV. Energies are now being increased to multiple GeV (e.g. 4.25 GeV in [79]), but these energies are higher than needed for AI. Hence the principle challenges in applying LPA for photon sources are control of beam quality and operability. Several laboratories worldwide now routinely produce electron beams with few-percent energy spread in the half-GeV range using LPA [78, 80–82]. In the past, the required laser power to generate a given electron energy was significantly higher than those indicated by theory and too high for transportable applications. Recently, precise control of laser temporal pulse shape and detailed laser phase-front control via use of a deformable mirror has resulted in high quality laser focal spots over the full focal depth of the laser [60], providing a more uniform and controlled accelerator structure. This enabled acceleration approaching theoretical predictions, including for example 0.2 GeV beams using 10 TW of power, 0.4 J energy, and 40 fs duration, focused to 10 μm spot sizes. Transverse emittance of the beams, a measure of focusability, has been shown to be at the 0.1 μm level [83, 84]. This level is comparable to high quality conventional linacs [85]. These energies are suitable to produce photons up to 9 MeV, e.g. for photofission or for Z discrimination in radiography. MeV energies suitable for bremsstrahlung photon sources are also available from LPA, and require much lower laser energy which provides a path to miniaturized sources that can be easily positioned around targets.

Experiments have demonstrated that Thomson scattering from LPAs [86] obeys the well-understood formulae for dependence on beam parameters, indicating that tuning such sources requires the same methods as for those based on conventional accelerators. A simple setup can be achieved by placing a foil to back-reflect the drive laser onto the electron beam [87] but this gives limited control over the scattering laser parameters. A laser split from the LPA driver [88] allows control of pointing and intensity but not pulse length, and has been used to produce MeV photons. The optimal laser parameters for scattering are in the range of ps duration, quite different from the LPA drive laser, motivating independent laser systems. For this reason early experiments at MeV energies resulted in broad spectra at the 50% bandwidth level and limited yield due to low electron beam quality and lack of independent scattering laser control. Recent experiments have also started to narrow energy spread, particularly at sub-MeV energies, and demonstrated control over photon energy [89–91].

Development of LPAs towards narrower energy spread and high stability is important for a photon source, and requires control of particle injection. The simplest LPAs operate by driving the plasma wave to very high amplitude, where the wave breaks to inject and accelerate particles in pure hydrogen or helium plasmas that provide homogeneity, but this nonlinear process is highly sensitive. Several methods of injection control have been demonstrated by altering the plasma composition or structure, or using an additional laser. Doping the plasma with percent-level amounts of nitrogen introduces states that ionize near the laser peak intensity, yielding highly stable injection [92, 93]. Stable operation over many thousands of shots and months of operation has been demonstrated [94], but with relatively broad energy spread. A localized plasma density down-ramp in the direction of laser propagation can trigger injection by changing the structure period [95, 96]. This technique has produced stable electron beams with control over energy in the 0.1–0.5 GeV range and with few percent energy spreads appropriate to AI photon sources [97, 98]. Controlling injection using colliding laser pulses [99] has delivered the narrowest energy spreads ($<1.4\%$ FWHM) observed in LPAs, as well as tuning to allow photon energy selection [60, 80]. Examples of such beams are shown in Fig. 5.34 Recent experiments also indicate that slice energy spread in LPAs can be less than 1% [100], which with removal of correlation could enable sub-percent photon bandwidth sources. This provides a path towards photon energy spreads suitable for NRF methods. Charge for these methods is typically in the range of 10^8 e-/shot.

Simulations indicate that beams from state of the art LPAs are sufficient to produce low energy spread MeV photon beams via Thomson scattering (Fig. 5.35). Energy spread is dominated by electron beam divergence, and is at the 10–20%

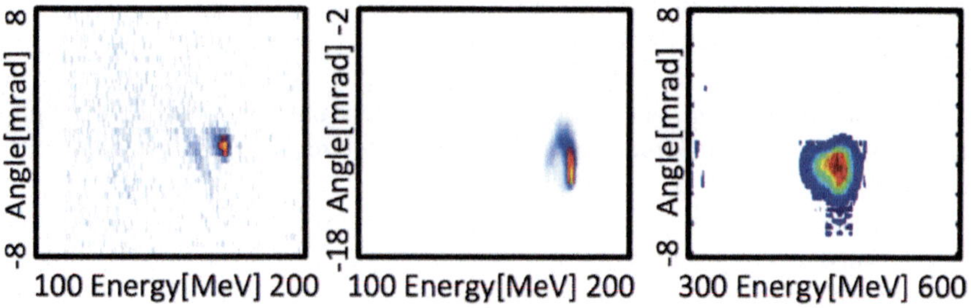

Fig. 5.34 LPA experiments produce high quality electron beams: colliding pulse injection (left) or self injection at 0.2–0.6 GeV (right two plots) (Reprinted from Ref. [60])

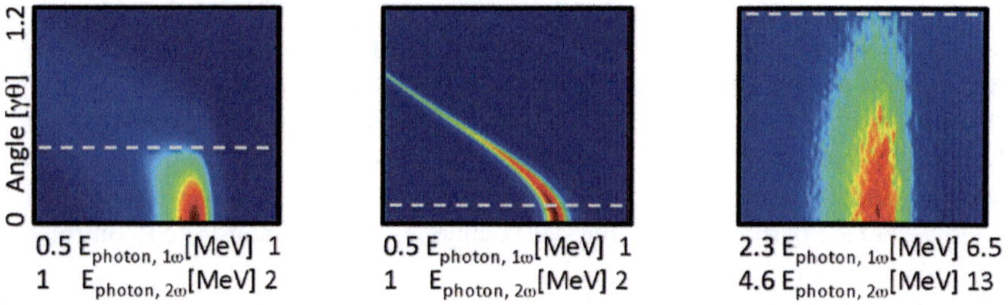

Fig. 5.35 Photon spectra simulated for the electron beams of high quality LPA and an independent scatter laser. Left: 200 MeV electron beams produce photon energy of 0.85 MeV with a scattering laser of 1ω (800 nm), or 1.7 MeV with 2ω (400 nm), both in 10% FWHM bandwidth set by beam divergence. Center: few percent bandwidth could be produced by expanding the same electron beam to reduce its divergence. Right: 460-MeV electron beams produce 4.7 MeV or 9 MeV photons using 800 or 400 nm scattering lasers, respectively, at bandwidth of 25% FWHM set by divergence. As above, beam expansion can reduce bandwidth. Collimator angles for each beam are illustrated with dashed lines, indicating how for narrower bandwidth a narrower collimator is required (Reprinted from Ref. [60])

level if scattering is done at the accelerator exit. To reduce bandwidth from the 10% level down to the limit imposed by electron beam energy spread, at or below the percent level for the lowest energy-spread LPAs, refocusing of the electron beam is needed. Conventional quadrupole lenses require meter-scale space, and this long drift space can increase beam length (which can limit deceleration) and emittance (which degrades photon energy spread). Techniques demonstrated for compact, deceleration-compatible manipulation of divergence, using tailored plasma targets or axial plasma currents to form plasma lenses [94, 101], would need to be used.

Achieving high photon yield requires new approaches using LPAs. For typical bunch charges, yield at or above one photon per electron is desirable to meet application flux targets, much higher than is used in conventional Thomson sources. This conversion yield is important both to reduce the electron current required for a given photon flux (which drives both laser and shielding size) and because LPA pulse structure does not accommodate the scattering laser pulse recirculation used on conventional linacs. Conventionally, however, achieving one photon per electron would require tens of Joules of scattering laser energy. This is an order of magnitude more laser energy than the LPA itself, which would be too large for transportable applications and not available at high repetition rates. However, most of this scattering laser energy does not interact with the electrons, because matching the long pulse length to the focal depth requires a focal spot size much larger than that of the electron beam. There is no gain from increasing the electron beam spot size because few of the laser photons are scattered. Hence, realizing high yield with lower scattering laser energies then requires that either the laser focal depth must be increased while keeping its radius small, or that the required interaction length must be shortened. The scattering laser focal depth could be extended using plasma channel like those used to guide the drive laser in LPA experiments over >10 focal depths at guided spot sizes as small as a few μm. This would reduce required scattering laser energy to the Joule class, similar to the LPA drive laser [102]. A parabolic plasma density profile with minimum density on axis forms a guiding fiber for the laser pulse which can withstand intensities beyond those required for Thomson scattering. For very narrow bandwidth photon sources, it may be desirable to use a hollow plasma channel [103, 104] to remove focusing forces on the electron beam while preserving laser focusing [102]. Further reduction in scatter laser energy may be possible by shaping the laser pulse in amplitude and frequency (chirp), which compensates for nonlinear effects on photon energy spread. Analytic and numerical solutions have been derived [105–108] indicating this enables higher interaction intensity and shorter interaction length. These approaches provide a path to yield of >1 photon/electron to reduce required electron current, while using laser energy comparable to the LPA drive laser energy. This mitigates both scattering laser and shielding contributions to MPS size.

To reduce secondary radiation from the high-energy electron beam, which in conventional Thomson MPSs requires heavy shielding that severely limits transportability, the LPA structure can be used to decelerate the beam after photon production. The high gradient of the LPA means this adds negligibly (cm-scale) to source size. The simplest implementation exploits the fact that the LPA structure travels slower than the speed of light: hence a injected particle beam first accelerates, then dephases at a defined distance, slipping into the decelerating phase and losing energy as shown in Fig. 5.36. Photons would be produced by scattering at the point where the electron beam reaches maximum energy. This approach is suitable for photon sources at the 10% level of energy spread. Simulations indicate deceleration of 85% or more is achievable [102]. For energy spread at the percent level, as noted above the scattering should be conducted after refocusing of the electron beam and in a different plasma structure than is used for acceleration. This motivates use of a multi-segment plasma structure. Two experiments indicate that deceleration is feasible via this latter technique. The beam generated in a first LPA structure was injected into a second plasma structure powered by a separate laser pulse and timing of the drive laser of the second stage controlled the phasing of the electron bunch in the second structure. Deceleration was achieved by appropriately timing the drive laser, and as expected acceleration was observed in the opposite

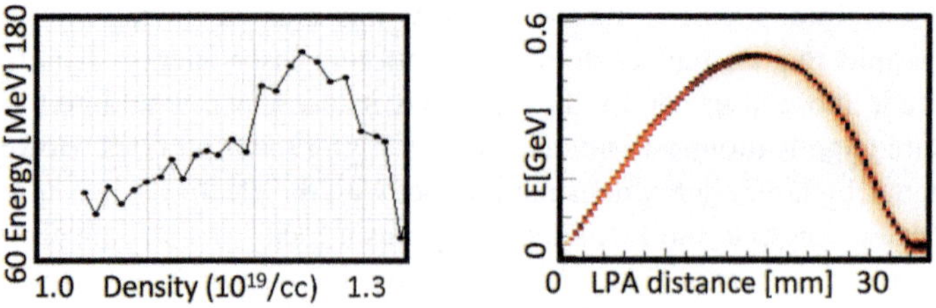

Fig. 5.36 (left) Colliding pulse experiments demonstrate control of LPA energy, including deceleration of the beam, by adjusting plasma density, with peak energy reached at the predicted density of $1.25 \times 10^{19}/\text{cm}^3$. (right) Simulations show that controlled deceleration can bring the beam down to 10% or less of initial energy, reducing shielding requirements

phaser [94]. In a single plasma structure deceleration could only be inferred because electron energy is observed only at the end of the plasma. With injection location locked by colliding laser pulses, the dephasing length was adjusted by changing the plasma density. As density increased the energy increases until a maximum value, where the plasma length approximately matches the dephasing length. At higher densities, the observed decrease of beam energy versus density indicates that the beam has decelerated. Simulations indicate that higher deceleration efficiency can be achieved by adjusting the plasma density and profile with more than 90% being possible [109], strongly reducing shielding requirements.

The principal characteristics of laser plasma Thomson sources are similar to those from conventional linacs, with three exceptions: pulse length, pulse structure, and rastering methods. Pulse duration will be fs, compared to ps for conventional systems but this difference is not known to be important to AI applications. Pulse structure will consist of evenly spaced pulses, rather than the bursts typical of conventional linacs. For rastering of laser-plasma based sources, due to the compact (1–2 m) size of the system, the whole accelerator and scattering system could be rotated to steer the beam, or alternatively the accelerator can be steered by steering the laser using high speed optical scanning mirrors.

To meet AI needs for photon flux, development lasers at kHz repetition rates and above will be crucial. Current lasers appropriate to drive MPSs are available in trailerable form (http://www.amplitude-technologies.com/) but allow only 10 Hz, and in a few larger systems 100 Hz. A development path has been identified to kHz, high efficiency lasers with pulse parameters suitable for LPAs and a program to realize such lasers is in progress [110]. Fiber lasers are very efficient and can operate at high repetition rates and high powers. Coherent combination of many fiber channels is required to address MPS application pulse energy requirements. Alternatively, high-efficiency pumping of solid state media is also under development. An LPA facility at kHz is a near-term step. For optimal Thomson source performance, modifications will be made to the scattering laser including ps pulse lengths and shaped pulses to minimize bandwidth and maximize yield.

Use of LPA-based Thomson MPSs for AI applications has the potential to reduce the size of the electron accelerator, scattering, and shielding systems. This is made possible by orders of magnitude higher accelerating gradients. The largest system component would be the laser (both LPA and scattering systems), which currently is at the scale of a standard cargo container for photon energies of 2–15 MeV and is dropping rapidly in size as lasers develop, such that smaller systems are likely to be realistic in the near future. Energy spreads at the ten percent level can been accessed on the experiments described above using direct scattering from the LPA electron beam, which is adequate for photofission and radiography applications. For NRF applications, development of narrower energy spread, at or below the 2% level, via precise control of the electron beam will require refocusing of the electron beam from LPAs or reduction of emittance. To enable use of LPAs, experiments must move from proofs of principle to demonstrate the combination of LPA, controlled scattering and deceleration described in this chapter. Operation must then be demonstrated over many days, and operation without requirement for tuning by a scientist. Laser development is required in parallel to move from the state of the art 10-Hz level to higher rates of 1 kHz in the near term and to potentially 100 kHz or more over time.

5.4.4 Laser-Driven Neutron Sources

Neutrons in many cases offer complementary properties to photons for AI and other applications. While intense, directional, high energy and pulsed neutron sources have been available from accelerators, large-scale facilities are needed to combine those characteristics. AI applications have typically used broad-angle sources from D-D or D-T reactions. Laser-driven accelerators offer new possibilities to produce compact, directional neutron sources from ion acceleration or using a monoenergetic photon source.

Laser plasma acceleration can produce beams of multi-MeV ions such as deuterons from relatively compact systems, and these beams can be used to generate neutron beams via interactions in a low-Z converter such as Be. Historically, such experiments utilized a sheath acceleration mechanism [111] in relatively thick targets opaque to the laser pulse [112], which produced limited ion energies that in turn limited performance. New regimes of ion acceleration now offer improved neutron yield. One recent experiment [113] utilized an 80 J, 600 fs laser in the regime where the target (initially 700 nm thick) becomes transparent to the laser beam during acceleration due to the laser's relativistic intensity [114]. This process produced deuterons up to 170 MeV. Directing this beam into a Be converter produced neutrons with yield of 10^{10} n/sr/shot, in a spectrum peaked at 10 MeV and extending up to 100 MeV. The neutrons were emitted from a mm-scale spot, into a forward cone of approximately 50° opening angle. As in photon sources, such a directional beam better matches with many targets of interest, and reduces shielding required in broad-angle sources to eliminate undesired particles. The beams were used for proof of principle radiography [113] and AI experiments that verified they can be applied as predicted by simulations. For a 100 J laser system based on the results at PHELIX and TRIDENT simulations indicate yields can reach 10^{12} neutrons/shot using optimized target/converter designs. Lasers at this scale are under development at 10 Hz rates, which could enable yields of 10^{13}/s in a forward directed beam. Use of more compact lasers is also of interest. Scaling from achieved performance [115] indicates that a Joule-class ps laser at kHz repetition rate could produce a fluence within the emission cone comparable to an isotropic source of 10^{11}–10^{12}/s. Joule-class lasers at kHz repetition rate are being developed as noted above, and a single such laser could be used to drive both a MPS and neutron source. Either approach could enable a compact and even mobile laser-driven neutron source in the near future. Higher performance in the future may be enabled by exploiting the radiation pressure acceleration regime in which the whole target is accelerated by light pressure [116]. Exploiting this regime requires very high laser temporal contrast.

Neutron beams can alternatively be produced from photon beams using γ-n reactions in high-Z targets. Recent experiments used a laser-produced bremsstrahlung beam from an electron beam having a 10 MeV Maxwellian distribution, produced from a 90 J, 150 fs laser pulse [117]. Bremsstrahlung and neutron production were accomplished in a Cu target, yielding of order 10^9 neutrons per pulse, with close to isotropic emission. Yields a few times higher were achieved in W and U converters. Utilizing a monoenergetic photon beam as described above, the bremsstrahlung step is removed and all photons could be delivered at the appropriate energy for γ-n reactions, increasing relative yield and enabling control of neutron emission [62]. By tuning the MPS energy or by inserting and removing the converter such a MPS-based neutron source can be turned on only when needed, while in other cases photons only can be used with no neutrons produced, reducing radiation concerns. Cases of interest could include use of a high-energy photon beam, above 10 MeV, to produce neutrons within an object of interest, potentially enabling novel AI methods. Because neutron production can be turned on and off by changing the photon beam energy, this could be done after an initial lower-energy scan using photons only, which could be used for example to exclude the presence of stowaway persons.

5.4.5 Conclusion

By bringing to compact facilities high performance probe sources until now available only at large fixed scientific research facilities, laser driven methods have the potential to significantly advance the state of the art in AI and related security applications. Laser scattering produces monoenergetic photon beams with narrow emission angle, while laser-driven acceleration techniques enable compact systems. Sources of nearly monoenergetic photons can hence be produced compactly by combining laser-plasma acceleration, control of scattering, and deceleration to limit shielding needs. Applications assessments indicate that such beams could enable strong reductions of dose required (in the range of one to two orders of magnitude in many cases) for AI together with improved signal strength and contrast to enable finer discrimination of materials [56]. The same acceleration technology can also be used to produce miniature bremsstrahlung sources that are easily positionable, addressing applications where size is a key driver. Directional neutrons can also be produced via a separate laser-plasma acceleration process, but in many cases can use similar lasers to those used for monoenergetic photon sources. Neutrons have complementary sensing properties to photons in many AI cases. The ability to use a single laser driver could enable a *switchable* source producing the particle needed on demand, further increasing performance.

Acknowledgements The Sect. 5.4 was supported by the U.S. Dept. of Energy National Nuclear Security administration Defense Nuclear Nonproliferation R and D (NA-22), and by the Office of Science Office of High Energy Physics, under Contract No. DE-AC02-05CH11231 to Lawrence Berkeley National Laboratory. The simulations used the computational facilities at the National Energy Research Scientific Computing Center, a DOE Office of Science User Facility supported by the Office of Science of the U.S. Department of Energy under Contract No. DE-AC02-05CH11231. We would like to acknowledge contributions by Markus Roth, Hai-En Tsai, Jeroen van Tilborg, Sven Steinke, Csaba Toth, Jean-Luc Vay, Carl Schroeder, Eric Esarey, Kei Nakamura, Bernhard Ludewigt, Brian Quiter, John Valentine, Marie-Anne Descalle, Matt Kinalw, David Chichester, Glen Warren, Cameron Miller, Sara Pozzi, and Wim Leemans.

References

1. W. Bertozzi, W. Franklin, S. Korbly, R.J. Ledoux, R. Niyazov, D.R. Swenson, A. Klimenko, Int. J. Modern Phys. A **26**(10–11), 1713 (2011). https://doi.org/10.1142/S0217751X11053122. International Conference on Fixed Field Alternating Gradient Accelerators (FFAG 09), Batavia, Sep 21–25, 2009
2. A. Jason, H. Miyadera, P. Turchi, SNM detection by active interrogation. LA-UR-10-04398, Los Alamos National Laboratory (2010)
3. E.I. Simakov, H.L. Andrews, M.J. Herman, K.M. Hubbard, E. Weis, vol. 1812 (2017), p. 060010. http://aip.scitation.org/doi/abs/10.1063/1.4975877
4. M. Roth, D. Jung, K. Falk, N. Guler, O. Deppert, M. Devlin, A. Favalli, J. Fernandez, D. Gautier, M. Geissel, R. Haight, C.E. Hamilton, B.M. Hegelich, R.P. Johnson, F. Merrill, G. Schaumann, K. Schoenberg, M. Schollmeier, T. Shimada, T. Taddeucci, J.L. Tybo, F. Wagner, S.A. Wender, C.H. Wilde, G.A. Wurden, Phys. Rev. Lett. **110**, 044802 (2013). https://link.aps.org/doi/10.1103/PhysRevLett.110.044802
5. I. Hofmann, Phys. Rev. ST Accel. Beams **16**(4), 041302 (2013). https://doi.org/10.1103/PhysRevSTAB.16.041302
6. M. Schollmeier, S. Becker, M. Geißel, K.A. Flippo, A. Blažević, S.A. Gaillard, D.C. Gautier, F. Grüner, K. Harres, M. Kimmel, F. Nürnberg, P. Rambo, U. Schramm, J. Schreiber, J. Schütrumpf, J. Schwarz, N.A. Tahir, B. Atherton, D. Habs, B.M. Hegelich, M. Roth, Phys. Rev. Lett. **101**, 055004 (2008). https://link.aps.org/doi/10.1103/PhysRevLett.101.055004
7. J.C. Fernandez, in *FESAC Workshop on Scientific Opportunities in High Energy Density Laboratory Plasma Physics*, Washington, DC, 2008
8. V.A. Dolgashev, AIP Conf. Proc. **1507**(1), 76 (2012). http://aip.scitation.org/doi/abs/10.1063/1.4773679

9. R.A. Marsh, F. Albert, S. Anderson, C. Barty, D. Gibson, F. Hartemann, S. Wu, Ultracompact accelerator technology for a next-generation gamma-ray source, in *Proceedings of the 2012 International Particle Accelerator Conference (IPAC 2012)*, New Orleans, 20-25 May 2012, pp. 3190–3194
10. O. Williams, *Basic Theory of Inverse Compton (Thomson) Scattering and Conceptual Design of X-Ray Source using the PEGASUS Linac at UCLA and a Tabletop Terawatt Laser*. Physics 199 (UCLA Particle Beam Physics Lab, UCLA Department of Physics and Astronomy, Winter 2003)
11. T. Wangler, *RF Linear Accelerators*. Physics Textbook (Wiley, New York, 2008). https://books.google.com/books?id=5EO4_nyndUMC
12. C. Tang, H. Chen, Q. Jin, T. Du, C. Cheng, D. Tong, Y. Lin, Y. Liu, in *International Topical Meeting on Nuclear Research Applications and Utilization of Accelerators* (International Atomic Energy Agency, Vienna, 2009). INIS-XA-09N0647
13. D.C. Faircloth, in *Proceedings, CAS - CERN Accelerator School: Ion Sources: Senec, Slovakia, May 29–June 8, 2012*, pp. 285–310 (2013). https://doi.org/10.5170/CERN-2013-007.285. https://inspirehep.net/record/1288541/files/arXiv:1404.0944.pdf
14. M. Reiser, *Theory and Design of Charged Particle Beams*. Wiley Series in Beam Physics and Accelerator Technology (Wiley, New York, 2008). https://books.google.com/books?id=tqtnXnhUuvIC
15. T. Gozani, J. Stevenson, M.J. King, Nucl. Instrum. Methods Phys. Res. Sect. A **652**(1), 334 (2011). http://dx.doi.org/10.1016/j.nima.2011.01.029. http://www.sciencedirect.com/science/article/pii/S0168900211000805. Symposium on Radiation Measurements and Applications (SORMA) XII 2010
16. J. Wang, X-band accelerator structures R&D at SLAC. SLAC/LLNL Discussion (2011)
17. M.S. Livingston, J.P. Blewett, Iris-loaded linac structures, in *Particle Accelerators* (McGraw-Hill Book Company, New York, 1962), p. 324
18. J.K. Sekutowicz, in *Proceedings, CAS - CERN Accelerator School: RF for Accelerators*, Ebeltoft, 8–17 Jun 2010 (2012). https://inspirehep.net/record/1084441/files/arXiv:1201.2598.pdf
19. E. Colby. Klystrons. US Particle Accelerator School, Vanderbilt University (2010)
20. M. Vretenar, A. Dallocchio, V. Dimov, M. Garlaschè, A. Grudiev, A. Lombardi, S. Mathot, E. Montesinos, M. Timmins, in *27th Linear Accelerator Conference* (Geneva, 2014), p. THPP040. 4 p. https://cds.cern.ch/record/2062619
21. S. Jolly, An intra-pulse fast feedback system for a future linear collider. Ph.D. thesis, Exeter College, University of Oxford, Oxford, 2003
22. Computer Simulation Technology (CST), www.cst.com
23. S. Ramo, J. Whinnery, T. Van Duzer, *Fields and Waves in Communication Electronics* (Wiley, New York, 1994). https://books.google.com/books?id=lIFkQgAACAAJ
24. G. Ciovati, in *Proceedings, CAS - CERN Accelerator School : Course on Superconductivity for Accelerators (CAS 2013)*, Erice, April 24–May 4, 2013 (2014), pp. 57–75. https://doi.org/10.5170/CERN-2014-005.57. https://inspirehep.net/record/1342040/files/arXiv:1501.07398.pdf
25. A. Chao, M. Tigner, *Handbook of Accelerator Physics and Engineering* (World Scientific, Singapore, 1999). https://books.google.com/books?id=Z3J4SjftF1YC
26. C. Boulware, Introduction to the physics of high-quality electron beams. PITZ Physics Lecture for Summer Students (2007)
27. G.J. Caporaso, eConf C00082, WE101 (2000) [658(2000)]
28. S. Sampayan, G. Caporaso, Y. Chen, S. Falabella, G. Guethlein, J. Harris, S. Hawkins, C. Holmes, M. Krogh, S. Nelson, W. Nunnally, A. Paul, B. Poole, M. Rhodes, D. Sanders, K. Selenes, K. Shaklee, S. Sitaraman, J. Sullivan, L. Wang, J. Watson, Nucl. Instrum. Methods Phys. Res. Sect. B **261**(1), 281 (2007). http://dx.doi.org/10.1016/j.nimb.2007.04.244. http://www.sciencedirect.com/science/article/pii/S0168583X07007380. The Application of Accelerators in Research and Industry

29. S. Boucher, R.B. Agustsson, P. Frigola, A.Y. Murokh, M. Ruelas, F.H. O'Shea, J.B. Rosenzweig, G. Travish, Conf. Proc. C0806233, TUPP150 (2008)
30. H.L. Luo, Y.C. Xu, X.Q. Wang, H.L. Xu, Chin. Phys. **C37**, 097001 (2013). https://doi.org/10.1088/1674-1137/37/9/097001
31. J.M. Potter, D. Schwellenbach, A. Meidinger, **525**, 178–183 (2013). http://aip.scitation.org/doi/abs/10.1063/1.4802315
32. D.E. Nagle, E.A. Knapp, B.C. Knapp, Rev. Sci. Instrum. **38**(11), 1583 (1967). http://dx.doi.org/10.1063/1.1720608
33. G.E. Dale, H.C. Kirbie, W.B. Haynes, C.E. Heath, T.A. Lopez, F.P. Romero, R.M. Wheat, in *2005 IEEE Pulsed Power Conference* (2005), pp. 1211–1214. https://doi.org/10.1109/PPC.2005.300572
34. W. Leemans, in *Proceedings of IPAC*, New Orleans, 2012
35. T.B. Zhang, J.L. Hirshfield, T.C. Marshall, B. Hafizi, Phys. Rev. E **56**, 4647 (1997). https://link.aps.org/doi/10.1103/PhysRevE.56.4647
36. E.A. Peralta, K. Soong, R.J. England, E.R. Colby, Z. Wu, B. Montazeri, C. McGuinness, J. McNeur, K.J. Leedle, D. Walz, E.B. Sozer, B. Cowan, B. Schwartz, G. Travish, R.L. Byer, Nature **503**(7474), 91 (2013). http://dx.doi.org/10.1038/nature12664
37. M. Uesaka, H. Kobayashi, *Compact Neutron Sources for Energy and Security* (World Scientific Publishing Co, Singapore, 2015), pp. 181–207. https://doi.org/10.1142/9789813108905_0010
38. A. Murata, S. Ikeda, N. Hayashizaki, Nucl. Instrum. Methods Phys. Res. B **406**, 260 (2017). http://dx.doi.org/10.1016/j.nimb.2016.12.024. http://www.sciencedirect.com/science/article/pii/S0168583X16305626. Proceedings of the 12th European Conference on Accelerators in Applied Research and Technology (ECAART12)
39. D.L. Chichester, J.D. Simpson, Compact neutron generators. The Industrial Physicist, 22-25 (December 2003/January 2004)
40. M. Rickey, R. Smythe, Nucl. Instrum. Methods **18**, 66 (1962). http://dx.doi.org/10.1016/S0029-554X(62)80010-X. http://www.sciencedirect.com/science/article/pii/S0029554X6280010X
41. R. Lanza, in *IAEA RCM, Neutron Based Techniques for the Detection of Illicit Materials and Explosives*, Johannesburg, 2009
42. I.M. Kapchinskij, I.M. Teplyakov, Prib. Tekh. Eksp **4**, 17 (1970)
43. R.H. Stokes, K.R. Crandall, J.E. Stovall, D.A. Swenson, IEEE Trans. Nucl. Sci. **26**(3), 3469 (1979). https://doi.org/10.1109/TNS.1979.4330069
44. B. Koubek, J. Schmidt, A. Schempp, L. Groening, in *Proceedings of IPAC 2011*, San Sebastian, 2011, pp. 2568–2570
45. T.J.T. Kwan, R.E. Morgado, T.S.F. Wang, B. Vodolaga, V. Terekhin, L.M. Onischenko, S.B. Vorozhtsov, E.V. Samsonov, A.S. Vorozhtsov, Y.G. Alenitsky, E.E. Perpelkin, A.A. Glazov, D.L. Novikov, V. Parkhomchuk, V. Reva, V. Vostrikov, V.A. Mashinin, S.N. Fedotov, S.A. Minayev, Rev. Sci. Instrum. **81**(10), 103304 (2010). http://dx.doi.org/10.1063/1.3488354
46. R. Duperrier, R. Ferdinand, J.M. Lagniel, N. Pichoff, *Proceedings of the 20th International Linear Accelerator Conference (Linac 2000)*, Monterey, 21-25 August 2000, pp. 839–842
47. J.D. Schneider, in *Proceedings of the 1999 Particle Accelerator Conference (Cat. No.99CH36366)*, vol. 1 (1999), pp. 503–507. https://doi.org/10.1109/PAC.1999.795744
48. M. Kaushik, L. Joshi, J. Electromagnetic Waves Appl. **29**(15), 2027 (2015). http://dx.doi.org/10.1080/09205071.2015.1074874
49. M. Heron, A miniature accelerator to treat cancer. https://home.cern/about/updates/2015/07/miniature-accelerator-treat-cancer; http://accelconf.web.cern.ch/AccelConf/linac2016/talks/th1a06_talk.pdf
50. R. Morgado, G. Arnone, C. Cappiello, Technical Report LA-UR-96-1206, Los Alamos National Laboratory (1996)

51. P. Kerr, M. Rowland, D. Dietrich, W. Stoeffl, B. Wheeler, L. Nakae, D. Howard, C. Hagmann, J. Newby, R. Porter, Nucl. Instrum. Methods Phys. Res. B **261**(1), 347 (2007). http://dx.doi.org/10.1016/j.nimb.2007.04.190. http://www.sciencedirect.com/science/article/pii/S0168583X07009202. The Application of Accelerators in Research and Industry
52. J.W. Staples, M.D. Hoff, J.W. Kwan, D. Li, B.A. Ludewigt, A. Ratti, S.P. Virostek, R.P.W. Lbnl, in *Proceedings of LINAC 2006*, Knoxville, 2006, pp. 253–255
53. C.L. Morris, Active interrogation using energetic protons. Technical Report LA-UR-10-04680, Los Alamos National Laboratory (2010)
54. R.W. Garnett, in *Proceedings of 28th Linear Accelerator Conference (LINAC16)*, East Lansing, 2016
55. R.C. Runkle, D.L. Chichester, S.J. Thompson, Nucl. Instrum. Methods Phys. Res., Sect. A **663**(1), 75 (2012)
56. C. Geddes, B. Ludewigt, J. Valentine, B. Quiter, M.A. Descalle, G. Warren, M. Kinlaw, S. Thompson, D. Chichester, C. Miller, S. Pozzi, Impact of monoenergetic photon sources on nonproliferation applications. Technical Report, DNN R & D Final Project Report (2017). http://www.osti.gov/scitech/servlets/purl/1376659
57. W. Leemans, E. Esarey, Phys. Today **62**(3), 44 (2009). https://doi.org/10.1063/1.3099645
58. E. Esarey, C.B. Schroeder, W.P. Leemans, Rev. Mod. Phys. **81**, 1229 (2009)
59. W.P. Leemans, E. Esarey, J. van Tilborg, P.A. Michel, C.B. Schroeder, C. Toth, C.G.R. Geddes, B.A. Shadwick, IEEE Trans. Plasma Sci. **33**(1), 8 (2005)
60. C. Geddes, S. Rykovanov, N. Matlis, S. Steinke, J.L. Vay, E. Esarey, B. Ludewigt, K. Nakamura, B. Quiter, C. Schroeder, C. Toth, W. Leemans, Nucl. Instrum. Methods Phys. Res. B **350**, 116 (2015)
61. V. Litvinenko, J. Madey, Nucl. Instrum. Methods Phys. Res. Sect. A **375**(1–3), 580 (1996)
62. O. Tesileanu, D. Ursescu, R. Dabu, N.V. Zamfir, J. Phys. Conf. Ser. **420**, 1–8 (2013)
63. R. Hajima, T. Hayakawa, T. Shizuma, C. Angell, R. Nagai, N. Nishimori, M. Sawamura, S. Matsuba, A. Kosuge, M. Mori, M. Seya, Eur. Phys. J. Spec. Top. **223**, 1229 (2014)
64. W.P. Leemans, R.W. Schoenlein, P. Volfbeyn, A.H. Chin, T.E. Glover, P. Balling, M. Zolotorev, K.J. Kim, S. Chattopadhyay, C.V. Shank, Phys. Rev. Lett. **77**(20), 4182 (1996)
65. A. Jochmann et al., Phys. Rev. Lett. **111**, 114803 (2013)
66. J.M. Hall, V.A. Semenov, F. Albert, C.P.J. Barty, in *Proceedings of 52nd Annual Meeting of the Institute of Nuclear Material Management*, Palm Desert, 2011
67. A. Gehring et al., Proc. SPIE **9215** (2014)
68. Y. Zhang et al., in *Proceedings of IEEE NSS* (2016)
69. C.R. Hatcher et al., Nucl. Inst. Methods **14**, 337 (1961)
70. L. Campbell, J. Fast, R. Wittman, R. Abrams, A. Afanasev, C. Ankenbrandt, K. Beard, G. Flanagan, R. Johnson, C. Yoshikawa, M. Popovic, Monoenergetic photon beams based on positron annihilation. Technical Report, Pacific Northwest National Laboratory, Richland, PNNL-22882 (2013). http://www.osti.gov/scitech/servlets/purl/1376659
71. T. Taddeucci, R. Sheffield, Neutron and gamma-ray production with low-energy beams. Technical Report, Los Alamos National Laboratory Report LA-UR-07-2724 (2010)
72. T. Kwan et al., Rev. Sci. Instrum. **81**, 1033034 (2010)
73. B. O'Day, Z. Hartwig, R. Lanza, A. Danagoulian, Nucl. Instrum. Methods A **832**, 68 (2016)
74. F. Albert, S.G. Anderson, D.J. Gibson, C.A. Hagmann, M.S. Johnson, M. Messerly, V. Semenov, M.Y. Shverdin, B. Rusnak, A.M. Tremaine, F.V. Hartemann, C.W. Siders, D.P. McNabb, C.P.J. Barty, Phys. Rev. Spec. Top. - Accel. Beams **13**, 070704 (2010)
75. C.G.R. Geddes, C. Toth, J. van Tilborg, E. Esarey, C.B. Schroeder, D. Bruhwiler, C. Nieter, J. Cary, W.P. Leemans, Nature **431**(7008), 538 (2004)
76. J. Faure, Y. Glinec, A. Pukhov, S. Kiselev, S. Gordienko, E. Lefebvre, J.P. Rousseau, F. Burgy, V. Malka, Nature **431**(7008), 541 (2004)
77. S.P.D. Mangles, C.D. Murphy, Z. Najmudin, A.G.R. Thomas, J.L. Collier, A.E. Dangor, E.J. Divall, P.S. Foster, J.G. Gallacher, C.J. Hooker, D.A. Jaroszynski, A.J. Langley, W.B. Mori, P.A. Norreys, F.S. Tsung, R. Viskup, B.R. Walton, K. Krushelnick, Nature **431**(7008), 535 (2004)

78. W.P. Leemans, B. Nagler, A.J. Gonsalves, C. Tóth, K. Nakamura, C.G.R. Geddes, E. Esarey, C.B. Schroeder, S.M. Hooker, Nat. Phys. **2**, 696 (2006)
79. W.P. Leemans, A.J. Gonsalves, H.S. Mao, K. Nakamura, C. Benedetti, C.B. Schroeder, C. Tóth, J. Daniels, D.E. Mittelberger, S.S. Bulanov, J.L. Vay, C.G.R. Geddes, E. Esarey, Phys. Rev. Lett. **113**, 245002 (2014)
80. J. Faure, C. Rechatin, A. Norlin, A. Lifschitz, Y. Glinec, V. Malka, Nature **444**(7120), 737 (2006)
81. T.P.A. Ibbotson, N. Bourgeois, T.P. Rowlands-Rees, L.S. Caballero, S.I. Bajlekov, P.A. Walker, S. Kneip, S.P.D. Mangles, S.R. Nagel, C.A.J. Palmer, N. Delerue, G. Doucas, D. Urner, O. Chekhlov, R.J. Clarke, E. Divall, K. Ertel, P.S. Foster, S.J. Hawkes, C.J. Hooker, B. Parry, P.P. Rajeev, M.J.V. Streeter, S.M. Hooker, Phys. Rev. ST Accel. Beams **13**, 031301 (2010)
82. S. Banerjee, S.Y. Kalmykov, N.D. Powers, G. Golovin, V. Ramanathan, N.J. Cunningham, K.J. Brown, S. Chen, I. Ghebregziabher, B.A. Shadwick, D.P. Umstadter, Phys. Rev. Spec. Top. Accel Beams **16**, 031302 (2013)
83. G.R. Plateau, C.G.R. Geddes, D.B. Thorn, M. Chen, C. Benedetti, E. Esarey, A.J. Gonsalves, N.H. Matlis, K. Nakamura, C.B. Schroeder, S. Shiraishi, T. Sokollik, J. van Tilborg, Cs. Tóth, S. Trotsenko, T.S. Kim, M. Battaglia, T. Stöhlker, W.P. Leemans, Phys. Rev. Lett. **109**, 064802 (2012)
84. R. Weingartner, S. Raith, A. Popp, S. Chou, J. Wenz, K. Khrennikov, M. Heigoldt, A.R. Maier, N. Kajumba, M. Fuchs, B. Zeitler, F. Krausz, S. Karsch, F. Grüner, Phys. Rev. ST Accel. Beams **15**, 111302 (2012). http://link.aps.org/doi/10.1103/PhysRevSTAB.15.111302
85. Y. Ding, A. Brachmann, F.J. Decker, D. Dowell, P. Emma, J. Frisch, S. Gilevich, G. Hays, P. Hering, Z. Huang, R. Iverson, H. Loos, A. Miahnahri, H.D. Nuhn, D. Ratner, J. Turner, J. Welch, W. White, J. Wu, Phys. Rev. Lett. **102**, 254801 (2009)
86. H. Schwoerer, B. Liesfeld, H.P. Schlenvoigt, K.U. Amthor, R. Sauerbrey, Phys. Rev. Lett. **96**(1), 014802 (2006)
87. K. Phuoc et al., Nat. Photon. **6**, 308 (2012)
88. S. Chen et al., Phys. Rev. Lett. **110**, 155003 (2013)
89. K. Khrennikov, J. Wenz, A. Buck, J. Xu, M. Heigoldt, L. Veisz, S. Karsch, Phys. Rev. Lett. **114**, 195003 (2015)
90. H.E. Tsai, X. Wang1, J.M. Shaw, Z. Li, A.V. Arefiev, X. Zhang, R. Zgadzaj, W. Henderson, V. Khudik, G. Shvets, M.C. Downer, Phys. Plasmas **22**, 023106 (2015)
91. G. Golovin, S. Banerjee, C. Liu, S. Chen, J. Zhang, B. Zhao, P. Zhang, M. Veale, M. Wilson, P. Seller, D. Umstadter, Nat. Sci. Rep. **6**, 24622 (2016)
92. M. Chen, Z.M. Sheng, Y.Y. Ma, J. Zhang, J. Appl. Phys. **99**(5), 056109 (2006)
93. C. McGuffey, A.G.R. Thomas, W. Schumaker, T. Matsuoka, V. Chvykov, F.J. Dollar, G. Kalintchenko, V. Yanovsky, A. Maksimchuk, K. Krushelnick, V.Y. Bychenkov, I.V. Glazyrin, A.V. Karpeev, Phys. Rev. Lett. **104**(2), 025004 (2010)
94. S. Steinke, J. van Tilborg, C. Benedetti, C.G.R. Geddes, C.B. Schroeder, J. Daniels, K.K. Swanson, A.J. Gonsalves, K. Nakamura, N.H. Matlis, B.H. Shaw, E. Esarey, W.P. Leemans, Nature **530**, 190 (2016)
95. S. Bulanov, N. Naumova, F. Pegoraro, J. Sakai, Phys. Rev. E **58**(5), R5257 (1998)
96. C.G.R. Geddes, K. Nakamura, G.R. Plateau, Cs. Tóth, E. Cormier-Michel, E. Esarey, C.B. Schroeder, J.R. Cary, W.P. Leemans, Phys. Rev. Lett. **100**(21), 215004 (2008)
97. A.J. Gonsalves, K. Nakamura, C. Lin, D. Panasenko, S. Shiraishi, T. Sokollik, C. Benedetti, C.B. Schroeder, C.G.R. Geddes, J. van Tilborg, J. Osterhoff, E. Esarey, C. Toth, W.P. Leemans, Nat. Phys. **7**, 862 (2011)
98. J. Faure, C. Rechatin, O. Lundh, L. Ammoura, V. Malka, Phys. Plasmas **17**(8), 083107 (2010)
99. E. Esarey, R.F. Hubbard, W.P. Leemans, A. Ting, P. Sprangle, Phys. Rev. Lett. **79**(14), 2682 (1997)
100. C. Lin, J. van Tilborg, K. Nakamura, A.J. Gonsalves, N.H. Matlis, T. Sokollik, S. Shiraishi, J. Osterhoff, C. Benedetti, C.B. Schroeder, C. Tóth, E. Esarey, W.P. Leemans, Phys. Rev. Lett. **108**, 094801 (2012)

101. C. Thaury, E. Guillaume, A. Doppl, R. Lehe, A. Lifschitz, K.T. Phuoc, J. Gautier, J.P. Goddet, A. Tafzi, A. Flacco, F. Tissandier, S. Sebban, A. Rousse, V. Malka, Nat. Commun. **6**, 6860 (2015)
102. S.G. Rykovanov, C.G.R. Geddes, J.L. Vay, C.B. Schroeder, E. Esarey, W.P. Leemans, J. Phys. B **47**, 234013 (2014)
103. T.C. Chiou, T. Katsouleas, C. Decker, W.B. Mori, G. Shvets, J.S. Wurtele, Phys. Plasmas **2**, 310 (1995)
104. C.B. Schroeder, J.S. Wurtele, D.H. Whittum, Phys. Rev. Lett. **82**(6), 1177 (1999)
105. F. Hartemann, A. Troha, N. Luhmann, Z. Toffano, Phys. Rev. E. Stat. Phys. Plasmas Fluids Relat. Interdiscip. Top. **54**(3), 2956 (1996)
106. I. Ghebregziabher, B.A. Shadwick, D. Umstadter, Phys. Rev. Spec. Top. Accel. Beams **16(3)**, 030705 (2013)
107. S.G. Rykovanov, C.G.R. Geddes, C.B. Schroeder, E. Esarey, W.P. Leemans, Phys. Rev. Accel. Beams **19**, 030701 (2016). http://link.aps.org/doi/10.1103/PhysRevAccelBeams.19.030701
108. B. Terzic, K. Deitrick, A.S. Hofler, G. Krafft, Phys. Rev. Lett. **112(7)**, 074801 (2014)
109. A. Bonatto, C.B. Schroeder, J.L. Vay, C.G.R. Geddes, C. Benedetti, E. Esarey, W.P. Leemans, Phys. Plasmas **22**, 083106 (2015)
110. W. Leemans et al., ICFA beam dynamics newsletter **56** (2011)
111. S.P. Hatchett, C.G. Brown, T.E. Cowan, E.A. Henry, J.S. Johnson, M.H. Key, J.A. Koch, A.B. Langdon, B.F. Lasinski, R.W. Lee, A.J. Mackinnon, D.M. Pennington, M.D. Perry, T.W. Phillips, M. Roth, T.C. Sangster, M.S. Singh, R.A. Snavely, M.A. Stoyer, S.C. Wilks, K. Yasuike, Phys. Plasmas **7**(5), 2076 (2000)
112. D.P. Higginson et al., Phys. Plasmas **17**, 100701 (2010)
113. M. Roth et al., Phys. Rev. Lett. **110**, 044802 (2013)
114. L. Yin, B.J. Albright, B.M. Hegelich, J.C. Fernandez, Laser Part. Beams **24**, 291 (2006)
115. C. Zulick, F. Dollar, V. Chvykov, J. Davis, G. Kalinchenko, A. Maksimchuk, G.M. Petrov, A. Raymond, A.G.R. Thomas, L. Willingale, V. Yanovsky, K. Krushelnick, Appl. Phys. Lett. **102**(12), 124101 (2013). http://dx.doi.org/10.1063/1.4795723
116. T. Esirkepov et al., Phys. Rev. Lett. **92**, 175003 (2004)
117. I. Pomerantz, E. McCary, A.R. Meadows, A. Arefiev, A.C. Bernstein, C. Chester, J. Cortez, M.E. Donovan, G. Dyer, E.W. Gaul, D. Hamilton, D. Kuk, A.C. Lestrade, C. Wang, T. Ditmire, B.M. Hegelich, Phys. Rev. Lett. **113**, 184801 (2014). https://link.aps.org/doi/10.1103/PhysRevLett.113.184801

Chapter 6
Detectors in Active Interrogation

Sara A. Pozzi, Anna S. Erickson, and Igor Jovanovic

Abstract Similar to passive measurements, active interrogation relies on the use of radiation detectors to measure and isolate the characteristic signatures of sought materials. While almost all advances in radiation detectors used in passive measurements can also improve the performance of active interrogation systems, there are some key differences in the desired characteristics between the two types of measurements that affect the choice and optimization of the detectors. This chapter brings up those distinctions while also providing a broad introduction to various detector types that could find use in active interrogation.

6.1 General Characteristics of the Detectors for AI

The selection of radiation detectors for active interrogation (AI) depends on the application and the interrogating sources. In the context of AI, the detectors can be classified into those that detect photons, neutrons, and those that can detect both types of radiation. Based on their operating principles and materials, they may be classified into gas, scintillation, semiconductor, Cherenkov detectors, as well as those that operate on less traditional principles such as bubble detectors. The source of the interrogating radiation does not always dictate the selection of the detectors; rather, the desired outcome is used as a guiding principle for the detector system design. For example, spatial imaging nearly always requires photon detectors to

S. A. Pozzi
University of Michigan, Ann Arbor, MI, USA
e-mail: pozzisa@umich.edu; ijov@umich.edu

A. S. Erickson (✉)
Georgia Institute of Technology, Atlanta, GA, USA
e-mail: erickson@gatech.edu

I. Jovanovic
Department of Nuclear Engineering and Radiological Sciences, University of Michigan, Ann Arbor, MI, USA

I. Jovanovic, A. S. Erickson (eds.), *Active Interrogation in Nuclear Security*, Advanced Sciences and Technologies for Security Applications,
https://doi.org/10.1007/978-3-319-74467-4_6

produce transmission or backscatter images or induction, while the confirmation of SNM presence via delayed signatures generally relies on neutron or neutron-gamma detectors.

A significant advantage of AI detection over passive methods when attempting to detect special nuclear material (SNM) is a more abundant, energetic and time-dependent penetrating radiation feedback in the form of both γ rays and neutrons. Regardless of whether a fissile or a fissionable nucleus is interrogated, the fission or photofission process proceeds through the same steps through a timeline as illustrated in Fig. 2.1. The signature of an isotope can consist of prompt neutrons, prompt γ rays, delayed neutrons, and delayed γ rays. After the interrogating particles (photons or neutrons) initiate fission events, prompt neutrons and γ rays are released. Depending on the nucleus, there are on average two to three prompt neutrons per fission and about eight photons. In 0.1–100 s following the prompt radiation release, delayed particles are emitted. The number of delayed emitted neutrons is two orders of magnitude lower than that of the prompt release. However, the number of delayed γ rays is much greater and thus the detection of delayed γ radiation has been considered an important goal in AI applications. It is important to note that photons of sufficiently high energy can induce photoneutron emission in non-fissile materials, for example in structural components of an object that is being interrogated. However, interrogation of non-fissile materials results almost exclusively in *prompt* emission of neutron. Delayed radiation is characteristic only of fissile or fissionable materials, which is the reason for the majority of the AI methods to rely on detecting delayed neutrons or γ rays. Considering these two delayed signature types, material identification via unique signature detection and analysis is generally performed with delayed neutrons [1]. Even though delayed γ rays are more abundant, their signatures typically lack material attribution characteristics and suffer from background and activation.

Delayed neutrons can be detected with a number of options, including gas and scintillating counters. Due to their limited emission, the detectors are generally required to cover large areas and be able to effectively reject the background associated with photons. In AI applications beyond material identification, the primary application of detectors is imaging, which is most commonly performed with scintillators. The scintillator characteristics of primary interest include the rise and decay time as well as its stability with respect to temperature and radiation exposure. AI applications routinely expose radiation detectors to challenging operating conditions. One of the largest vulnerabilities of detectors is damage due to harsh radiation environments, often consisting of mixed radiation fields, caused by the interrogating beams [2]. Commercially available detectors are generally designed to operate in laboratory settings, where the environment is stable against temperature fluctuations and radiation intensity is significantly lower. In addition to efficiency and radiation damage considerations, detectors in AI imaging applications are also selected based on the desired spatial resolution and crosstalk effects between detector pixels. In this discussion, crosstalk refers to transmission photons scattering from one detector pixel and creating a signal in a neighboring pixel.

6.2 Gaseous Detectors

Gaseous ionization counters were among the earliest types of instrumentation developed for radiation detection. Gas-based detectors have long proven to be a reliable and efficient tool for detecting and quantifying the presence of neutrons in AI. These detectors are especially robust in cases where the AI source is a photon source (i.e. bremsstrahlung) because they exhibit little sensitivity to photons. Thermal neutron detectors are useful in applications where measuring the kinetic energy of fast neutrons is irrelevant and one is mainly interested in neutron counting.

6.2.1 Interaction Mechanisms in Gaseous Detectors

Gaseous detectors rely on ionization of the gas by the incident radiation and subsequent transport of the electrons and ion pairs to the electrodes. In case of neutron detection, gas counters are filled with gases or lined with a thin converter layer material containing isotopes with a high thermal-neutron capture cross sections, for example ^{3}He gas, ^{10}B in a form of gas (BF_3) or as part of inner lining, or ^{235}U-based lining in fission chambers. Other examples of thermal neutron detector material include ^{6}Li and capture-gated detectors using ^{157}Gd or Cd, although these isotopes are commonly used in scintillating detectors rather than gas-based counters. As illustrated in Fig. 6.1, the neutron absorption cross section exhibits a $1/v$ behavior (reciprocal of neutron velocity) over the thermal energy range.

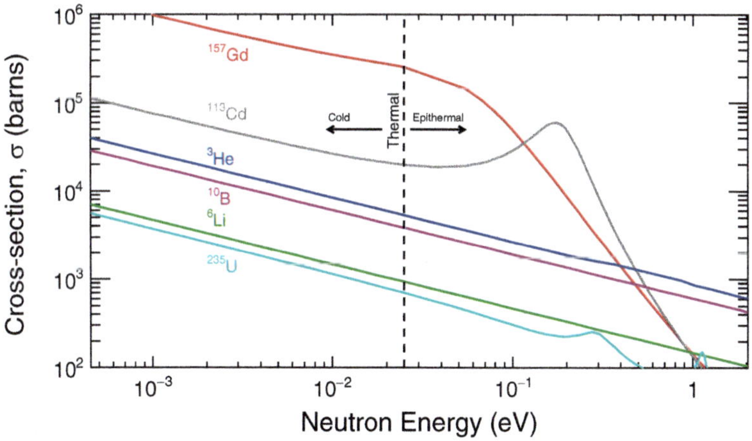

Fig. 6.1 Cross sections as functions of incident neutron energy for the most common nuclides used in thermal neutron detectors [4, 5]

When thermal neutrons are captured, the result is the generation of charged reaction products, such as protons, α particles, or fission fragments, which are emitted in correlated (antiparallel) directions and have a total kinetic energy equal to the Q-value of the reaction. If the Q-value is sufficiently high, the charged particles can be detected with a proportional counter. If the reaction products deposit their entire energy in the proportional counter, monoenergetic pulses consistent with the Q-value are measured. In reality, not every neutron interaction, especially those near the detector wall, result in full energy deposition of the reaction products. A *wall effect* continuum therefore is frequently observed, corresponding to one of the charged particles from the reaction escaping the detector without depositing its entire energy. For ^{10}B and ^{3}He-based gas proportional counters, the Q-values are sufficiently high that a sizable gap in energy exists between γ-ray pulses and neutron pulses suffering from wall effects. Therefore, an energy threshold can be readily chosen to simultaneously achieve good γ-ray signal rejection and high neutron detection efficiency.

$$\mathrm{n} + {}^{3}\mathrm{He} \rightarrow \mathrm{p} + {}^{3}\mathrm{H}, \quad Q = 0.764\,\mathrm{MeV}, \tag{6.1}$$

$$\begin{aligned} \mathrm{n} + {}^{10}\mathrm{B} \rightarrow & \alpha + {}^{7}\mathrm{Li}, \quad Q = 2.792\,\mathrm{MeV}, \\ & \alpha + {}^{7}\mathrm{Li}^{*}, \quad Q = 2.310\,\mathrm{MeV}. \end{aligned} \tag{6.2}$$

Recoil electrons from γ-ray interactions in the detector wall can be discriminated based on their lower signal amplitudes. These properties result in a highly efficient thermal neutron detector which is relatively insensitive to γ-ray radiation [3, 4].

6.2.2 ^{3}He Gaseous Tubes

^{3}He has long been the medium of choice for neutron detection in gaseous detectors [6]. Tubes of pressurized inert ^{3}He gas are operated as proportional counters. Since the cross section of ^{3}He for thermal neutron absorption is 5330 barns, the detectors are typically embedded in a polyethylene moderator to thermalize fast neutrons and increase the probability of reaction. The efficiency of ^{3}He neutron detection systems primarily depends on two parameters: optimization of the surrounding polyethylene moderator and fill pressure of the tubes [7]. A thermal neutron detection efficiency of 77% is quoted for a 2.5-cm diameter 4-atm ^{3}He proportional counter [3]. On the other hand, detection of fast neutrons requires an optimization of tube diameter, fill pressure, and certain amount of polyethylene in order to achieve efficiencies between 5% and 12% [7]. For coincidence counters, often with a dozen or more ^{3}He tubes in a closely packed configuration, neutron efficiencies of the order of 20% are achievable [3]. The efficiency for detecting correlated neutrons in coincidence, however, is significantly lower. The probability of detecting k neutrons in coincidence assuming n correlated neutrons were emitted, for example in a spontaneous fission, is a function of n, k, and the system neutron efficiency, ϵ:

$$P(n, k) = \frac{n!}{(n-k)!k!}\epsilon^k(1-\epsilon)^{n-k}. \tag{6.3}$$

For example, ^{238}U and ^{240}Pu spontaneous fissions both emit 2–3 neutrons on average per fission. The probability of detecting two neutrons in coincidence with a coincidence counter with neutron efficiency of 20% would therefore be on the order of only 4%.

In an AI scenario, SNM in a cargo container might be interrogated with an external neutron source [8, 9]. Fast neutrons from the induced fissions are detected in a ^3He array embedded in polyethylene. Between birth from fission and detection in ^3He, the neutrons will undergo many elastic scattering interactions in both the cargo container fill material and in the polyethylene surrounding the ^3He tubes. In the polyethylene, neutrons may scatter or be absorbed or leak out of the detector module entirely. After every fission event, the neutron population $N(t)$ decreases exponentially as a function of time with a detector-dependent time constant, τ, commonly referred to as the die-away time:

$$N(t) = N(0)\exp(-t/\tau). \tag{6.4}$$

Typical die-away times for thermal neutron detectors are on the order of 30–100 μs [3, 10, 11].

When actively interrogating a cargo container, the fill material may also contribute significantly to the thermalization of neutrons, leading to additional induced fissions. As a result, longer fission chains would increase the die-away time measured with the neutron detection systems. In order to counteract this effect, a Cd liner is often included in the detection system. The Cd liner absorbs neutrons thermalized in the environment, but allows fast neutrons to pass through to the ^3He module, where they are subsequently thermalized in the polyethylene and undergo detection.

Even though ^3He-based detectors have been considered the *gold standard* of thermal neutron counting due to their superior γ-ray rejection capabilities, the scarcity of supply of ^3He and the resulting high price have driven the effort to develop suitable alternatives. It is worth noting that ^3He is a byproduct of the decay of federal tritium stockpiles and sold in annual auctions. Due to the cessation of large-scale tritium production towards the end of the Cold War, production of new ^3He has been declining, and a spike in ^3He prices has driven interest in alternative thermal neutron detector materials [12–14].

6.2.3 ^{10}B-Based Gaseous Detectors

Even though the cross section for thermal neutron capture for ^{10}B (3840 b) is smaller than for ^3He, the associated Q-value of the reaction is sufficiently high, justifying a widespread use of detectors based on boron, with two major types being BF_3 gas and boron-lined proportional counters. BF_3 proportional counters are gas-filled tubes similar to ^{3}He tubes in their operation. However, unlike inert ^{3}He, BF_3 is a toxic material requiring stricter fill pressure limits imposed on the manufacturing process of the tubes [3, 15]. BF_3 gas is also known to dissociate in high γ-ray rate environments [16]. Boron-lined proportional counters are standard proportional counters with an inert fill gas, but the addition of a thin boron layer deposited on the detector wall interior makes it possible to detect thermal neutrons. BF_3 tubes exhibit wall effects similar to ^{3}He proportional tubes. In boron-lined proportional tubes all neutron interactions occur near the wall, so that one of the charged particle reaction products always enters the wall without depositing its energy in the gas.

The smaller cross section and reduced gas fill pressure result in a lower thermal neutron detection efficiency for BF_3 detectors as compared to ^{3}He counters, though this is compensated partially by the use of larger tube diameters. For example, a 5-cm diameter, 1.18-atm BF_3 tube has a thermal neutron efficiency of 46% versus the 77% efficiency quoted previously for a standard ^{3}He tube [3]. The efficiency of boron-lined proportional tubes is limited by the thickness of the boron coating. In order for a thermal neutron to reach the proportional tube fill gas, the boron coating thickness is limited to 0.2 mg/cm^2 [3]. Due to its lower operating voltage and lower fill pressure, boron-lined proportional counters exhibit lower γ-ray sensitivity, but also lower thermal neutron detection efficiency ($\sim$10%) [3]. Like other thermal neutron detectors, boron-based detectors exhibit similar die-away times as ^{3}He-based detection systems.

6.2.4 Applications of Gaseous Detectors to Fission Signatures

In typical AI scenarios, fission signatures are often characterized using thermal neutron detectors. In this approach, the information associated with the initial fission neutron energy is lost. However, in list mode counting the user has a pulse train of thermal neutron events to analyze. A histogram of the total number of neutrons in a user-defined time window results in a multiplicity distribution, which allows for coincidence counting. Figure 6.2 shows an example of a pulse train of detected neutrons and a user-defined time window. The pulse train can be stored as list mode

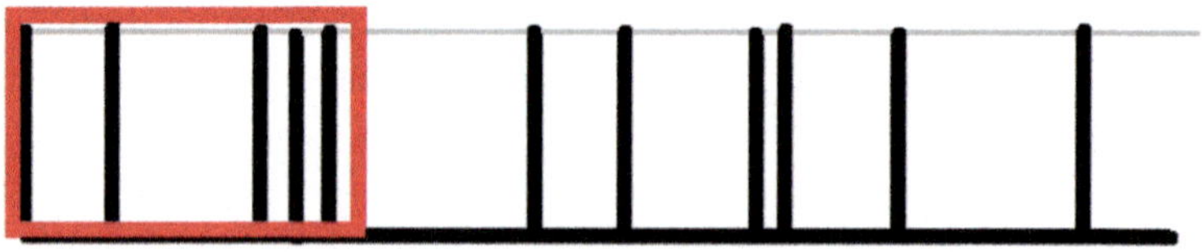

Fig. 6.2 Representation of neutron pulse train where the horizontal axis represents time, each vertical line represents a detected neutron, and the red box is the time window used for creation of the multiplicity distributions

data and analyzed offline. Alternatively, it can be analyzed online with a coincidence circuit. In this case, when a neutron is detected, a gate with some fixed time width is opened. For a non-paralyzable coincidence circuit, any additional neutron signal that occurs during the open gate from a previously detected neutron would be lost. A paralyzable coincidence circuit results in the open gate extended by new incoming signals. If the count rates are sufficiently high, a paralyzable coincidence circuit would become completely blind to neutrons since its gate would be open for infinitely long time [3]. Alternatively, pulses can be read into a shift register coincidence circuit to avoid the dead time issues associated with other coincidence circuits. In a coincidence shift register, a segment of the pulse train is stored in memory, allowing for all of the detected neutrons in the pulse train to be analyzed without dead time losses. This is typically done with a gate of length G, in which real coincidence (R) and accidental coincidence (A) signals are measured as $(R + A)$. Then, a second gate of length G is open with some long delay D, for which only accidentals A should be present [3].

6.2.4.1 Prompt and Delayed Neutron Counting (with Moderation)

Using list mode data, the time between detected neutron events within the user-defined time window can be histogrammed to create a Rossi-Alpha distribution (see Fig. 6.3). Given sufficient moderator in the environment, Eq. (6.4) must be modified to include a slow component due to the presence of longer fission chains. Equation (6.5) shows that detected neutrons are made up of both the detector geometry-dependent fast component (die-away time τ_{fast}), and the slow component (die-away time τ_{slow}) resulting from a longer neutron lifetime in the source and scattering media. Both fast and slow components exist together with a constant

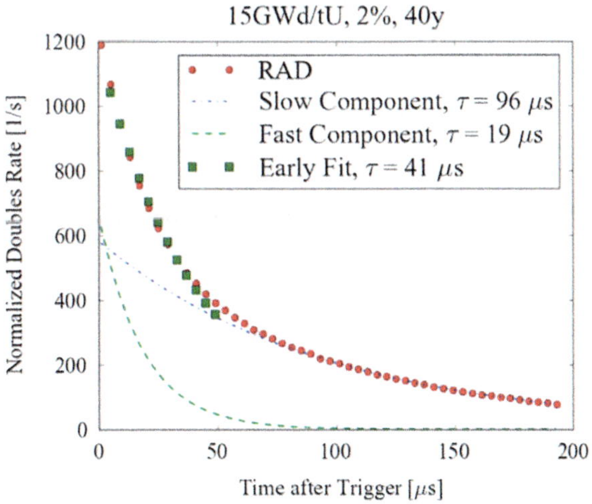

Fig. 6.3 Rossi-Alpha distribution for spent fuel differential die-away self-interrogation. Distribution contains fast and slow components. Reproduced from Ref. [10]

background of accidentals A. The division into a slow and fast region assumes the presence of a cadmium wrapper on the detector module, so that only fast neutrons from first or induced fissions enter the detector module, thermalize, and score [10].

$$N(t) = N_0 \left(\exp(-t/\tau_{fast}) + \exp(-t/\tau_{slow})\right) + A. \tag{6.5}$$

An early and a late gate can be used to gain information from the sample such as plutonium effective mass and multiplication, but using only the early die-away time, as done in differential die-away self-interrogation, is more effective [10, 17].

6.2.4.2 Multiplicity

In neutron multiplicity counting, a signal frequency distribution is created. The number of time windows from the pulse train with one, two, three, and so forth detected neutron events are recorded. Factorial moments are formed and used to compute the effective number of singles, doubles, and triples. These measured quantities are then related in equations that are functions of both the instrument parameters, such as neutron detection efficiency and die-away time, and sample parameters, such as spontaneous fission rate, sample self-multiplication, and (α,n) rate. Unknown parameters, such as special nuclear material sample mass, can be determined by using these equations [18–20].

Common applications include well counters in which a fissile sample is placed in a cavity surrounded by rings of ^{3}He tubes embedded in polyethylene [21, 22]. An AmLi source is used to induce additional fissions in measurements of uranium samples [23].

6.2.4.3 Differential Die-Away Self-Interrogation (DDSI)

Differential die-away self-interrogation is a technique developed to compute multiplication and plutonium content in spent fuel. The neutrons from spontaneous fission and (α,n) reactions within the spent fuel are used to self-interrogate the spent fuel. Neutron coincidence counting is implemented with ^{3}He detector assemblies. The entire Rossi-Alpha distribution, as opposed to an early and a late gate, are used to characterize the measured spent fuel [10, 24, 25].

6.3 Scintillation Detectors

Photon detectors in AI generally rely on scintillation. In scintillation detectors the deposited energy of the photon is converted into an electron, which deposits its energy, a fraction of which is converted into scintillation light. Thus, the selection of detectors generally starts with high density material for efficient photon conversion

Table 6.1 Selected scintillators and their relevant properties for AI applications

Property	LYSO	BGO	CsI	LaBr	NaI	EJ-309	$CdWO_4$	EJ-299-34	Stilbene
Density (g/cm^3)	7.15	7.13	4.51	5.10	3.67	0.96	7.90	1.08	1.15
Decay time (ns)	42	300	980	30	250	3.5	12,000	13, 35, 270	4.50
Intrinsic radioactivity	Yes	No	No	Yes	No	No	No	No	No
Hygroscopic	No	No	No	Yes	Yes	N/A	No	No	Yes
Luminosity (photons/MeV)	24,000	8000	55,000	75,000	45,000	12,300	7800	8600	8700
Energy resolution (@ 661 keV)	7.1%	12%	5.7%	2.9%	7.0%	>20%	>20%	>20%	>20%

into an electron. This is followed by a requirement for high efficiency for conversion of excitation energy to fluorescent radiation. The scintillators are further chosen such that they are transparent to their own fluorescence in order to avoid light self-absorption. The spectral characteristics of the emitted fluorescent radiation are further matched with a proper photosensors, for example photomultiplier tubes (PMT) or Si-based photomultipliers (SiPM). Table 6.1 summarizes some of the important characteristics of scintillators relevant for AI applications.

6.3.1 Scintillation Mechanism in Organic and Inorganic Scintillators

Organic and inorganic scintillators both rely on incident particles exciting atoms and molecules to produce a signal via luminescence. However, the scintillation mechanism associated with each type relies on principally different processes. The luminescence process in organic materials is a molecular process which does not require a lattice structure. On the other hand, light production in inorganic crystals is a result of band structure. The differences in scintillation light production mechanisms influence critical properties of these detectors in AI applications. In particular, the following characteristics influence the detector choice:

- *Time of response*: in AI applications, the speed of detector response influences the ability of the system to process high count rates without significant dead time. The most important characteristic of a scintillator in this context is its fluorescence decay time. The shorter the decay the time, the lower the effect of pile-up.
- *Light output*: above a certain minimum particle energy, scintillators tend to behave in a relatively linear fashion when charged particle energy is converted into detectable light photons. This property allows for energy spectroscopy and particle characterization.

- *Pulse shape discrimination*: pulse time profile can be dependent on the particle characteristics (charge, mass, energy), as they affect the stopping power in the medium. The pulse shape discrimination (PSD) technique is based on exploitation of this dependence in order to separate neutron and γ ray signatures on an event-by-event basis.

6.3.1.1 Inorganic Scintillators

Inorganic scintillators are characterized by a lattice structure that has a direct effect on luminescence centers responsible for production of light. A representative example of inorganic scintillators are the alkali halides, either in a pure form or containing small amounts of impurities that can also serve as activators. For example, one of the most widely used inorganic scintillators is NaI(Tl), where Tl is added as the impurity activator. Non-alkali examples include BGO ($Bi_4Ge_3O_12$) crystals, ZnO(Ag), and $CdWO_4$.

Regardless of their chemical composition, inorganic scintillators rely on electron-hole formation in the band structure, as illustrated in Fig. 6.4. Upon ionizing radiation interaction in the crystal, two important processes can result in the emission of light: ionization of valence band electron into the conduction band, resulting in the creation of a free electron and a free hole, and creation of exciton. Exciton is a state where the electron and the hole remain bound, but can freely migrate through the crystal as a pair. Impurity atoms create additional local energy

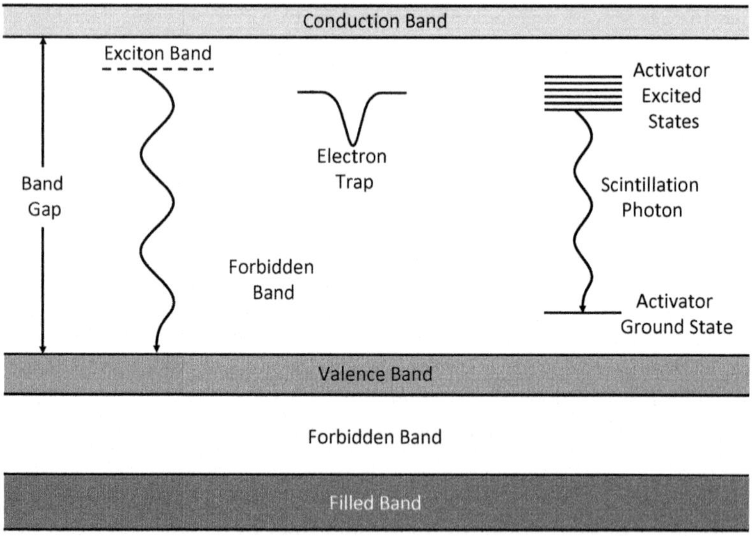

Fig. 6.4 Band structure of an alkali halide scintillator with impurities. Impurity levels can aid in the electron drop to the valence band or can trap electrons, resulting in a metastable state

levels in the forbidden band. A migrating hole can ionize the impurity atom. If an electron drops to the impurity energy level, it can de-excite to the valence band with an emission of radiation, known as fluorescence. If, on the other hand, the electron loses its energy via non-radiation route (thermal or vibrational), this energy is lost for detection, a process also known as *quenching*.

6.3.1.2 Organic Scintillators

While inorganic scintillators rely on band structure characteristic of crystals (and have been described as *semiconductors* for that reason by material science community), the scintillation mechanism in organic media is molecular in nature. It is common to encounter a reference to organic scintillators as *molecular glass* materials. The organic scintillators are generally hydrocarbon compounds composed on benzene ring structures with non-aromatic additions and substitutions. Unlike inorganic crystals, organic scintillators feature excitation and emission spectra that are practically independent of the physical state of the material (solid, liquid or gas). The main reason for this behavior is the nature of the scintillation process, which happens at a molecular level without interaction with neighboring molecules. Scintillation light is produced by free valence electrons occupying π-orbitals in molecules and not associated with a particular atom.

At room temperature, most of the electrons are in their ground state, S_0, as shown in Fig. 6.5 for a typical organic scintillator. Incident radiation populates S_1, excited

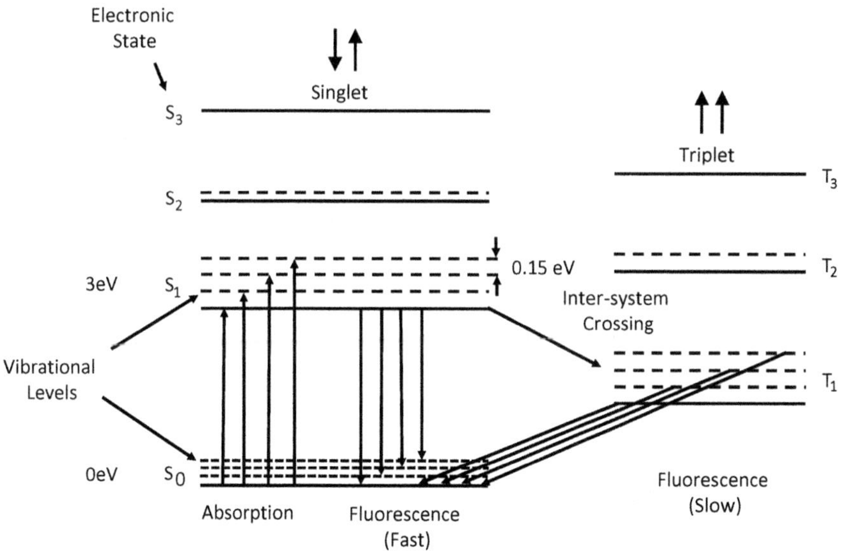

Fig. 6.5 Energy level diagram for organic scintillators. S_0 refers to the ground state of the molecule. Triplet states are indicated with T

singlet state shown along with fine structure corresponding to vibrational states of the molecule. The singlet state decays almost immediately (less than 10 ps) to either the ground state emitting light or to the adjacent triplet levels. It is important to note that transition from triplet levels to S_0 is possible, although forbidden by multipole selection rules, causing decay through interaction with another excited molecule in T_1 state:

$$T_1 + T_1 \rightarrow S_1 + S_0 + \text{phonons}. \tag{6.6}$$

S_1 states can then decay through fluorescence. The delay introduced by $T_1 + T_1$ integration causes the slow component of scintillation light. Formation mechanism of singlet and triplet states is the basis for particle discrimination based on their pulse shape. Since the excited triplet state formation is highly dependent on the ionization energy loss of the particle passing through the scintillator, it is possible to differentiate between electrons and ions. The effect is more pronounced in liquid and organic crystal scintillators, for example stilbene, than for plastic scintillator materials, in which timing of singlet and triplet excited states are similar.

6.3.2 Gamma Detection with Scintillators

Gamma-ray detection relies on the conversion of the photon to an electron via photoelectric effect, Compton scattering, or pair production. The density of a scintillator coupled with the effective atomic number of its constituents plays a significant role in photon detection applications. In this regard, inorganic scintillators outperform organic scintillators. In addition to γ-ray conversion, the efficiency with which radiation energy is converted to optical photons is essential to achieve sufficiently high energy resolution. Comparing different scintillators, inorganic materials outperform organic materials, resulting in the widespread use of inorganics for γ-ray spectroscopy.

6.3.3 Fast Neutron Detection with Organic Scintillators

In organic scintillators, γ rays primarily interact with electrons through Compton scattering and neutrons interact with hydrogen through elastic scattering. This results in different venues for pulse formation, making it possible to separately identify collected pulses as γ rays or neutrons [26]. Figure 6.6a shows two pulses from stilbene, an organic crystal scintillator, identified as electron and proton pulses. In pulses of the same amplitude, proton pulses contain more light in their tail region than electron pulses. This difference can be used to identify an unknown pulse as generated by either a proton or an electron [27]. Charge integration is a

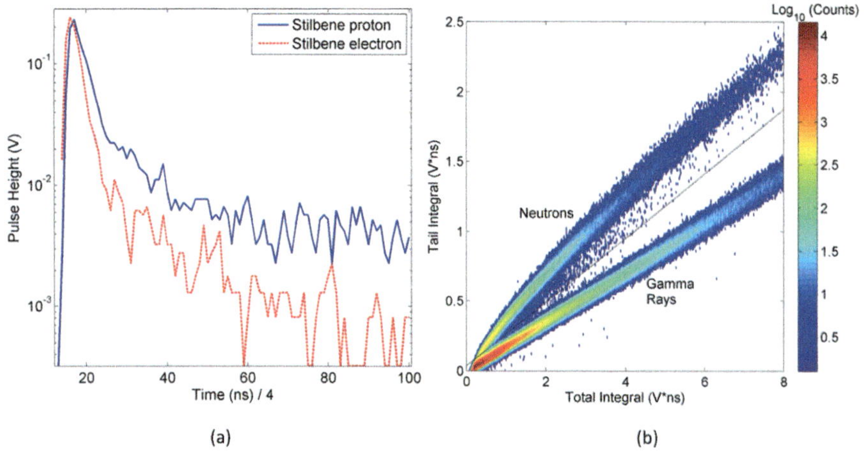

Fig. 6.6 (**a**) 250-mV pulses from a stilbene scintillator highlighting the difference between the γ ray (electron) and neutron (proton) interactions; (**b**) Integration of each pulse through both their tail region and full pulse length yields the PSD distribution for ^{252}Cf, emitter of both γ rays and neutrons

common technique that quantifies the difference in the tail by integrating both the full pulse length and the pulse tail [28]. Figure 6.6b shows the PSD distribution for a cylindrical stilbene scintillator with diameter and length of 5 cm coupled to an Electron Tubes photomultiplier tube (model number 9214B). The stilbene detector measured neutrons and γ rays from a ^{252}Cf spontaneous fission source. Using charge integration PSD, pulses can be visually separated and identified as neutrons or γ rays.

Organic scintillators come in three different varieties. Plastic scintillators are historically the least expensive organic scintillators and can easily be scaled to larger volumes as dictated by applications. Traditional plastic compositions such as EJ-200 lack PSD capability, but newer compositions such as EJ-299-33 have been developed and are PSD-capable [29]. Liquid scintillators can be easily scaled to large volumes and exhibit very good PSD. In field applications, organic liquid detectors are avoided due to environmental risks from potential leaks, although some compositions such as EJ-309 are not classified as hazardous. Organic crystals, such as stilbene and anthracene, are difficult to scale to large sizes and are generally the most expensive, but offer significant performance advantages such as excellent PSD in stilbene and high light output in anthracene [30, 31]. Recent developments in crystal growth methods for stilbene are lowering the cost and improving the commercialization of stilbene [30].

The application of PSD-capable organic scintillators to AI scenarios of SNM poses significant technical challenges. In these applications, we frequently need to measure neutron in the presence of numerous γ rays (generated by the interrogating

Table 6.2 Intrinsic neutron efficiency and γ-ray misclassification rate for different organic scintillators

Detector type	Manufacturer	$E_n(\%)$	MR_γ
Stilbene	Inrad optics	15.1	2.9×10^{-6}
Liquid	Eljen (EJ-309)	12.9	1.60×10^{-5}
Plastic	Radiation monitoring devices (BB3-5)	7.1	1.93×10^{-6}

source or by the SNM). In these cases, high fluxes of γ rays create pileup, producing measured signals that would be classified as neutrons when using charge integration PSD. Neutron detectors are evaluated by their intrinsic neutron efficiency E_n and γ-ray misclassification rate MR_γ. E_n is defined as the number of neutron counts (after background subtraction) divided by the number of neutrons incident on the detector. The MR_γ is defined as the number of neutron counts (after background subtraction) when the detector is exposed to a pure gamma ray field (as generated by a Cs-137 source, for example). A high E_n and a low MR_γ are desirable for AI applications. These values are both dependent on the choice of the PSD curve, which is used to classify an event as a neutron or a photon (see Fig. 6.6b: neutron pulses lie above, and photon pulses below the PSD curve). A more conservative PSD curve will produce lower MR_γ values at the expense of the E_n. A pileup rejection algorithm was used to demonstrate that organic scintillators can function in environments of 1000 incident γ rays per neutron without using lead. Table 6.2 shows the intrinsic neutron efficiency and γ-ray misclassification rate for stilbene, liquid, and plastic scintillators when making measurements in this environment [32].

6.4 Hybrid and Other Detector Types

6.4.1 ^{4}He Detectors

High-pressure ^{4}He tubes are gas scintillators sensitive to fast neutrons [33, 34]. The detection mechanism is based on neutron scattering: an energetic neutron scatters off a ^{4}He nucleus and sets it in motion. This ^{4}He recoil ionizes the gas and induces light in the extreme ultraviolet region. A wavelength shifter is used to shift the wavelengths to the visible range. Silicon photomultipliers or conventional photomultiplier tubes can be used to detect and amplify this light. This type of detector is very insensitive to γ rays, making it useful for AI applications. Drawbacks of this approach include low detection efficiency and the need for high pressurization of the gas.

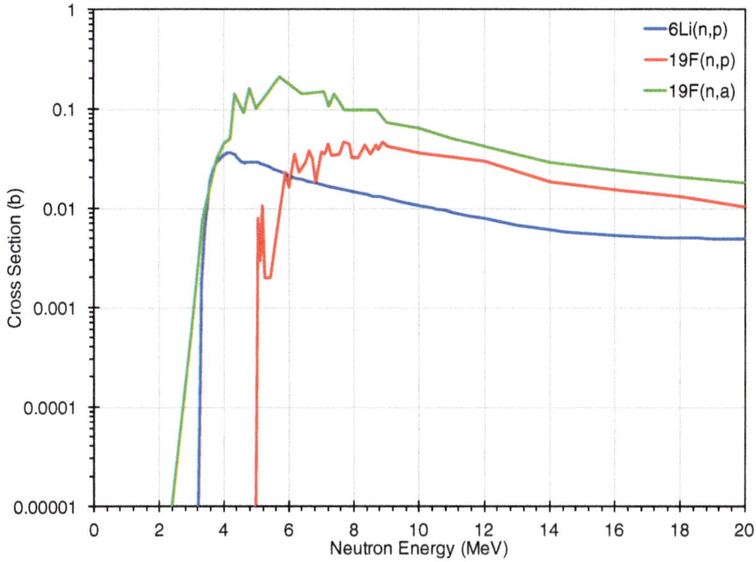

Fig. 6.7 ENDF/B-VII cross-section data for selected reactions relevant to threshold activated detectors: ^{6}Li(n,p), ^{19}F(n,p), and ^{19}F(n,α) [4, 5]

6.4.2 Threshold Activated Detectors

Threshold activated detectors (TADs) contain nuclides in the detection medium that exhibit detectable reactions only above a distinct energy threshold. For example, (n,p) reactions on ^{6}Li and ^{19}F only occur at neutron energies above 3.18 and 4.25 MeV, respectively; ^{19}F also has an (n,α) reaction with a threshold of 1.6 MeV. Figure 6.7 shows the ENDF/B-VII cross-section data for these reactions. The activation products that are produced during irradiation are then counted by the detection medium as they decay. Recently, these types of detectors have gained consideration in photon-based AI because high-energy neutrons are a reliable characteristic signature of fission [35]. TADs excel in this application due to their insensitivity to photons and environmentally scattered neutrons.

6.4.3 Superheated Emulsions and Bubble Detectors

Superheated emulsions (SE) consist of metastable liquid droplets dispersed in a viscous, inert matrix. SE radiation detection mechanism relies on the same basic principle as that of bubble chambers used in high-energy physics [36]. Upon interaction with ionizing radiation, droplets undergo a phase transition from liquid to vapor, producing visible bubbles. The first application of superheated emulsions to practical neutron detection was the superheated drop detector (SDD), introduced by Apfel in the late 1970s [37]. Ing and Birnboim later developed the bubble-damage polymer detector (BD) [38]. The main difference between the SDDs and BDs is the matrix, which is a viscoelastic water-based gel in the case of SDDs, and a polymeric mixture in the BDs. ISO and ANSI included both detector versions in their records [39, 40], under the standard denomination of superheated emulsions.

6.4.3.1 Neutron Interaction Mechanisms

Detection of fast neutrons in SE relies on the neutron scattering reaction with the nuclei of metastable halocarbons or fluorocarbons liquids, whose boiling point is between −5 and 5 °C. At room temperature, these compounds are kept in the superheated liquid state due to the emulsification in the inert host medium and are free of heterogeneous nucleation sites, which would otherwise trigger the liquid evaporation. Charged recoil ions can deposit an amount of energy sufficient to initiate the vaporization. If the initial vapor cavities reach a radius that makes them thermodynamically unstable, i.e. the *critical radius* [41], they continue to grow irreversibly until the entire liquid inside the droplet evaporates into a visible vapor bubble. Materials typically used in superheated emulsions for fast neutron detection are octafluorocyclobutane (C-318) and dichlorotetrafluoroethane (R-114), which are insensitive to photons (Fig. 6.8) [42] and suitable for the detection of charged particles with a linear energy transfer of hundreds of keV per μm, such as fast neutron recoils.

Fig. 6.8 Computed tomography images of an SE detector based on octafluorocyclobutane after photon (1 Gy of 6 MV X rays, left) and neutron irradiation (2.5 mSv of Am-Be, right). Adapted from Ref. [42]

6.4.3.2 Application of Superheated Emulsions for the Detection of Special Nuclear Materials in AI

SE detectors exhibit a threshold-like response as a function of neutron energy. The energy threshold of each liquid depends on the operating pressure and temperature. As the temperature increases, the degree of superheat of the emulsion decreases and a lower amount of energy is required for the liquid to vaporize. Conversely, the higher the pressure, the higher the neutron energy threshold. In AI scenarios, it is possible to control the emulsion temperature and pressure and set the desired energy threshold to detect only high-energy induced fission neutrons. The system is intrinsically insensitive to the interrogation beam: either neutrons below the chosen energy threshold or X rays, as well as to low-energy neutrons produced by (γ,n) or (n,xn) reactions in the scanned object.

The application of SE in AI requires to promptly count evaporated bubbles in large detector modules and is enabled by some recent developments in bubble readout technology [43, 44]. An optical readout technique can be used to relate the amount of light scattered by evaporated bubbles to their density, in real time. This approach achieved a 1-σ readout uncertainty of 6% with 1000 evaporated bubbles on a test volume of 150 mL (6 cm detector diameter). It should be noted that bubbles do not travel inside the matrix once evaporated, which guarantees a steady readout, and they can be condensed to liquid state by applying to the emulsions a pressure higher than the vapor tension of the liquid droplets. This annealing procedure is typically performed in a few seconds by applying an external pressure of approximately 20 atmospheres [42].

6.4.4 *Interaction Mechanisms in Hybrid Detectors*

In hybrid detectors, multiple detection mechanisms are present in the same detection volume. For example, a detector could be sensitive to both thermal neutrons and have good γ-ray detection capabilities. In capture-gated detectors [45], a fast neutron produces a scattering pulse and a capture pulse within the same active volume. These detectors typically consist of organic scintillators doped with a neutron absorbing medium such as boron, cadmium, or lithium.

6.4.5 *Cs_2LiYCl_6 (CLYC)*

CLYC is an inorganic scintillator (Fig. 6.9a) that combines γ-ray detection and spectroscopy capabilities with neutron detection capabilities [46]. Lithium enriched in ^{6}Li is used to capture low-energy neutrons. These neutron captures produce high-energy pulses that can be easily distinguished from the γ-ray pulses. When

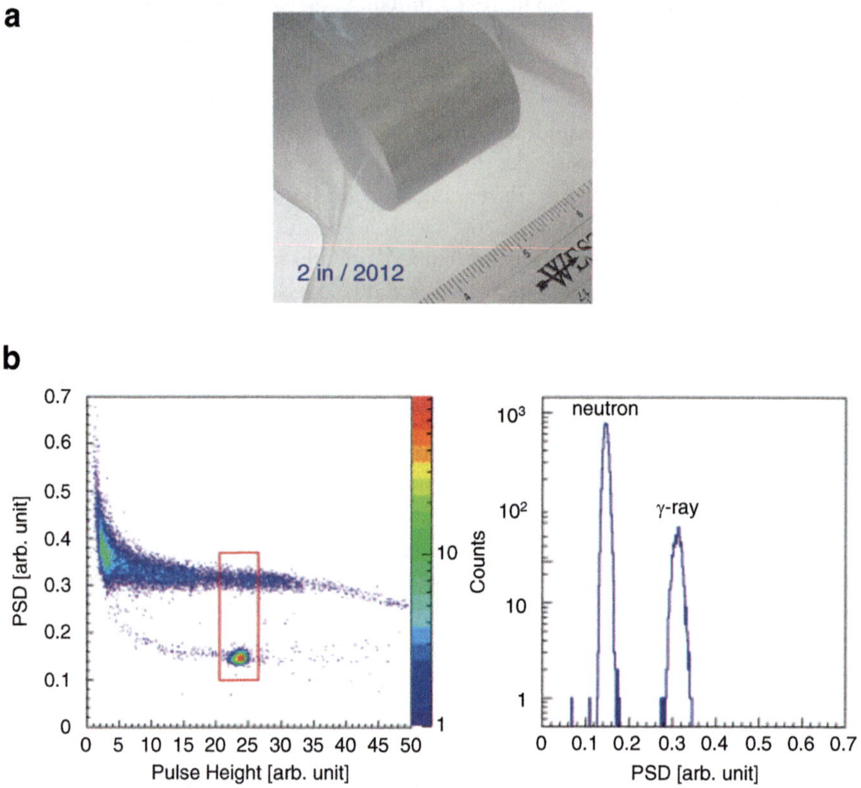

Fig. 6.9 (**a**) Photograph of a CLYC crystal grown in 2012 at Radiation Monitoring Devices, Inc. Reproduced from Ref. [46]. (**b**) Pulse-shape analysis of CLYC measuring γ rays and thermal neutrons. Reproduced from Ref. [47]

pulse shape analysis is applied, CLYC has shown the capability for pulse shape discrimination (Fig. 6.9b), separating γ-ray pulses from neutron pulses [47].

6.4.6 Capture-Gated Organic Scintillators

Most common neutron detectors presented in this chapter provide little information on the incident energy of a neutron interacting in the detector. Thermal neutron detectors require a fast neutron to lose most of its energy in the moderator material surrounding the thermal neutron detector. Many fast neutron detectors are too small, and a fast neutron deposits only a portion of its energy before escaping the detector volume. In many applications, including AI, a neutron spectrometer may be helpful to distinguish different neutron sources based upon their differing energy spectra. Capture-gated neutron spectrometers have been developed to address this need. These detectors operate as follows: a sufficiently large organic scintillator (solid plastic, or liquid) allows fast neutrons to undergo scattering interactions to reach thermal energies. A rapid sequence of proton recoils created by the fast neutron interactions occurs, producing a single pulse that is correlated to the energy deposited by the fast neutron. The next step involves discriminating pulses from full energy depositions from pulses generated by neutrons that ultimately escape the detector. A material with high thermal neutron capture cross section, such as ^{10}B, ^{6}Li, or Gd is incorporated in the design. The presence of a distinctive delayed signal produced by the thermalized neutron capture products verifies that a preceding proton recoil pulse corresponded to a (nearly) full energy deposition by a fast neutron.

Examples of capture-gated neutron detectors include liquid scintillators doped with ^{10}B [48–51], as well as plastic scintillators incorporating ^{10}B [52–54]. Other examples include the use of ^{6}Li glass [55–57], Gd [58], as well as borated crystals [59]. Capture-gated, intrinsic neutron efficiencies are typically of the order of a few percent. The efficiency is strongly dependent upon the detector design, including how the thermal neutron capture material is incorporated and distributed. Intrinsic neutron efficiencies of approximately 1% have been reported for ^{10}B loaded liquid scintillators and neutrons with energies between 1 and 4 MeV [51].

Organic scintillators may be capable of PSD, which can be one method for distinguishing thermalized neutron capture product interactions from scattered neutrons and γ rays. An example of this PSD is shown in Fig. 6.10 for ^{6}Li glass-/plastic scintillation detector measuring PuBe. This sandwich design detector consists of four disks of BC-408 plastic scintillator separated by thin layers of ^{6}Li glass. The capture pulses from neutron capture on ^{6}Li are clearly distinguishable from scattering pulses of neutrons elastically scattering in the plastic scintillator.

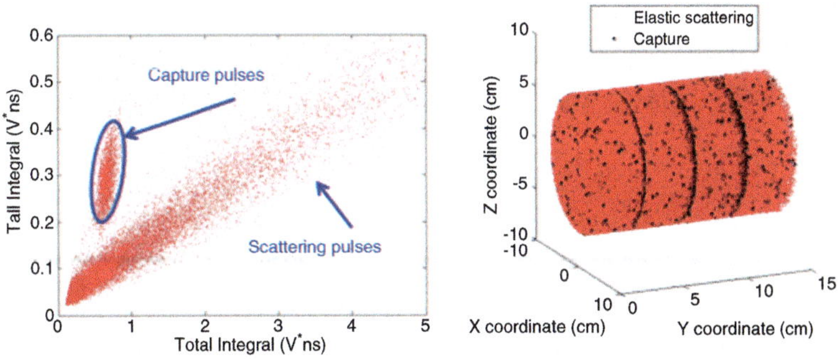

Fig. 6.10 Left: PSD for ^{6}Li glass-/plastic scintillation detector measuring PuBe. Right: MCNP-PoliMi simulation tracking neutron captures on ^{6}Li and elastic scattering in ^{6}Li glass-/plastic scintillation detector for PuBe source at 30 cm source to detector distance and two inches of Pb shielding. Reproduced from Ref. [60]

6.4.7 Heterogeneous Composite Scintillators

An emerging detector type is the *heterogeneous composite scintillator*, which relies on a combination of two or more distinct scintillators to realize special properties, such as particle discrimination or scaling to large volume. In such scintillators, materials are optically coupled, such that fluorescence from one scintillator can propagate into another. It is further necessary to minimize the overlap between emission and absorption bands of multiple scintillator materials as well as maintain the quality of material interface to ensure unimpeded, low-loss propagation of light to the photosensor.

To realize PSD in such systems, two or more materials are chosen with distinct fluorescence decay times. If the geometry of the heterogeneous scintillator is designed such that interactions of different particles deposit their energy predominantly in different scintillators [61], the resulting pulse shapes can by very distinct and indicate the particle type with very high level of discrimination [62]. In practice, this may be achieved by choosing a scintillator in which neutron capture occurs (for example, ^{6}Li-containing scintillating glass [63] or a ^{6}Li-containing ZnS(Ag) [57]) to be relatively thin, such that heavy charged particles from neutron capture are fully contained, whereas high-energy electrons produced in γ-ray interactions are not. A practical implementation of this design with ^{6}Li-doped glass rods embedded in a scintillating polymer that achieves exquisite neutron/γ discrimination is shown in Fig. 6.11 and has been shown to be capable of neutron capture spectroscopy with relatively high resolution [64, 65]. It has recently been shown that it is possible to develop detectors that exhibit *triple PSD* properties by combining scintillating glass and PSD materials such as the EJ-299 scintillators [66]. Such detectors can be used to simultaneously measure neutrons by recoil and capture, while also detecting γ rays. Unlike homogeneous composite scintillators in which the concentration of

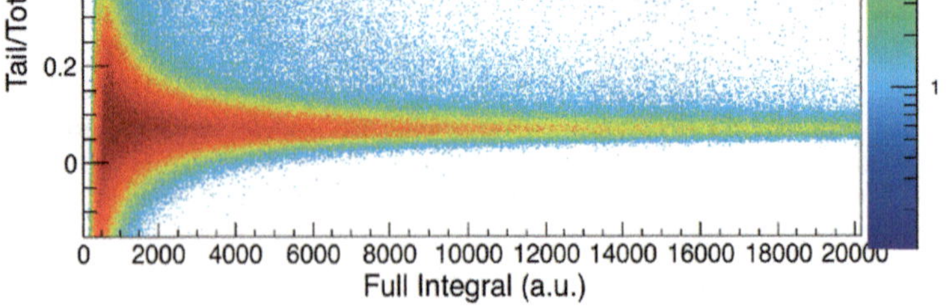

Fig. 6.11 PSD characteristic of a ^{6}Li-glass/scintillating polymer detector (inset)

the neutron capture agent dissolved in a polymer is limited, they can support much higher concentration of neutron capture agents since these agents can be placed into other materials, such as scintillating glass.

6.4.8 Cherenkov Detectors

Detectors based on Cherenkov radiation represent an example of nontraditional counters for nuclear security applications. Common in high energy physics, these detectors feature an inherent energy threshold and directionality of emitted light due its electromagnetic nature as opposed to scintillators. Cherenkov radiation was first observed and described by A.P. Cherenkov in 1932 during his graduate studies [67]. In 1937, Tamm and Frank developed the original formal classical theory of Cherenkov radiation [68].

The phenomenon of Cherenkov radiation is best known as blue light emitted in nuclear reactor cores and spent fuel pool. The glow is the response of matter when a charged particle travels faster than the phase velocity of light in that medium. As a result, Cherenkov radiation significantly differs from scintillators because of directionality, concentration of light emission in the ultraviolet spectrum, and energy threshold. Cherenkov light is emitted in the direction of the charged particle in a form of a cone. The medium and the particle velocity define the parameters of the cone, in particular the emission angle. This property is often used in high energy physics applications in order to distinguish between particle types, but it is rather underutilized in nuclear security applications. In active interrogation applications, the threshold nature of Cherenkov radiation is quite useful when the lower energy radiation is not of importance. In radiography of SNM in particular, presence of lower-energy part of the interrogating photon spectrum may result in noisy image, making energy threshold an attractive feature of the detectors. The velocity threshold for charged particles can be described as the speed of light normalized by the refractive coefficient of the medium:

$$v > c/n(\omega), \tag{6.7}$$

where n is the refractive index of the medium and ω is angular frequency of the photon.

The threshold feature also aids with rejection of background if the detectors are operated near an active interrogation source. However, one major issue with Cherenkov detectors is relatively small light yield as compared to scintillators. Frank and Tamm derived their Cherenkov light yield per unit path per unit wavelength interval as

$$\frac{d^2N}{dxd\lambda} = \frac{2\pi\alpha}{\lambda^2}\left(1 - \frac{1}{\beta^2 n^2}\right) dxd\lambda, \tag{6.8}$$

where α is the fine structure constant, and β is the normalized particle velocity. In this equation, it the refraction coefficient is assumed to be constant with photon wavelength which is a rather accurate assumption for materials that would be used in active interrogation applications, for example water or glass. Integrating the equation over the sensitivity region of a typical PMT (400–700 nm), one gets the generation of photons per cm path in water as

$$\frac{dN}{dx} \approx 490 \sin^2 \theta. \tag{6.9}$$

Gamma-ray detection is possible with Cherenkov counters because of their interaction with electrons, light charged particles whose energy threshold in water is only 262 keV. Sowerby [69] provided a detailed discussion of threshold energies for γ rays as a function of refractive indices of radiators, a summary of which is provided in Fig. 6.12. On the other hand, neutron detection relies on proton recoil or generation of relatively heavy, as compared to electrons, charged particles. The Cherenkov energy threshold for these reactions is far above practical kinetic energies that are observed in active interrogation applications. The low-level light output makes the detector design a critical consideration. While Cherenkov radiator media, such as water and glass, are relatively cheap and thus can be scaled to large volumes in order to improve detection efficiency, light collection may limit

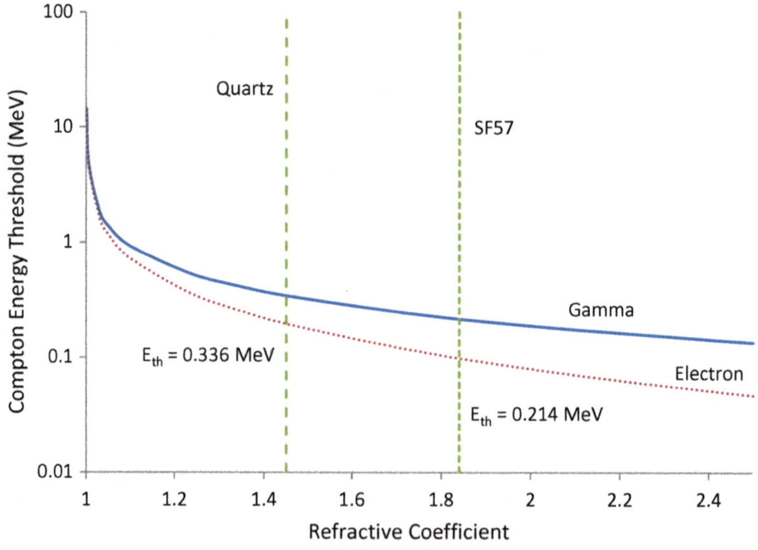

Fig. 6.12 Electron energy threshold for Cherenkov light production. Gamma line corresponds to Compton scattering of a photon with a maximum energy transfer

the choice of medium. For example, water absorption length has a strong effect on Cherenkov detector efficiency since the absorption length tends to overlap with the PMT quantum efficiency curve [70, 71]. The choice of reflector also affects the efficiency of the counters, especially in case of solid detectors. Finally, consideration of refractive index can play a significant role in the design: higher refractive index corresponds to denser materials, which can be more efficient in γ-ray conversion into electrons, but significantly decrease the energy threshold resulting in more noise. This consideration becomes especially important when the spectroscopic capabilities of Cherenkov radiation are used for transmission imaging of SNM. Figure 6.12 compares threshold energies for electrons and photons resulting in Compton electrons (maximum energy transfer is shown) for two materials capable of Cherenkov radiation generation: SF57 leaded glass and High-purity quartz (SiO_2) glass. SF57 has a higher density (3.53 g/cm^3) as compared to quartz (2.2 g/cm^3), resulting in a lower Cherenkov energy threshold and higher efficiency of photon conversion due to large amount of lead, both of which are important from the viewpoint of spectroscopy. However, the higher energy threshold of quartz is preferable to avoid contribution from low energy background (including environmental produces by the accelerator.) The threshold of leaded glass also presents a problem for pair production events. The maximum energy transferred to a Compton electron from a 0.511 MeV annihilation photon is 0.314 MeV. This is below the kinetic energy threshold of quartz, but can radiate Cherenkov in the leaded glass. Signal resulting from this in the leaded glass is indistinguishable from the photoelectric signal reducing the ability to use spectral features, according to the interaction cross sections shown in Fig. 6.13. Photons of 0.511 MeV are frequently emitted from activation products and other reactions that will only compound the problem.

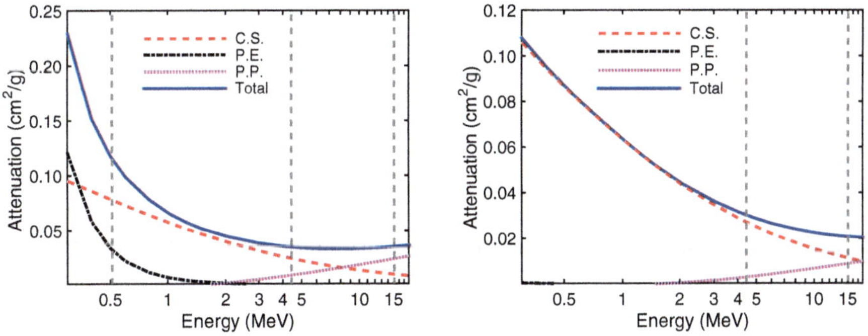

Fig. 6.13 SF57 leaded glass (left) and High-purity quartz (SiO_2) glass (right) cross sections

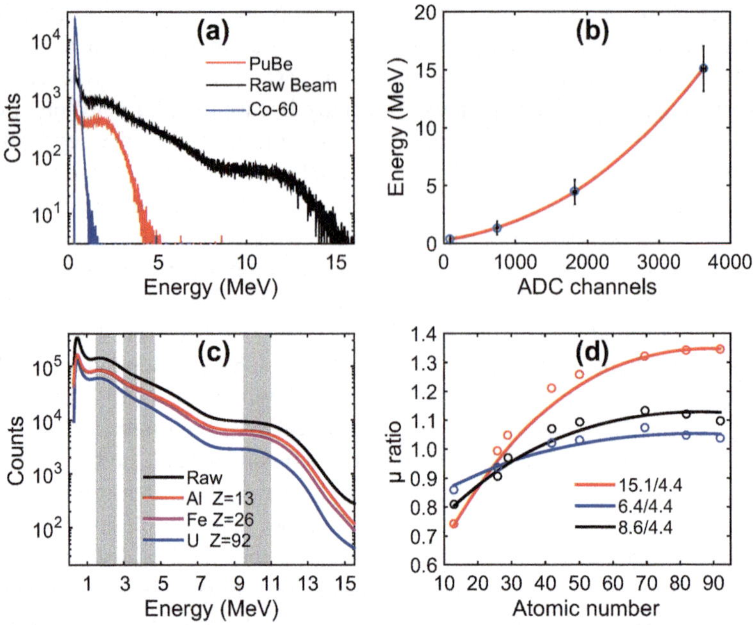

Fig. 6.14 Spectroscopic capabilities of Cherenkov detectors. (**a**) Response of quartz Cherenkov detectors to various γ-ray sources, where *Raw Beam* refers to 4.4 and 15.1 MeV γ-ray lines and well as some less prominent γ rays from $^{11}B(d,n\gamma)^{12}C$ reaction. (**b**) Energy calibration of Cherenkov detectors. (**c**) Transmission properties of various materials. (**d**) Reconstruction of Z_{eff} based on relative transmission. Adapted from Ref. [75]

6.4.9 Imaging with Arrays of Cherenkov Detectors

Cherenkov detectors have a long history of use in high-energy physics, including imaging applications [72]. Recently, quartz-type Cherenkov detectors have been applied to a small-scale imaging of SNM using a source emitting well-resolved γ rays [73]. Figure 6.14 illustrates the response of Cherenkov detectors to γ rays generated with $^{11}B(d,n\gamma)^{12}C$ reaction. In this study, the 3-MeV deuteron beam impinges on a 2-mm thick natural boron target consisting of 19.9% ^{10}B and 80.1% ^{11}B. The result of this reaction is the emission of copious γ rays at 4.4 and 15.1 MeV as well as neutrons and other less-intense γ rays. The detectors were capable of crude energy spectroscopy [74], which allowed to resolve 4.4 and 15.1 MeV γ-ray lines sufficiently well. As a result, the ratio of relative attenuation coefficients was used to reconstruct spatially-dependent Z_{eff} of the material using Beer-Lambert law:

$$T(E, Z) = \frac{I(E, Z)}{I_0(E)} = e^{\frac{-\mu(E,Z)}{\rho}\xi} \tag{6.10}$$

where E is the energy of the photon, Z is the atomic number of the material, I_0 is the initial intensity of the photon (raw beam), I is the intensity of the photon after

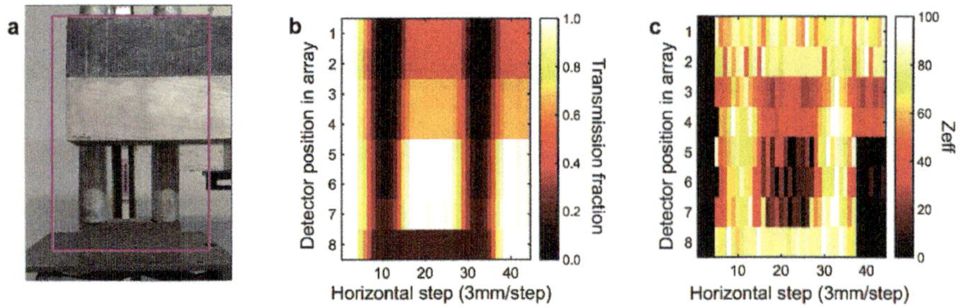

Fig. 6.15 Photon radiography with an array of Cherenkov detectors. (**a**) An object composed of various materials, including lead, aluminum, tungsten and depleted uranium. (**b**) Relative transmission image reconstructed with integrated spectrum. (**c**) Z_{eff} reconstruction based on 15.1/4.4 MeV photon penetration and correlation with the atomic number. Adapted from Ref. [73]

penetrating through the material Z, $\mu(E, Z)/\rho$ is the mass attenuation coefficient of the photons with energies E and ξ is the areal density of the material Z.

In Fig. 6.15, the application of Cherenkov detectors to radiography of composite materials shows that for well-resolved photons, for example 15.1 and 4.4 MeV, are capable of atomic number reconstruction based on relative attenuation [73]. Z_{eff} reconstruction is based on the ratio of energy-dependent effective mass attenuation coefficients that is independent of the areal density of the interrogated material, ξ, as shown in Eq. (6.11). More details can be found in Rose [73, 75].

$$R(E_1, E_2, Z) = \frac{\ln T(E_1, Z)}{\ln T(E_2, Z)} = \frac{\mu(E_1, Z)}{\mu(E_2, Z)} \tag{6.11}$$

6.5 Applications to Fission Signatures

6.5.1 *Photofission Time of Flight*

AI techniques are well established for identifying concealed SNM using delayed neutron detection. The presence of delayed neutrons uniquely signifies the presence of fissionable material. Uranium or plutonium, for example, generate a signal after interrogation has ceased, which is due to delayed neutron emission and subsequent fission reactions. Non-fissionable materials, however, do not emit neutrons when the AI has ceased, and therefore produce no signal. Delayed neutron emission is rare compared to prompt neutron emission and the delayed neutrons have a lower average energy (approximately 500 keV compared to 2 MeV). Therefore, detection of the more abundant, higher-energy, prompt neutrons is desirable; however, the AI source provides an intense background during the prompt regime.

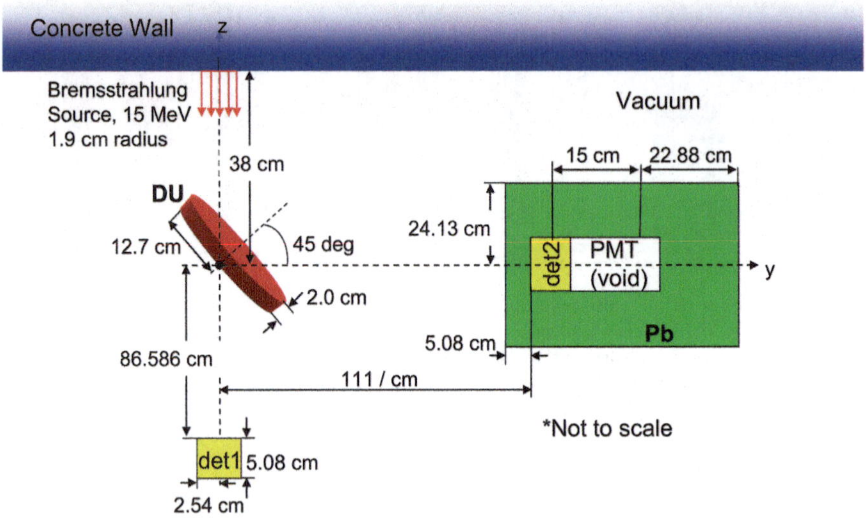

Fig. 6.16 Experimental setup for time-of-flight measurements of photoneutrons produced in depleted uranium; experiments were performed at the Idaho Accelerator Center. Reproduced from Ref. [76]

A correlation experiment using time of flight is one way to access the information from the prompt neutrons emitted in fission. In a proof-of-concept experiment, time-of-flight correlation measurements were made with two cylindrical plastic scintillation detectors. Figure 6.16 shows a schematic of the experimental setup at the Idaho Accelerator Center with a depleted uranium (DU) target. Some of the photons from the accelerator beam pass through the target and are used to start the clock for the time-of-flight measurement. The neutrons and photons produced by reactions in the DU target are then detected in the stop detector, located a large distance away. The length of the flight path can be used to convert the measured times of flight into energy; the longer the flight path, the better the energy resolution. In these experiments, the stop detector was placed at two different distances: 111 and 201 cm. Figure 6.17 shows the time-of-flight distributions resulting from bremsstrahlung interrogation of DU. The signature is characterized by two main features: a sharp initial peak followed by a smooth distribution. The initial peak corresponds to the arrival of the prompt γ rays at the stop detector; all photons arrive at essentially the same time since they all travel at the same speed, c, and start from approximately the same point. The subsequent distribution is the arrival of the photo-neutrons at the stop detector. These secondary neutrons originate from three reactions: (γ,n), (γ,2n), and (γ,fission).

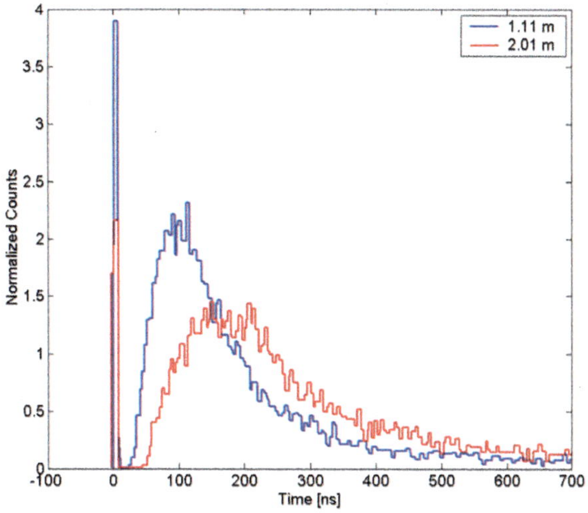

Fig. 6.17 Idaho Accelerator Center measurement results for bremsstrahlung interrogation of a depleted uranium target; data is shown for two different stop-detector distances. Reproduced from Ref. [76]

6.5.2 Detection of Delayed Neutron Emission Profile

An attractive method to enhance the detection of unique signatures of SNM in AI is the detection of the unique temporal profile of delayed neutrons, which is governed by the dynamics of decay of delayed neutron precursors. This method goes beyond the more usual observation of an excess of delayed neutrons integrated over some period after the AI beam is turned off to incorporate the observation of its time profile and thus better discriminate this relatively signature from background. The exact profile of the emitted delayed neutrons is further affected by the time profile of the AI source. An AI source is usually operated over a period of time, such that the delayed neutron precursor population gradually builds up. For a continuous source, this population may be detected once the AI beam is turned off and follows a time profile [77]:

$$R_d(t) = B + C \sum_{i=1}^{n} \epsilon_i Y_i (\exp(t_b/\tau_i) - 1) \exp(-t/\tau_i), \tag{6.12}$$

where $R_d(t)$ is the detected decaying delayed neutron rate, B is the constant neutron background, C is a scaling constant, index i is the group number, n is the number of delayed neutron groups, ϵ_i is the detector efficiency for group i, Y_i is the β-delayed neutron yield per fission for group i, and t_b is the period over which the AI beam was turned on. A typical situation is that the AI beam consists of series of low-duty-cycle micropulses, and in this case it may also be possible to observe the buildup of delayed neutron emission between micropulses. The buildup profile can be described as [77]:

$$R_b(t) = B + C \sum_{i=1}^{n} Y_i \epsilon_i \left(1 - \exp(-t/\tau_i)\right). \tag{6.13}$$

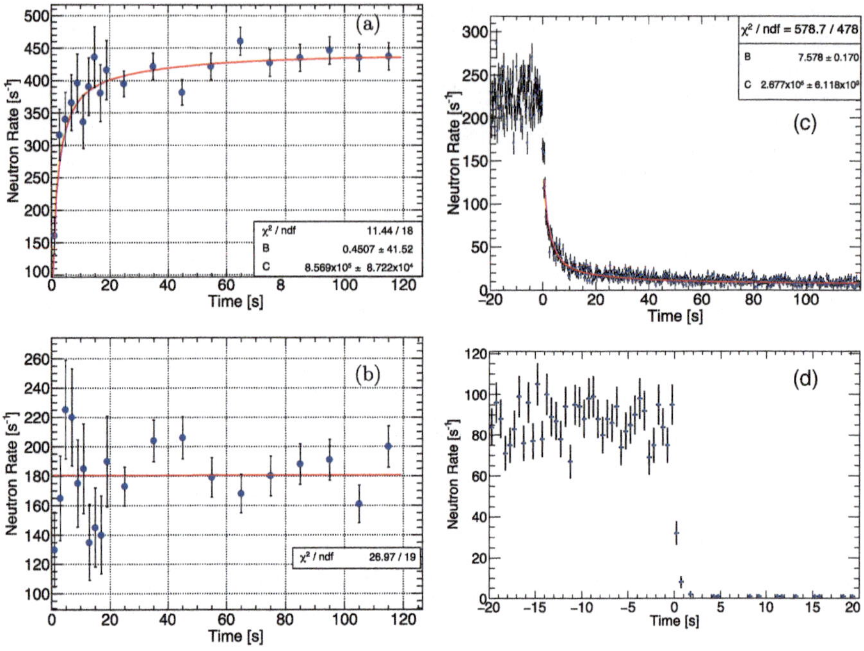

Fig. 6.18 β-delayed neutrons observed during the AI beam incidence on natural uranium in (**a**) buildup and (**c**) decay. A comparison of the null result with tungsten is shown in (**b**) buildup and (**d**) decay. The fit to nuclear data is shown in red. Adapted from [77]

Both of these characteristic profiles have been recorded in AI experiments in which most fissions were induced by fast neutrons and it has been shown that they exhibit excellent agreement with the time profiles calculated from the available nuclear data (Fig. 6.18).

An emerging special type of source has been demonstrated which produces a powerful burst of neutrons on the order of a nanosecond long by accelerating a large number of ions using an intense laser pulse. Such source has $t_b \approx 0$ and has recently been used to detect a clear signal of delayed neutrons [78].

While delayed neutrons have traditionally been detected by counters that employ ^{3}He tubes [79], it has been recently shown that other, ^{3}He-free detectors may be used to accomplish this goal. In the recent work [73] the novel composite heterogeneous scintillators [63] have been shown to be capable of detecting the relatively low-energy neutrons through captures on ^{6}Li in a complex AI radiation environment immediately after the AI beam is turned off. It has subsequently shown that delayed neutrons can be detected as a buildup between accelerator pulses and also through proton recoil if the material is not significantly shielded to thermalize the delayed neutrons [77]. Ideally, detectors that are highly discriminating in both recoil and capture mode (*triple PSD* detectors [66]) may be able to accomplish the goal of detecting delayed neutrons with higher efficiency, and other simple sandwich-type designs may be scaled to large, container-like areas for delayed neutron screening [80].

6.5.3 Neutron Interrogation with Imaging

Radiation imaging techniques provide an increased ability to detect and characterize nuclear material such as HEU. Localization of neutrons improves the measured signal-to-background, which lowers the minimum detectable activity compared to a technique using count rates. The dual-particle imager (DPI) is a device ideally suited for this application because it combines localization capabilities with spectroscopy [81, 82]. Neutron sensitivity is achieved by using two arrays containing organic liquid scintillators to form a neutron scatter camera [83–86]. A Compton camera is formed by using the front array of organic liquid scintillators as a scatter plane and liquid scintillators and NaI(Tl) scintillators in the back array as the absorption plane.

The DPI detects a neutron by looking for neutron counts that occur in coincidence between the front and back liquid scintillator arrays. The energy of the incident neutron can then be calculated using the energy deposited in the first liquid scintillator, E_1:

$$E_0 = E_1 + E_{TOF}. \tag{6.14}$$

E_{TOF} is the energy remaining after interaction in the first detector and is calculated by using the time of flight between interactions in the two detectors. The scattering angle of the neutron, θ_n, relative to the axis between both detectors is calculated using

$$\theta_n = \cos^{-1} \sqrt{E_1/E_0}. \tag{6.15}$$

The calculated angle θ_n defines a conical surface of possible origins for the detected neutron. Then a superposition of many cones localizes the detected particles, which is known as backprojection. More advanced image reconstruction methods use statistical techniques to improve image resolution and signal-to-noise compared to the result achieved with backprojection. Two such methods are maximum-likelihood expectation maximization and stochastic origin ensembles (SOE) [87–90].

Experiments using a 13.7-kg sphere of HEU and a portable neutron generator were conducted to demonstrate performance of the DPI in an AI setting. The neutron generator was placed adjacent to the HEU sphere and outfitted with a DT target to produce 14.1-MeV neutrons. The generator was pulsed at a rate of 300 Hz and emitted approximately 6×10^7 neutrons/s. Because the DPI received direct shine from the neutron generator, a digital veto was applied in post-processing to remove neutrons detected during a generator pulse.

Figure 6.19 shows a comparison of three experiments: no HEU present, bare HEU, and the HEU moderated with 3.8 cm of high density polyethylene (HDPE). In all three images, reconstructed using the SOE method, the scale is given in counts per second with a red box overlaid to show the position of the neutron generator and a green box showing the location of the HEU sphere. When the HEU sample is

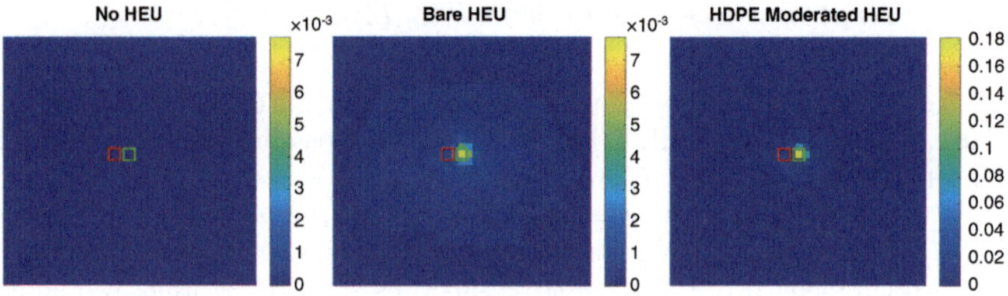

Fig. 6.19 Reconstructed neutron images localizing a 13.7-kg HEU sphere. The overlaid red box shows the location of the DT neutron generator and the green box shows the location of the HEU sphere. When no HEU is present the image is blank. Reproduced from Ref. [91]

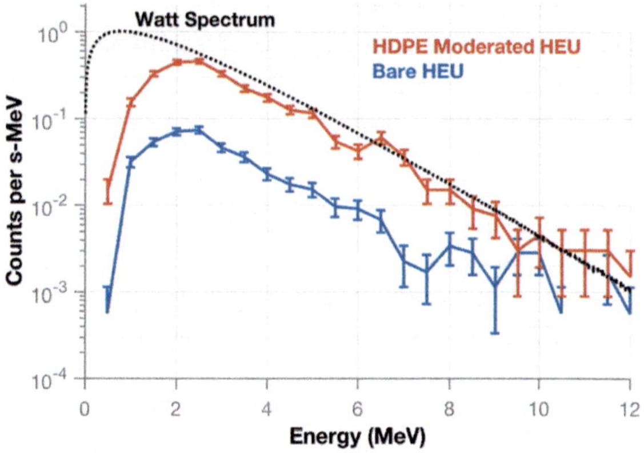

Fig. 6.20 Reconstructed neutron spectra for bare and HDPE moderated HEU. The spectral shape follows the expected Watt spectrum for neutron energies greater than 2.5 MeV. Reproduced from Ref. [91]

present, the DPI correctly localizes the emitted neutrons. With no HEU present, a blank image appears, which means that the 14.1-MeV neutrons from the DT neutron generator are successfully vetoed.

The HDPE moderator increased the detected count rate and produced an image with less noise surrounding the hot spot. The increased count rate occurs because the HDPE reflected emitted neutrons back into the sphere, which induced more fission events. The k_{eff} of the HDPE moderated HEU was 0.764 compared to 0.649 for the bare sphere.[1]

The fast neutron spectrum created from induced fission is a signature that can be used to characterize interrogated material. Figure 6.20 shows the reconstructed neutron spectrum measured by the DPI for the bare and HDPE moderated HEU. The plot also shows a Watt spectrum, which is the expected spectral shape. Both spectra follow the shape of the Watt spectrum from 2.5 MeV through 10 MeV. At

[1] k_{eff} values calculated using MCNP6 KCODE.

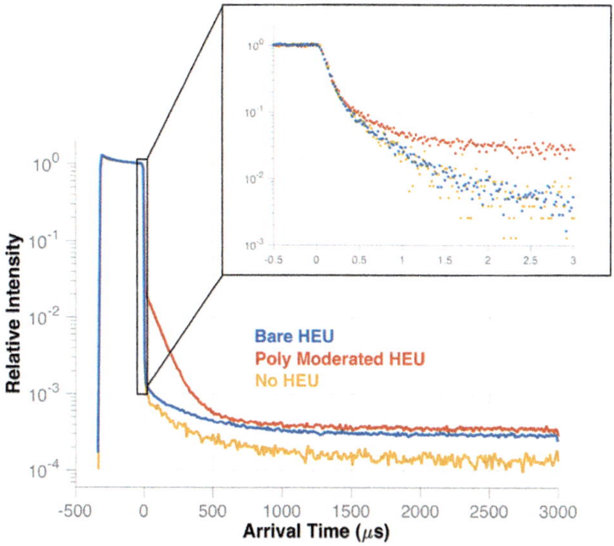

Fig. 6.21 A comparison of the neutron die-away for bare HEU, HDPE moderated HEU, and when no HEU was present. Both the first 3 μs and the full 3000 μs show that a sample with a larger multiplication will produce a slower neutron die-away. Reproduced from Ref. [91]

energies below 2.5 MeV, detector threshold effects limit detection efficiency, which causes the discrepancy between the Watt spectrum and reconstructed spectra at low energies.

Neutron die-away is another signature measured from the interrogated HEU. A sample with a larger multiplication has a slower die-away of the neutron population because the fission chains created in the material extend longer in time. To create a plot of the die-away, the detection times for neutrons detected in the 16 front-plane liquid scintillators are histogrammed based on the arrival time after the end of the generator pulse. For each spectrum, the last time bin of the pulse is set to a value of one.

Figure 6.21 shows a comparison of the normalized spectra for different configurations. Neutrons detected from direct shine of the neutron generator are seen at the negative time values. It is clear from both the plot zoomed in to the first 3 μs and the plot showing the entire 3000 μs that the neutron die-away is slower for the HDPE moderated HEU than for the bare sample. However, in the first 3 μs, there is no discernable difference between the die-away for the bare HEU sample and when no HEU was present. However, over the course of the entire pulse structure, these cases can be discriminated, with the bare HEU producing a slower die-away than the no sample present case.

6.5.4 Neutron Multiplicity Counting

Neutron coincidence and multiplicity counting are typically used in nuclear safeguards and waste management to quantify the mass of spontaneous fission or fissile nuclear materials, such as plutonium or uranium, in a variety of samples, including metal samples, fuel pellets, oxides and fluorides.

While radiation emitted by plutonium isotopes may be detected by high-sensitivity devices in passive mode, radiation signatures from highly-enriched ^{235}U are virtually impossible to detect with passive measurements. In fact, in this case the emission consists of a low yield of fission neutrons and γ rays and low-energy γ rays from the radioactive decay of the actinides, which are easily attenuated by the sample itself and surrounding materials. For these reasons, AI techniques, whereby an external neutron source is typically used to trigger fission reactions, are typically used to perform non-destructive assay of uranium samples [92].

Historically, since the late 1970s [93, 94], AI has relied on ^{3}He proportional counters, arranged in a well-shaped assembly. The active well coincidence counter (AWCC), originally developed at Los Alamos National Laboratory [94], is an instrument used to assay both low and high-enriched uranium samples and includes 42 ^{3}He tubes embedded in a polyethylene collar and two AmLi interrogation sources. The counting cavity is typically surrounded by a cadmium liner, which prevents thermal neutrons to be back-scattered by the surrounding polyethylene and induce further fissions in the sample. By removing the cadmium sleeve, the system is operated in thermal mode, as opposed to the fast mode with cadmium in place, and the fission rate noticeably increases. Thermal mode operation is suitable for low-enriched, small samples. Other AI devices based on ^{3}He detectors are the uranium neutron coincidence collar [3], used to estimate the fissile content in boiling-water-reactor and pressurized-water-reactor fuel assemblies and the ^{252}Cf neutron shuffler [93], optimized for 55-gallons waste barrels.

Neutron coincidence counting can be used to quantify the ^{235}U mass in a sample by measuring the coincidence rate of neutron doubles. Calibration curves, which relate the neutron doubles rate to the ^{235}U mass, are needed in this case. This approach yields accurate results if the calibration samples have a similar composition and geometry as the material to be assayed. However, calibration curves are not always available and are prone to bias error when the standard is not representative of the assayed sample. Several counting methods have been developed to overcome this issue [95, 96]. For example, an active mode multiplicity analysis was developed which does not rely on a calibration curve. This approach is similar to the one used for passive assays [18], as it relates measured neutron double and triplets counts to the sample multiplication and mass. The singles rate typically is not included in active mode multiplicity analysis, because it is dominated by the AmLi neutron emission, whereas the *coupling effect* due to fissions induced in the sample by the interrogation sources must be taken into account [96].

In an attempt to replace ^{3}He tubes in active coincidence and multiplicity counters, several detectors have been considered [97], including boron trifluoride

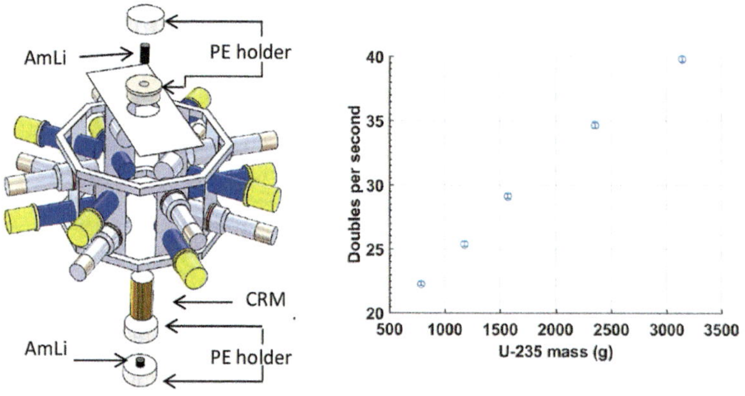

Fig. 6.22 Three-dimensional view of a fast-neutron multiplicity counter (left), including stilbene and EJ-309 liquid scintillators while measuring a sample of uranium certified reference material (CRM) [102] in active mode and detected doubles for CRM with ^{235}U mass in the 0.25–4 kg range and 93% enrichment (right)

(BF_3) proportional counters, ^{10}B lined proportional counters, ^{10}B doped plastic scintillators, ^{6}Li loaded optical fiber scintillators and $^{6}LiF/ZnS(Ag)$ based scintillators [98]. These detectors are sensitive to thermal neutrons, so the overall performances are expected to resemble those achieved by ^{3}He based systems. A different approach relies on organic scintillators, which are sensitive to fast neutrons [99, 100]. Fast-neutron multiplicity counters (Fig. 6.22) do not require moderation and feature a coincidence gate of the same order of magnitude of the neutron lifetime within the system, i.e. tens of nanoseconds. The coincidence window is approximately three orders of magnitude shorter than that used in thermal systems. The measurement is thus less affected by accidental counts, which are the main source of uncertainty [101].

Acknowledgements We gratefully acknowledge the contributions to this chapter of the students and research staff in the primary authors' research groups. This work is based in part upon funding by the Consortium for Verification Technology under Department of Energy National Nuclear Security Administration award number DE-NA0002534, the Department of Defense, Defense Threat Reduction Agency under contract number HDTRA117C0046, and the Department of Homeland Security, Domestic Nuclear Detection Office, Academic Research Initiative award number 2016-DN-077-ARI106.

References

1. T. Gozani, IEEE Trans. Nucl. Sci. **56**(3), 736 (2009). https://doi.org/10.1109/TNS.2009.2015309
2. P.E. Vanier, L. Forman, D.R. Norman, AIP Conf. Proc. **1099**(1), 583 (2009). http://aip.scitation.org/doi/abs/10.1063/1.3120104

3. D. Reilly, N. Ensslin, H.J. Smith, Passive nondestructive assay of nuclear materials. Technical Report, LA-UR-90-732, Los Alamos National Laboratory (1991)
4. M. Chadwick, M. Herman, P. Obložinský, M. Dunn, Y. Danon, A. Kahler, D. Smith, B. Pritychenko, G. Arbanas, R. Arcilla, R. Brewer, D. Brown, R. Capote, A. Carlson, Y. Cho, H. Derrien, K. Guber, G. Hale, S. Hoblit, S. Holloway, T. Johnson, T. Kawano, B. Kiedrowski, H. Kim, S. Kunieda, N. Larson, L. Leal, J. Lestone, R. Little, E. McCutchan, R. MacFarlane, M. MacInnes, C. Mattoon, R. McKnight, S. Mughabghab, G. Nobre, G. Palmiotti, A. Palumbo, M. Pigni, V. Pronyaev, R. Sayer, A. Sonzogni, N. Summers, P. Talou, I. Thompson, A. Trkov, R. Vogt, S. van der Marck, A. Wallner, M. White, D. Wiarda, P. Young, Nucl. Data Sheets **112**(12), 2887 (2011). http://dx.doi.org/10.1016/j.nds.2011.11.002. http://www.sciencedirect.com/science/article/pii/S009037521100113X. Special Issue on ENDF/B-VII.1 Library
5. Iaea, evaluated nuclear data file (endf). https://www-nds.iaea.org/exfor/endf.htm
6. W.R. Mills, R.L. Caldwell Jr., I.L. Morgan, Rev. Sci. Instrum. **33**(8), 866 (1962). http://dx.doi.org/10.1063/1.1717994
7. A. Tomanin, P. Peerani, G. Janssens-Maenhout, Nucl. Instrum. Methods Phys. Res. Sect. A **700**, 81 (2013). http://dx.doi.org/10.1016/j.nima.2012.10.002. http://www.sciencedirect.com/science/article/pii/S0168900212011254
8. E.H. Seabury, D.L. Chichester, in *2009 IEEE Nuclear Science Symposium Conference Record (NSS/MIC)* (2009), pp. 710–712. https://doi.org/10.1109/NSSMIC.2009.5402101
9. R.C. Runkle, D.L. Chichester, S.J. Thompson, Nucl. Instrum. Methods Phys. Res. Sect. A **663**(1), 75 (2012)
10. A.C. Trahan, Utilization of the differential die-away self-interrogation technique for characterization and verification of spent nuclear fuel. Ph.D. thesis, University of Michigan (2016)
11. K.A. Jordan, J. Vujic, T. Gozani, Nucl. Instrum. Methods Phys. Res. Sect. A **579**(1), 407 (2007). http://dx.doi.org/10.1016/j.nima.2007.04.089. http://www.sciencedirect.com/science/article/pii/S0168900207006638. Proceedings of the 11th Symposium on Radiation Measurements and Applications
12. D. Henzlova, R. Kouzes, R. McElroy, P. Peerani, M. Aspinall, K. Baird, A. Bakel, M. Borella, M. Bourne, L. Bourva, F. Cave, R. Chandra, D. Chernikova, S. Croft, G. Dermody, A. Dougan, J. Ely, E. Fanchini, P. Finocchiaro, V. Gavron, M. Kureta, K. Ianakiev, K. Ishiyama, T. Lee, C. Martin, K. McKinny, H. Menlove, C. Orton, A. Pappalardo, B. Pedersen, D. Peranteau, R. Plenteda, S. Pozzi, M. Schear, M. Seya, E. Siciliano, S. Stave, L. Sun, M. Swinhoe, H. Tagziria, S. Vaccaro, J. Takamine, A.L. Weber, T. Yamaguchi, H. Zhu, Current status of helium-3 alternative technologies for nuclear safeguards. Technical Report LA-UR–15-21201, Los Alamos National Laboratory (2015)
13. R.T. Kouzes, A.T. Lintereur, E.R. Siciliano, Nucl. Instrum. Methods Phys. Res. Sect. A **784**, 172 (2015). http://dx.doi.org/10.1016/j.nima.2014.10.046. http://www.sciencedirect.com/science/article/pii/S0168900214012030. Symposium on Radiation Measurements and Applications 2014 (SORMA XV)
14. R.T. Kouzes, J.H. Ely, L.E. Erikson, W.J. Kernan, A.T. Lintereur, E.R. Siciliano, D.L. Stephens, D.C. Stromswold, R.M.V. Ginhoven, M.L. Woodring, Nucl. Instrum. Methods Phys. Res. Sect. A **623**(3), 1035 (2010). http://dx.doi.org/10.1016/j.nima.2010.08.021. http://www.sciencedirect.com/science/article/pii/S016890021001795X
15. A. Lintereur, K. Conlin, J. Ely, L. Erikson, R. Kouzes, E. Siciliano, D. Stromswold, M. Woodring, Nucl. Instrum. Methods Phys. Res. Sect. A **652**(1), 347 (2011). http://dx.doi.org/10.1016/j.nima.2010.10.040. http://www.sciencedirect.com/science/article/pii/S0168900210022795. Symposium on Radiation Measurements and Applications (SORMA) XII 2010
16. A.J. Stokes, T.J. Meal, J.E. Myers, IEEE Trans. Nucl. Sci. **13**(1), 630 (1966). https://doi.org/10.1109/TNS.1966.4324024
17. M. Schear, H. Menlove, L. Evans, S. Tobin, S. Croft, Spent fuel characterization using the differential die-away self-interrogation technique. Technical Report LA-UR-11-03423, Los Alamos National Laboratory (2011)

18. N. Ensslin, W. Harker, M. Crick, D. Langner, M. Pickrell, J. Stewart, Application guide to neutron multiplicity counting. Technical Report LA-13422-M, Los Alamos National Laboratory (1998)
19. W. Hage, D. Cifarelli, Nucl. Instrum. Methods Phys. Res. Sect. A **236**(1), 165 (1985). http://dx.doi.org/10.1016/0168-9002(85)90142-1. http://www.sciencedirect.com/science/article/pii/0168900285901421
20. D. Cifarelli, W. Hage, Nucl. Instrum. Methods Phys. Res. Sect. A **251**(3), 550 (1986). http://dx.doi.org/10.1016/0168-9002(86)90651-0. http://www.sciencedirect.com/science/article/pii/0168900286906510
21. M.M. Ferrer, P. Peerani, M.R. Looman, L. Dechamp, Nucl. Instrum. Methods Phys. Res. Sect. A **574**(2), 297 (2007). http://dx.doi.org/10.1016/j.nima.2007.01.167. http://www.sciencedirect.com/science/article/pii/S0168900207001696
22. B. Goddard, S. Croft, Nucl. Instrum. Methods Phys. Res. Sect. A **712**, 147 (2013). http://dx.doi.org/10.1016/j.nima.2013.02.007. http://www.sciencedirect.com/science/article/pii/S0168900213001836
23. R. Weinmann-Smith, D. Beddingfield, A. Enqvist, M. Swinhoe, Nucl. Instrum. Methods Phys. Res. Sect. A **856**, 17 (2017). http://dx.doi.org/10.1016/j.nima.2017.02.083. http://www.sciencedirect.com/science/article/pii/S0168900217303042
24. A.C. Kaplan, V. Henzl, H.O. Menlove, M.T. Swinhoe, A.P. Belian, M. Flaska, S.A. Pozzi, Nucl. Instrum. Methods Phys. Res. Sect. A **757**, 20 (2014). http://dx.doi.org/10.1016/j.nima.2014.04.023. http://www.sciencedirect.com/science/article/pii/S0168900214004197
25. A.C. Kaplan, V. Henzl, H.O. Menlove, M.T. Swinhoe, A.P. Belian, M. Flaska, S.A. Pozzi, Nucl. Instrum. Methods Phys. Res. Sect. A **764**, 347 (2014). http://dx.doi.org/10.1016/j.nima.2014.08.003. http://www.sciencedirect.com/science/article/pii/S0168900214009231
26. F. Brooks, Nucl. Instrum. Methods **162**(1), 477 (1979). http://dx.doi.org/10.1016/0029-554X(79)90729-8. http://www.sciencedirect.com/science/article/pii/0029554X79907298
27. A. Comrie, A. Buffler, F. Smit, H. Wörtche, Nucl. Instrum. Methods Phys. Res. Sect. A **772**, 43 (2015). http://dx.doi.org/10.1016/j.nima.2014.10.058. http://www.sciencedirect.com/science/article/pii/S0168900214012157
28. T. Alharbi, Radiat. Phys. Chem. **106**, 50 (2015). http://dx.doi.org/10.1016/j.radphyschem.2014.06.031. http://www.sciencedirect.com/science/article/pii/S0969806X14002874
29. G.H. Bertrand, M. Hamel, S. Normand, F. Sguerra, Nucl. Instrum. Methods Phys. Res. Sect. A **776**, 114 (2015). http://dx.doi.org/10.1016/j.nima.2014.12.024. http://www.sciencedirect.com/science/article/pii/S016890021401465X
30. N. Zaitseva, A. Glenn, L. Carman, H.P. Martinez, R. Hatarik, H. Klapper, S. Payne, Nucl. Instrum. Methods Phys. Res. Sect. A **789**, 8 (2015). http://dx.doi.org/10.1016/j.nima.2015.03.090. http://www.sciencedirect.com/science/article/pii/S0168900215004635
31. T. Yanagida, K. Watanabe, Y. Fujimoto, Nucl. Instrum. Methods Phys. Res. Sect. A **784**, 111 (2015). https://doi.org/10.1016/j.nima.2014.12.031. http://www.sciencedirect.com/science/article/pii/S0168900214014739. Symposium on Radiation Measurements and Applications 2014 (SORMA XV)
32. M. Bourne, S. Clarke, N. Adamowicz, S. Pozzi, N. Zaitseva, L. Carman, Nucl. Instrum. Methods Phys. Res. Sect. A **806**, 348 (2016). http://dx.doi.org/10.1016/j.nima.2015.10.025. http://www.sciencedirect.com/science/article/pii/S0168900215012322
33. R. Chandra, G. Davatz, H. Friederich, U. Gendotti, D. Murer, J. Instrum. **7**, C03035 (2012). https://doi.org/10.1088/1748-0221/7/03/C03035
34. R. Jebali, J. Scherzinger, J. Annand, R. Chandra, G. Davatz, K. Fissum, H. Friederich, U. Gendotti, R. Hall-Wilton, E. Håkansson, K. Kanaki, M. Lundin, D. Murer, B. Nilsson, A. Rosborg, H. Svensson, Nucl. Instrum. Methods Phys. Res. Sect. A **794**, 102 (2015). http://dx.doi.org/10.1016/j.nima.2015.04.058. http://www.sciencedirect.com/science/article/pii/S0168900215005690
35. T. Gozani, J. Stevenson, M.J. King, Nucl. Instrum. Methods Phys. Res. Sect. A **652**(1), 334 (2011). http://dx.doi.org/10.1016/j.nima.2011.01.029. http://www.sciencedirect.com/science/article/pii/S0168900211000805. Symposium on Radiation Measurements and Applications (SORMA) XII 2010

36. D.A. Glaser, Phys. Rev. **87**, 665 (1952). 10.1103/PhysRev.87.665
37. R.E. Apfel, Nucl. Instrum. Methods **162**(1), 603 (1979). http://dx.doi.org/10.1016/0029-554X(79)90735-3. http://www.sciencedirect.com/science/article/pii/0029554X79907353
38. H. Ing, H. Birnboim, Nucl. Tracks Radiat. Meas. (1982) **8**(1), 285 (1984). http://dx.doi.org/10.1016/0735-245X(84)90106-6. http://www.sciencedirect.com/science/article/pii/0735245X84901066. Special Volume: Solid State Nuclear Track Detectors
39. Passive neutron dosimetry systems — Part 1: performance and test requirements for personal dosimetry. Technical Report ISO 21909-1, International Organization for Standardization (2015)
40. Personnel neutron dosimeters (neutron energies less than 20 mev). Technical Reports ANSI/HPS N13.52, American National Standards Institute (1999)
41. F. d'Errico, Nucl. Instrum. Methods Phys. Res. Sect. B **184**(1), 229 (2001). http://dx.doi.org/10.1016/S0168-583X(01)00730-3. http://www.sciencedirect.com/science/article/pii/S0168583X01007303. Advanced Topics in Solid State Dosimetry
42. F. d'Errico, A.D. Fulvio, Radiat. Meas. **46**(12), 1690 (2011). http://dx.doi.org/10.1016/j.radmeas.2011.10.017. http://www.sciencedirect.com/science/article/pii/S1350448711004999. Proceedings of the 16th Solid State Dosimetry Conference, September 19–24, Sydney
43. F. d'Errico, A.D. Fulvio, M. Maryañski, S. Selici, M. Torrigiani, Radiat. Meas. **43**(2), 432 (2008). http://dx.doi.org/10.1016/j.radmeas.2008.02.011. http://www.sciencedirect.com/science/article/pii/S1350448708000590. Proceedings of the 15th Solid State Dosimetry (SSD15)
44. E. Kaplan, J. Lemley, T. Tsang, L. Milian. Real-time self-networking radiation detector apparatus (2007). http://www.google.com.pg/patents/US7230250. US Patent 7,230,250
45. G. Knoll, *Radiation Detection and Measurement* (Wiley, New York, 2010). https://books.google.com/books?id=4vTJ7UDel5IC
46. J. Glodo, R. Hawrami, K. Shah, J. Cryst. Growth **379**, 73 (2013). http://dx.doi.org/10.1016/j.jcrysgro.2013.03.023. http://www.sciencedirect.com/science/article/pii/S0022024813002108. Compound Semiconductors and Scintillators for Radiation Detection Applications: A Special Tribute to the Research of Michael Schieber
47. N. D'Olympia, P. Chowdhury, C. Lister, J. Glodo, R. Hawrami, K. Shah, U. Shirwadkar, Nucl. Instrum. Methods Phys. Res. Sect. A **714**, 121 (2013). http://dx.doi.org/10.1016/j.nima.2013.02.043. http://www.sciencedirect.com/science/article/pii/S0168900213002349
48. T. Aoyama, K. Honda, C. Mori, K. Kudo, N. Takeda, Nucl. Instrum. Methods Phys. Res. Sect. A **333**(2), 492 (1993). http://dx.doi.org/10.1016/0168-9002(93)91197-U. http://www.sciencedirect.com/science/article/pii/016890029391197U
49. S.D. Jastaniah, P.J. Sellin, IEEE Trans. Nucl. Sci. **49**(4), 1824 (2002). https://doi.org/10.1109/TNS.2002.801674
50. M. Flaska, S.A. Pozzi, Nucl. Instrum. Methods Phys. Res. Sect. A **599**(2), 221 (2009). http://dx.doi.org/10.1016/j.nima.2008.10.030. http://www.sciencedirect.com/science/article/pii/S0168900208015465
51. S. Hunt, C. Iliadis, R. Longland, Nucl. Instrum. Methods Phys. Res. Sect. A **811**, 108 (2016). http://dx.doi.org/10.1016/j.nima.2015.12.001. http://www.sciencedirect.com/science/article/pii/S0168900215015508
52. W. Feldman, G. Auchampaugh, R. Byrd, Nucl. Instrum. Methods Phys. Res. Sect. A **306**(1), 350 (1991). http://dx.doi.org/10.1016/0168-9002(91)90342-N. http://www.sciencedirect.com/science/article/pii/016890029190342N
53. E. Kamykowski, Nucl. Instrum. Methods Phys. Res. Sect. A **317**(3), 559 (1992). http://dx.doi.org/10.1016/0168-9002(92)91002-Q. http://www.sciencedirect.com/science/article/pii/016890029291002Q
54. P. Holm, K. Peräjärvi, S. Ristkari, T. Siiskonen, H. Toivonen, Nucl. Instrum. Methods Phys. Res. Sect. A **751**, 48 (2014). http://dx.doi.org/10.1016/j.nima.2014.03.021. http://www.sciencedirect.com/science/article/pii/S0168900214003040

55. J.B. Czirr, G.L. Jensen, Nucl. Instrum. Methods Phys. Res. Sect. A **349**(2), 532 (1994). http://dx.doi.org/10.1016/0168-9002(94)91222-X. http://www.sciencedirect.com/science/article/pii/016890029491222X
56. G.L. Jensen, J. Wang, J.B. Czirr, Nucl. Instrum. Methods Phys. Res. Sect. A **333**(2), 474 (1993). http://dx.doi.org/10.1016/0168-9002(93)91195-S. http://www.sciencedirect.com/science/article/pii/016890029391195S
57. K. Wilhelm, J. Nattress, I. Jovanovic, Nucl. Instrum. Methods Phys. Res. Sect. A **842**, 54 (2017). http://dx.doi.org/10.1016/j.nima.2016.10.042. http://www.sciencedirect.com/science/article/pii/S0168900216310865
58. J. Czirr, Nucl. Instrum. Methods **108**(3), 613 (1973). http://dx.doi.org/10.1016/0029-554X(73)90549-1. http://www.sciencedirect.com/science/article/pii/0029554X73905491
59. J. Czirr, D.B. Merrill, D. Buehler, T.K. McKnight, J.L. Carroll, T. Abbott, E. Wilcox, Nucl. Instrum. Methods Phys. Res. Sect. A **476**(1), 309 (2002). http://dx.doi.org/10.1016/S0168-9002(01)01445-0. http://www.sciencedirect.com/science/article/pii/S0168900201014450. International Workshop on Neutron Field Spectrometry in Science, Technology and Radiation Protection
60. M. Flaska, S. Pozzi, J. Czirr, W. Ulbricht, in *Symposium on Radiation Measurements Applications*, Berkeley, California, 2008
61. M. Mayer, J. Nattress, C. Trivelpiece, I. Jovanovic, Nucl. Instrum. Methods Phys. Res. Sect. A **784**(Supplement C), 168 (2015). https://doi.org/10.1016/j.nima.2014.09.023. http://www.sciencedirect.com/science/article/pii/S0168900214010249. Symposium on Radiation Measurements and Applications 2014 (SORMA XV)
62. G. Rich, K. Kazkaz, H. Martinez, T. Gushue, Nucl. Instrum. Methods Phys. Res. Sect. A **794**(Supplement C), 15 (2015). https://doi.org/10.1016/j.nima.2015.05.004. http://www.sciencedirect.com/science/article/pii/S0168900215006130
63. M. Mayer, J. Nattress, V. Kukharev, A. Foster, A. Meddeb, C. Trivelpiece, Z. Ounaies, I. Jovanovic, Nucl. Instrum. Methods Phys. Res. Sect. A **785**(Supplement C), 117 (2015). https://doi.org/10.1016/j.nima.2015.03.014. http://www.sciencedirect.com/science/article/pii/S0168900215003083
64. J. Nattress, M. Mayer, A. Foster, A.B. Meddeb, C. Trivelpiece, Z. Ounaies, I. Jovanovic, IEEE Trans. Nucl. Sci. **63**(2), 1227 (2016). https://doi.org/10.1109/TNS.2016.2537761
65. T. Shi, J. Nattress, M. Mayer, M.W. Lin, I. Jovanovic, Nucl. Instrum. Methods Phys. Res. Sect. A **839**(Supplement C), 86 (2016). https://doi.org/10.1016/j.nima.2016.09.041. http://www.sciencedirect.com/science/article/pii/S0168900216309809
66. M. Sharma, J. Nattress, K. Wilhelm, I. Jovanovic, Nucl. Instrum. Methods Phys. Res. Sect. A **857**(Supplement C), 75 (2017). https://doi.org/10.1016/j.nima.2017.03.019. http://www.sciencedirect.com/science/article/pii/S016890021730356X
67. P. Cherenkov, Dokl. Akad. Nauk SSSR **2**, 451 (1934)
68. I. Frank, I. Tamm, Dokl. Akad. Nauk SSSR **14**, 109 (1937)
69. B. Sowerby, Nucl. Instrum. Meth. **97**, 145 (1971)
70. G.M. Hale, M.R. Querry, Appl. Opt. **12**(3), 555 (1973)
71. R.C. Smith, K.S. Baker, Appl. Opt. **20**(2), 177 (1981)
72. D. Websdale, Nucl. Instrum. Meth. A **595**(1), 12 (2008). http://dx.doi.org/10.1016/j.nima.2008.07.040. http://www.sciencedirect.com/science/article/pii/S0168900208009303
73. P.B. Rose, A.S. Erickson, M. Mayer, J. Nattress, I. Jovanovic, Sci. Rep. **6**, 24388 EP (2016). http://dx.doi.org/10.1038/srep24388
74. P. Rose Jr, A. Erickson, Nucl. Instrum. Methods A **799**, 99 (2015)
75. P.B. Rose Jr., A.S. Erickson, J. Appl. Phys. **120**(6), 064903 (2016). http://dx.doi.org/10.1063/1.4960778
76. S. Clarke, Analysis of prompt emissions following photoninterrogation of fissionable material. Ph.D. thesis, Purdue University (2007)
77. M. Mayer, J. Nattress, I. Jovanovic, Appl. Phys. Lett. **108**(26), 264102 (2016). https://doi.org/10.1063/1.4955051

78. A. Favalli, F. Aymond, J.S. Bridgewater, S. Croft, O. Deppert, M.J. Devlin, K. Falk, J.C. Fernandez, D.C. Gautier, M.A. Gonzales et al., Nuclear material detection by one-short-pulse-laser-driven neutron source. Technical Report, Los Alamos National Laboratory (LANL), Los Alamos, NM (US) (2015)
79. M.T. Kinlaw, A.W. Hunt, Appl. Phys. Lett. **86**(25), 254104 (2005). https://doi.org/10.1063/1.1953874
80. F. Sutanto, J. Nattress, I. Jovanovic, J. Appl. Phys. **122**(5), 054901 (2017). https://doi.org/10.1063/1.4986382
81. A. Poitrasson-Rivière, M.C. Hamel, J.K. Polack, M. Flaska, S.D. Clarke, S.A. Pozzi, Nucl. Instrum. Methods Phys. Res. Sect. A **760**, 40 (2014). http://dx.doi.org/10.1016/j.nima.2014.05.056. http://www.sciencedirect.com/science/article/pii/S0168900214005889
82. A. Poitrasson-Rivière, J.K. Polack, M.C. Hamel, D.D. Klemm, K. Ito, A.T. McSpaden, M. Flaska, S.D. Clarke, S.A. Pozzi, A. Tomanin, P. Peerani, Nucl. Instrum. Methods Phys. Res. Sect. A **797**, 278 (2015). https://doi.org/10.1016/j.nima.2015.06.045. http://www.sciencedirect.com/science/article/pii/S0168900215007950
83. D. Herzo, R. Koga, W. Millard, S. Moon, J. Ryan, R. Wilson, A. Zych, R. White, Nucl. Instrum. Methods **123**(3), 583 (1975). http://dx.doi.org/10.1016/0029-554X(75)90215-3. http://www.sciencedirect.com/science/article/pii/0029554X75902153
84. P.E. Vanier, L. Forman, in *IEEE Nuclear Science Symposium Conference Record, 2005*, vol. 1 (2005), vol. 1, pp. 116–119. http://dx.doi.org/10.1109/NSSMIC.2005.1596219
85. N. Mascarenhas, J. Brennan, K. Krenz, J. Lund, P. Marleau, J. Rasmussen, J. Ryan, J. Macri, in *2006 IEEE Nuclear Science Symposium Conference Record*, vol. 1, pp. 185–188 (2006). http://dx.doi.org/10.1109/NSSMIC.2006.356135
86. N. Mascarenhas, J. Brennan, K. Krenz, P. Marleau, S. Mrowka, IEEE Trans. Nucl. Sci. **56**(3), 1269 (2009). http://dx.doi.org/10.1109/TNS.2009.2016659
87. S.J. Wilderman, N.H. Clinthorne, J.A. Fessler, W.L. Rogers, in *1998 IEEE Nuclear Science Symposium Conference Record. 1998 IEEE Nuclear Science Symposium and Medical Imaging Conference (Cat. No.98CH36255)*, vol. 3, pp. 1716–1720 (1998). http://dx.doi.org/10.1109/NSSMIC.1998.773871
88. J. Polack, A maximum-likelihood approach for localizing and characterizing special nuclear material with a dual-particle imager. Ph.D. thesis, University of Michigan (2016)
89. M. Hamel, J. Polack, A. Poitrasson-Rivière, M. Flaska, S. Clarke, S. Pozzi, A. Tomanin, P. Peerani, Nucl. Instrum. Methods Phys. Res. Sect. A **810**, 120 (2016). http://dx.doi.org/10.1016/j.nima.2015.12.002. http://www.sciencedirect.com/science/article/pii/S016890021501551X
90. M. Hamel, J. Polack, A. Poitrasson-Rivière, S. Clarke, S. Pozzi, Nucl. Instrum. Methods Phys. Res. Sect. A **841**, 24 (2017). http://dx.doi.org/10.1016/j.nima.2016.10.004. http://www.sciencedirect.com/science/article/pii/S0168900216310233
91. M. Hamel, J. Polack, M. Ruch, M. Marcath, S. Clarke, S. Pozzi, Sci. Rep. (2017). http://dx.doi.org/10.1038/s41598-017-08253-x
92. S.D. Clarke, M. Flaska, S.A. Pozzi, in *2011 IEEE Nuclear Science Symposium Conference Record* (2011), pp. 331–335. http://dx.doi.org/10.1109/NSSMIC.2011.6154511
93. H.O. Menlove, T.W. Crane, Nucl. Instrum. Methods **152**(2), 549 (1978). http://dx.doi.org/10.1016/0029-554X(78)90057-5. http://www.sciencedirect.com/science/article/pii/0029554X78900575
94. H. Menlove, Description and operation manual for the active well coincidence counter. Technical Report LA-7823-M, Los Alamos Scientific Laboratory (1979)
95. B. Goddard, W. Charlton, P. Peerani, Nucl. Instrum. Methods Phys. Res. Sect. A **739**, 1 (2014). http://dx.doi.org/10.1016/j.nima.2013.11.101. http://www.sciencedirect.com/science/article/pii/S0168900213016665
96. N. Ensslin, W.H. Geist, M.S. Krick, M.M. Pickrell, Active neutron multiplicity counter. Technical Report LA-UR-07-1403, Los Alamos Scientific Laboratory (2007)

97. A.P. Simpson, S. Jones, M.J. Clapham, S.A. McElhaney, in *2011 IEEE Nuclear Science Symposium Conference Record* (2011), pp. 4853–4857. https://doi.org/10.1109/NSSMIC.2011.6152484
98. S. Stave, M. Bliss, R. Kouzes, A. Lintereur, S. Robinson, E. Siciliano, L. Wood, Nucl. Instrum. Methods Phys. Res., Sect. A **784**, 208 (2015). http://dx.doi.org/10.1016/j.nima.2015.01.039. http://www.sciencedirect.com/science/article/pii/S0168900215000704. Symposium on Radiation Measurements and Applications 2014 (SORMA XV)
99. A.D. Fulvio, T. Shin, T. Jordan, C. Sosa, M. Ruch, S. Clarke, D. Chichester, S. Pozzi, Nucl. Instrum. Methods Phys. Res. Sect. A **855**, 92 (2017). http://dx.doi.org/10.1016/j.nima.2017.02.082. http://www.sciencedirect.com/science/article/pii/S0168900217303066
100. J.L. Dolan, M. Flaska, A. Poitrasson-Riviere, A. Enqvist, P. Peerani, D.L. Chichester, S.A. Pozzi, Nucl. Instrum. Methods Phys. Res. Sect. A **763**, 565 (2014). http://dx.doi.org/10.1016/j.nima.2014.06.028. http://www.sciencedirect.com/science/article/pii/S0168900214007359
101. D.L. Chichester, S.J. Thompson, M.T. Kinlaw, J.T. Johnson, J.L. Dolan, M. Flaska, S.A. Pozzi, Nucl. Instrum. Methods Phys. Res. Sect. A **784**, 448 (2015). http://dx.doi.org/10.1016/j.nima.2014.09.027. http://www.sciencedirect.com/science/article/pii/S0168900214010286. Symposium on Radiation Measurements and Applications 2014 (SORMA XV)
102. E. Garner, L. Machlan, W. Shileds, Certification of uranium isotopic standard reference materials. Technical Report 260, National Bureau of Standards Special Publication (1971)

Chapter 7
Data Acquisition and Processing Systems

Mark Ellis

Abstract A data acquisition system is required to interpret the electrical signals produced by radiation detectors when a gamma ray, neutron or other radiation type interacts with it. The data acquisition system can take on many forms to match the wide variety of radiation detector types available and the different ways in which the signals from these detectors can be processed to give information on the detected radiation. First presented is a detailed overview of a variety of analog data acquisition topologies ranging from simple radiation event counters through to pulse shape discrimination. A comparison is then made with the digital counterparts of these data acquisition topologies, highlighting the relative merits and shortcomings of both analog and digital implementations. Next, the important factors relating to how data is processed, stored and transferred through different stages of a data acquisition chain is discussed and suggestions given on how to avoid bottle-necks and ensure balanced data flow throughout the system. Also discussed are the key challenges associated with the design and configuration of data acquisition systems for active interrogation (AI) environments such as dealing with the high event rates, capturing the time development of active radiation signatures, and techniques for working with pulsed interrogating sources.

7.1 Introduction

An electronic system is required to interpret the electrical signals produced by radiation detectors when a γ ray, neutron or other radiation type interacts with it. The common term for such electronic systems is *data acquisition* (or simply *DAQ*) and there are many different designs of such DAQ to match the wide variety of radiation detector types available and the different ways in which the data from these detectors can be processed to give information on the detected radiation. What

M. Ellis (✉)
Atomic Weapons Establishment, Aldermaston, Reading, Berkshire, UK
e-mail: mark.ellis@awe.co.uk

I. Jovanovic, A. S. Erickson (eds.), *Active Interrogation in Nuclear Security*, Advanced Sciences and Technologies for Security Applications,
https://doi.org/10.1007/978-3-319-74467-4_7

was once a relatively straight forward component of a radiation detection system, data acquisition has, in the last decade or so, grown into a significant area of rapid research and development in its own right. As radiation detectors have become more sophisticated in terms of complex detection modes and physical arrangement, so too have the data acquisition systems required to support them.

7.1.1 Gamma-Ray and Neutron Detection

Gamma rays are high energy photons (wavelengths typically less than 10 pm), which can interact directly with the electrons of an atom. Although a large number of possible interaction mechanisms are known for γ rays in matter, only three major types play an important role in radiation measurements: photoelectric absorption, Compton scattering and pair production [1]. In all cases, photon energy is transferred to electron energy, which can be directly or indirectly detected in an electronic circuit.

Neutrons are uncharged particles and so cannot be detected through interactions with the electric field from the electrons within an atom. Instead, neutrons are detected through interactions with the nuclei of atoms. There are two primary interaction types: absorption reactions, where the neutron is absorbed and charged particles (and photons) are emitted, and proton recoil reactions, where the neutron elastically scatters with the nuclei. Both of these reactions produce charged particles that deposit their energy within a detection medium that can be detected with an electronic circuit [1–3].

7.1.2 Overview of Data Acquisition

In practice, a data acquisition circuit will typically be required for each radiation detector element in a given system. When a radiation particle interacts with a detector element, a certain amount of charge is liberated. The result is a short duration (typically tens of nanoseconds up to several microseconds in duration) current pulse at the detector element output (the definition of current being charge per unit time).

The detector systems of interest for AI (and indeed for most neutron and γ-ray detector systems) will be configured in such a way that the amount of charge liberated is directly proportional to the radiation energy deposited. The function of the DAQ circuit is to first detect the radiation event and then measure certain properties such as the radiation energy. The current pulse seen at the detector output when, for example, a γ ray interacts with the detector will be similar to that shown in Fig. 7.1.

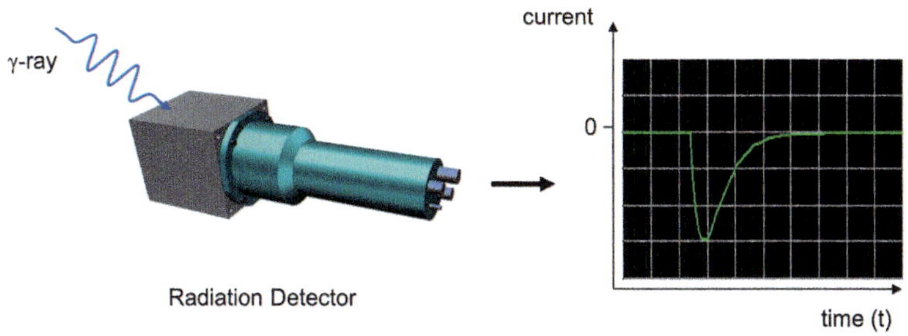

Fig. 7.1 Current pulse from detector resulting from a γ-ray interaction

The traditional method for measuring the particle energy or counting events within a given energy range has been the analog DAQ chain, which is a series of analog electronics modules configured to detect and record the events; analog data acquisition is described in Sect. 7.2. The last decade or so has seen the emergence and prevalence of digital data acquisition, which has brought about a step change in the way that radiation detector signals are acquired and processed, offering more flexibility in the design of radiation detection systems; digital data acquisition is introduced and detailed in Sect. 7.3. An important part of any data acquisition system is the way in which data is formatted, stored and transferred to the different stages of the data acquisition chain. The key areas related to data handling in both analog and digital DAQ systems are covered in Sect. 7.4. Lastly, Sect. 7.5 is dedicated to the challenges associated with the design and build of DAQ systems for AI.

7.2 Analog Data Acquisition

As described in Sect. 7.1, a radiation detector is designed to produce a short current pulse at its output when a radiation particle is detected. Each type of radiation detector (be it semiconductor-based, a scintillator, proportional counter etc.) will have its own way of creating this current pulse; which typically involves accelerating liberated electrons across a potential difference. An example of this pulse generating mechanism for a scintillation detector (such as a sodium iodide radiation detector), is shown in Fig. 7.2.

A typical photomultiplier tube (PMT) for a sodium iodide scintillator might contain 8 or 10 dynodes and offer an electron multiplication factor of the order of 10^6; a voltage bias of several hundred volts is typically required across the dynode chain in order to achieve this high multiplication factor. In a typical setup (as described above) one might expect to observe something in the region of 1 nC of charge at the output of the photomultiplier tube for a 662-keV γ ray that deposits all of its energy in the scintillator. This is a quantity of charge that can be readily measured by an electronic circuit (the data acquisition system). What has been

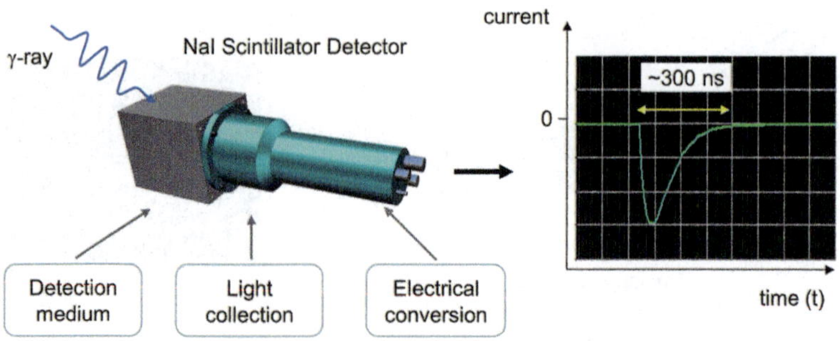

Fig. 7.2 Particle detection and current pulse generation in a scintillator detector

described here is the current pulse the current pulse generation process for a sodium iodide scintillator but the process would be identical for, say, a liquid scintillator on the detection of a neutron or γ ray. A semiconductor-based detector such as High-purity germanium (HPGe) or a proportional counter such as a ^{3}He tube would generate the current pulse via a different mechanism [1], but the general form of the current pulse will be the same as that shown in Fig. 7.2.

7.2.1 *Detection Mode*

There are three general modes of detector operation called pulse mode, current mode and mean square voltage mode. For AI (and similar applications) it is highly desirable to operate in *pulse mode* (as depicted in Fig. 7.2) as this is the only one of the three modes that preserves information on the amplitude and timing of individual radiation events [1]. For the remainder of this chapter, the focus will be on pulse mode detection since the vast majority of AI DAQ systems operate on this basis.

But there may be instances where the other modes may be necessary or sufficient. *Current mode* detection produces a continuous signal that is a time average of the individual current bursts and this average current depends on the product of the particle interaction rate and the charge per interaction [1]; current mode can be useful in some scenarios such as when the event rate is too high to get satisfactory results in pulse mode but where an indication of particle event rate is needed. *Mean square voltage* (MSV) mode is an adaptation of current mode where the direct-current (DC) component of the current mode signal is blocked and the signal of interest becomes the fluctuation of the current mode signal about its mean value. The MSV signal is proportional to the particle interaction event rate and, importantly, proportional to the square of the charge produced in each event. The MSV signal is thus useful when making measurements in mixed radiation fields where the charge produced from radiation events from one particle type is different from the charge produced from another particle type [1].

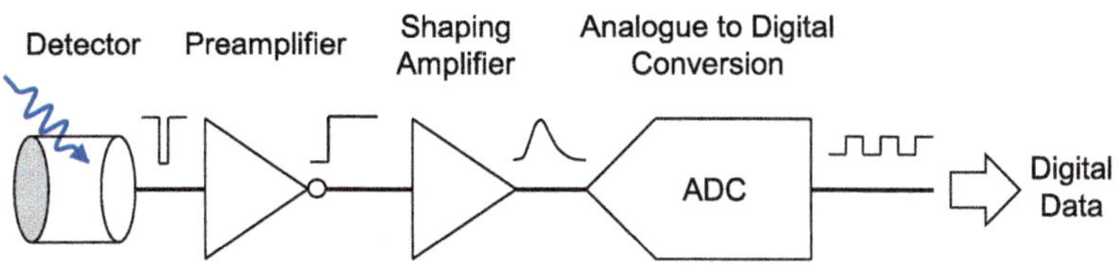

Fig. 7.3 Traditional analog data acquisition chain

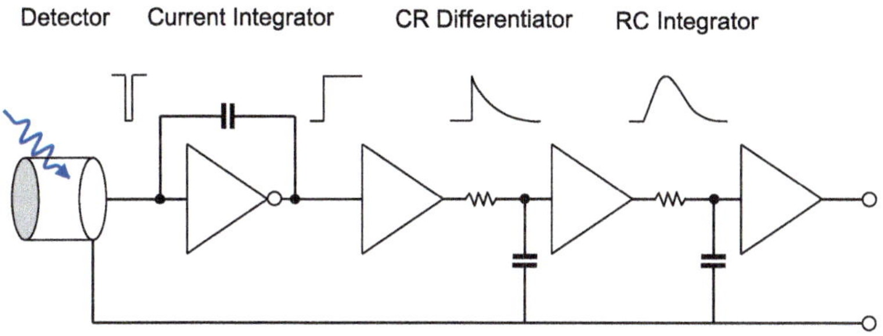

Fig. 7.4 Charge integration and pulse shaping

7.2.2 The Data Acquisition Chain

The traditional method for detecting and acquiring information on a radiation particle that interacts with a radiation detector is an analog DAQ chain. This has the steps of charge integration, pulse shaping and then detecting the peak height of the resulting waveform from which the energy deposited by the interacting particle can be inferred; see Fig. 7.3. This method has been the mainstay of neutron detection and γ-ray spectroscopy for many years. The data acquisition chain shown in Fig. 7.3 is, however, just one of a number of possible analog DAQ chains, all of which are described later in Sects. 7.2.4–7.2.6.

7.2.3 Charge Integration and Pulse Shaping

The preamplifier and pulse shaping amplifier stages shown in Fig. 7.3 are collectively referred to as charge integration and pulse shaping. This is illustrated in more detail in Fig. 7.4.

The charge output from the detector must be collected (integrated) for the full duration of the charge pulse; it is the integrated charge that is proportional to the particle energy. Hence the full and correct name for this stage in the chain is the *charge-sensitive preamplifier* (CSP). The CSP integrates the charge pulse and converts it to a voltage signal that can be accepted by the stage that follows. The CSP will usually have some gain to compensate for the (typically) low charge output from the detector.

The function of a pulse shaper is to simplify certain amplitude measurements, improve the Signal-to-Noise Ratio (SNR) of the pre-amplified signal and to bring the signal back to the baseline more rapidly, ready to accept the next pulse. The simplest implementation of this is a high-pass filter followed by low-pass filter (as shown in Fig. 7.4). The basic implementation in electronic components is a CR filter network (differentiator) followed by an RC filter network (integrator). Shaping amplifiers today will employ much more sophisticated shaping stages such as *n*th-order integrators, but the principle remains the same.

The peak of the shaped signal is smoother than the peak of the pre-amplified signal and this allows for a more accurate amplitude measurement. The SNR is improved by tailoring the frequency response of the shaping amplifier to favor the signal over the noise. The SNR improvement that can be achieved depends on the specific implementation of the shaping filter but in general an improvement in SNR comes at the expense of the duration of the shaped pulse. When configuring a DAQ system, it is typically a trade-off between noise reduction and maximizing the radiation count rate.

7.2.4 Peak Sensing Analog-to-Digital Conversion

The peak height of the pulse from the shaping amplifier is proportional to the energy of the interacting particle in the detector (as described in Sect. 7.2.3). A common implementation to capture this peak height value is to use a unit known as a *peak sensing analog-to-digital convertor* (ADC). The peak sensing ADC data acquisition chain is shown in Fig. 7.5. The first stage of the peak sensing ADC is to hold the peak value of the voltage pulse from the shaping amplifier. The second stage of the peak sensing ADC is a fast ADC, which converts this held voltage level into a digital number that can be stored on a computer or other digital storage device. Following a suitable calibration (establish the relationship between the recorded ADC values and deposited energy by using radiation sources of known energy), the digital number from the ADC can be scaled to give a value in units of energy.

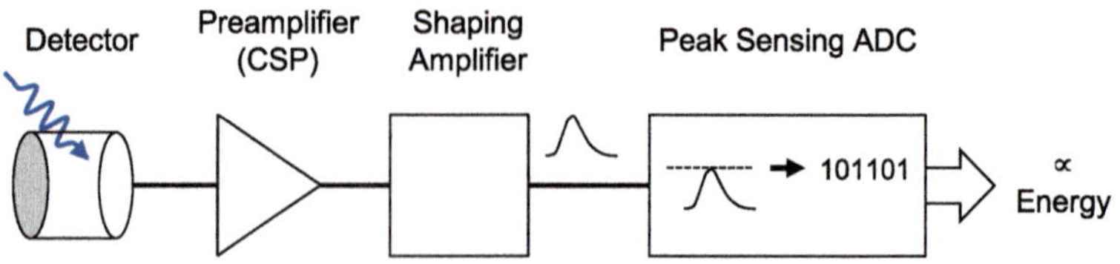

Fig. 7.5 Data acquisition chain based on a peak sensing ADC

The required resolution (number of digital bits used in the conversion) of the fast ADC will depend on the energy resolution of the radiation detector. A sodium iodide detector has an energy resolution in the region of 7% at 662 keV and so an 8-bit fast ADC is deemed sufficient (up to an energy range of, say, 7 MeV) to provide adequate energy value granularity in the digital domain to match the energy resolution that the sodium iodide detector is able to produce; less than 8 bits would be insufficient to faithfully capture the energy value of a detected particle and more than 8 bits would yield no additional information on the acquired energy value (but there may be a cost premium associated with a higher specification ADC). In contrast, a high-purity germanium (HPGe) detector has an energy resolution in the region of 0.3% (2 keV) at 662 keV and so a 14-bit fast ADC will most likely be necessary (depending on the energy range of interest) to provide adequate energy value granularity in the digital domain. For example, a 14-bit ADC will provide 16,384 (2^{14}) digital steps, which, spread over an energy range of, say, 7 MeV gives a digital step size of 0.43 keV. A digital step size of 0.43 keV is deemed sufficient compared to the HPGe energy resolution of 2 keV at 662 keV; but one needs to bear in mind that the required digital step size will be less than 2 keV for lower energy γ rays (since the energy resolution, in keV, of the HPGe detector will vary as the square root of the deposited energy [1]).

The ADC stage is usually followed by a multichannel analyzer to get a visual display of the detected radiation events (see Sect. 7.2.7).

7.2.5 *Charge-to-Digital Conversion*

The Charge-to-Digital Converter (commonly referred to as a QDC) is a unit designed to directly integrate the pulse of current from the detector to give a charge value (which is itself proportional to the energy deposited by the radiation particle). This QDC-based data acquisition chain is shown in Fig. 7.6. The QDC unit is made up of two stages; the first stage integrates the current pulse to give a single charge

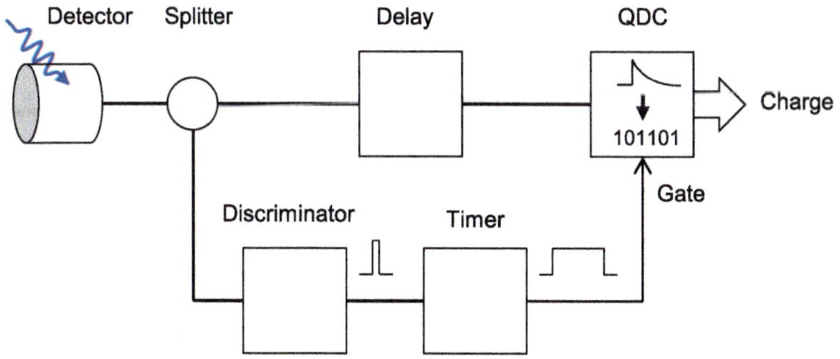

Fig. 7.6 Charge-to-digital conversion data acquisition chain

value, which is converted to a voltage level via a Charge-to-Amplitude Converter (QAC); in the second stage, a fast analogue-to-digital converter is used to convert this voltage level into a digital number that can be stored on a computer or other digital storage device.

The QDC requires a charge integration window (a time period over which to integrate the detector current pulse), which is provided by the timer. The start signal for the timer is provided by the discriminator (which can be either a leading edge triggered or a constant-fraction discriminator; see Sect. 7.3.4), which provides a timer start signal when an input pulse is detected. The delay unit (prior to the QDC) is necessary to align the current pulse with the charge integration window, as there is an inherent delay introduced by the discriminator relative to the current pulse. The notable difference between the QDC DAQ chain and the peak sensing ADC DAQ chain (described in Sect. 7.2.4) is the absence of a charge sensitive preamplifier, which can be particularly advantageous where radiation event timing is of particular importance or where the shape of the current pulse must be preserved (the charge sensitive preamplifier destroys pulse shape information and also modifies the pulse timing characteristics). Also, without the long decay time associated with the charge sensitive preamplifier (typically a decay time constant in the region of 50 μs), the undesirable effects of pulse pile-up (see Sect. 7.5.2) are also minimized. If the amount of charge per pulse produce by the detector is small then a current sensitive linear fast amplifier will be needed to amplify the current pulse to a level that can be readily accepted by the DAQ electronics; for most photomultiplier-based scintillator detectors, this pre-amplification is not required.

The QDC stage is usually followed by a multichannel analyzer (MCA) to get a visual display of the detected radiation events (see Sect. 7.2.7).

7.2.6 *Counter-Timer/Scaler*

Some applications or experiments only require a simple count of the radiation events within a certain energy range and the Counter-Timer (commonly referred to as a Scaler) DAQ chain (shown in Fig. 7.7) might be suitable and sufficient.

The CSP and shaping amplifier stages are exactly the same as that described in Sect. 7.2.4 for the peak sensing ADC DAQ chain but here the shaping amplifier output is fed to a *single channel analyzer* (SCA) unit. The SCA produces an output

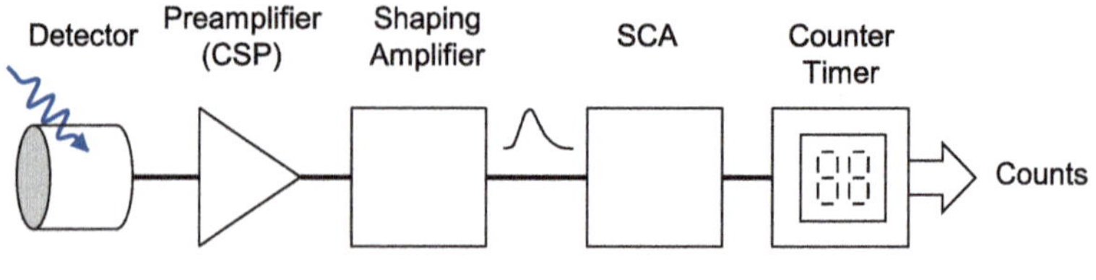

Fig. 7.7 Counter-Timer/Scalar data acquisition chain

logic pulse if the peak amplitude of the pulse from the shaping amplifier falls within a preconfigured pulse-height window. The lower range of the pulse-height window is set with a *lower level discriminator* (LLD) and the upper range set with an *upper level discriminator* (ULD); the LLD and ULD will typically be manual control knobs on the SCA unit. In its simplest form, the *counter-timer* will provide a visual readout of the number of detected events via an LED or similar display. Most likely the counter-timer would also have its own time-base, which allows it to display a count-rate.

When properly configured and energy calibrated, the counter-timer will increment each time a radiation particle, within a given energy range, is detected and will display either the absolute number of detected events or an event rate. It is not unusual for the SCA and counter-timer to be housed in the same unit.

A distinct advantage of the counter-timer DAQ chain is that it can be set up and run without the need for a host computer.

7.2.7 Multichannel Analyzer

The multichannel analyzer is a hardware unit that can be used at the back end of the data acquisition chain to visualize the energy (or indeed any quantity that can be represented as an analog voltage level) and frequency of radiation events. The MCA is connected to a computer or some other device that provides a visual display.

The MCA is primarily intended for use with a data acquisition chain of the type shown in Fig. 7.5, which uses a shaping amplifier followed by a peak sensing ADC but can be applied to many other types of DAQ chain. In a practical setup the shaping amplifier would be one hardware module and the MCA would be another hardware module; in this practical setup the MCA module encompasses the functionality of a Peak Sensing ADC unit but also has an interface to a computer. The interface to the computer can take on many forms; the most favored interface for modern systems is USB (Universal Serial Bus) but the MCA could, for example, be a computer add-in card that connects via PCIe (Peripheral Component Interface express). The MCA will be designed to work with a visualization and analysis program running in software on the host computer. The MCA setup is shown in Fig. 7.8.

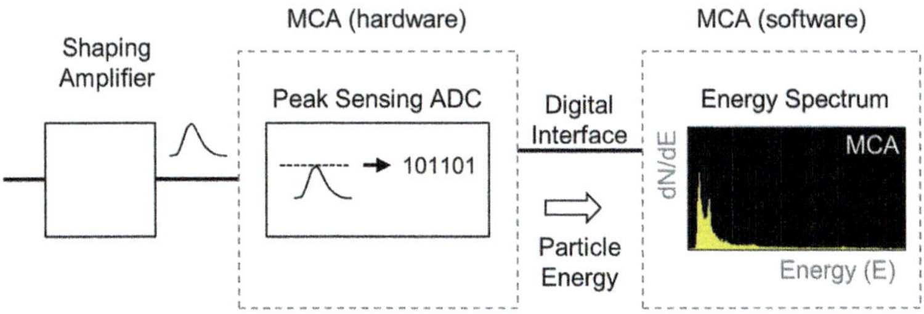

Fig. 7.8 Multichannel analyzer implementation

The displayed output of the MCA is a representation known as an *energy spectrum*, which is a histogram of particle energy versus the number of counts (N) per energy interval (often referred to as an energy *bin* or energy *channel*). The process is as follows: The MCA accepts individual pulse events from the shaping amplifier; the maximum amplitude of those pulses is converted to an energy value and those energy values are added one by one to the histogram (energy spectrum). So the energy spectrum is built up over time as more radiation events are acquired. An example energy spectrum can be seen towards the right of Fig. 7.8.

On completion of a data acquisition run, the energy spectrum can be saved as a simple histogram file so it can be displayed or post-processed in another software package. In a simple MCA implementation, information on the energy and time of arrival of each event is thrown away once it has been included in the energy spectrum; but in recent years MCA implementations have been adapted to run in a mode known as *time stamp list mode*, where some information on each radiation event is retained. The time stamp list mode is discussed later in Sect. 7.3.8.

7.2.8 Multichannel Scaler (MCS)

The *multichannel scaler* (MCS) is a hardware unit that can be used at the back end of the data acquisition chain to visualize the time profile of radiation events. The MCS is connected to a computer or some other device that provides a visual display. The MCS is primarily intended for use with a data acquisition chain of the type shown in Fig. 7.5, which uses a shaping amplifier followed by a Peak Sensing ADC. In a practical setup the shaping amplifier would be one hardware module and the MCS would be a separate hardware module; in this practical setup the MCS module encompasses the functionality of a peak sensing ADC unit but also has an interface to a computer. Similar to the MCA, the MCS interface to the computer can take on many forms; the most favored interface for modern systems is USB (universal serial bus) but the MCS could, for example, be a computer add-in card that connects via PCIe (peripheral component interface express). The MCS will be designed work with a visualization and analysis program running in software on the host computer. The MCS setup is shown in Fig. 7.9.

The displayed output of the MCS is a representation known as a *time spectrum*, which is a histogram of particle event times versus the number of counts (N) per time interval (often referred to as a time *bin*, time *channel* or dwell time). The MCS will have a set number of time channels, governed by the amount of storage memory in the unit; typically this will be a memory store addressed by a 16-bit bus, which equates to 65,536 (2^{16}) time channels. The dwell time can be chosen to suit the measurements being undertaken with a typical MCS unit having a dwell time range of nanoseconds up to many 100s of seconds. The overall experiment time is the product of the dwell time and the number of assigned channels, which could be from a few microseconds up to several years. An example time spectrum can be seen towards the right of Fig. 7.9.

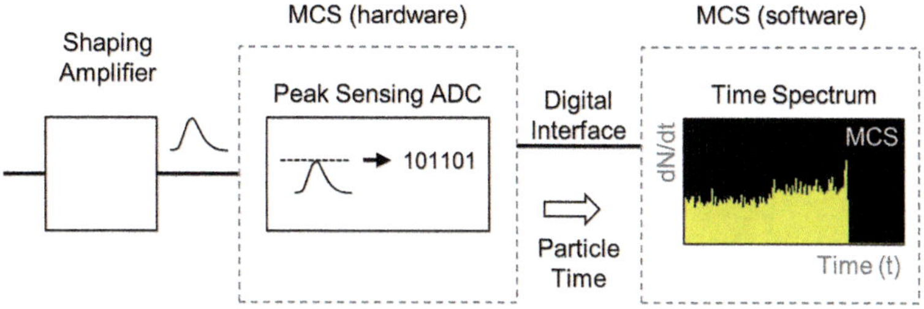

Fig. 7.9 Multichannel scaler implementation

On completion of a data acquisition run, the time spectrum can be saved as a simple histogram file so it can be displayed or post-processed in another software package.

Time spectra are particularly important for AI applications since we are often interested in the rate of decay of some induce signature. Adaptations of the MCS for AI are discussed later in Sect. 7.5.4.

7.2.9 *Triggering*

A trigger is the signal that tells the data acquisition electronics to start acquiring data. The trigger could come from an external source (for example, an associated particle tag signal from a neutron generator) or it could be generated from the radiation current pulse itself (known as self-triggering). When operating in a pulse detection mode (see Sect. 7.2.1), for the data acquisition topologies described in Sects. 7.2.3–7.2.8, it is necessary to have one trigger per radiation event, so self-triggering is the usual mode of operation. Although not explicitly stated or shown in Fig. 7.5, the peak sensing ADC can usually be configured to operate in either a self-trigger mode or a gated mode, where a gate signal (with a width that encompasses the pulse from the shaping amplifier) must be supplied to the peak sensing ADC unit. Whether or not the ADC is operated in a self-triggering or gated mode is very much down to the design and needs of the experiment or measurement. If it is possible to generate an accurate external trigger then this is normally the preferred trigger solution.

Particular attention is given to the self-trigger mode because the trigger scheme used can have a significant effect on the quality and accuracy of the data acquired. As the name suggests, when self-triggering, the trigger must be generated from the detector output signal itself. An example of where a self-trigger is necessary is the QDC data acquisition chain shown in Fig. 7.6. The two most common schemes for self-triggering are Leading Edge Discrimination (LED) and Constant Fraction Discrimination (CFD).

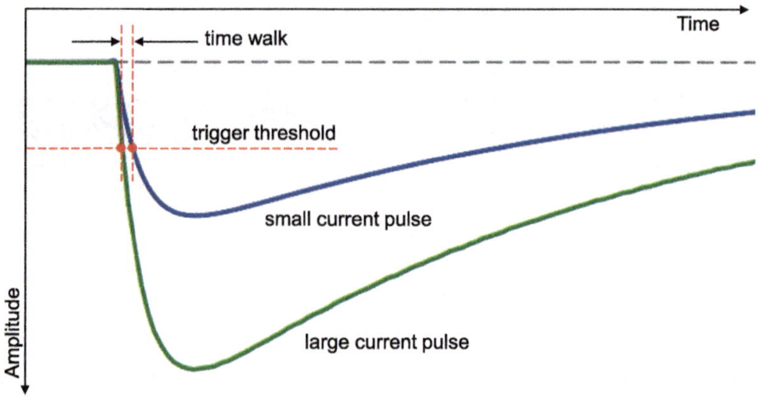

Fig. 7.10 Illustration of the leading edge trigger method and how the trigger time is amplitude dependent

In the LED method, a suitable amplitude threshold is set and a trigger is generated if the signal crosses this threshold. The threshold is typically set to a level that is sufficiently clear of any electrical or other noise in the system. The LED trigger method is reasonably straight forward to implement and is the mode typically used in oscilloscopes and many commercially available multichannel analyzers. But the LED method does have its drawbacks particularly if signal timing is of importance. Consider the case of two different current pulses from the detector, one of low amplitude and one of high amplitude, as shown in Fig. 7.10. Both pulses have the same leading edge time constant. With a fixed trigger threshold level it can be seen that the two current pulse signals cross the threshold at slightly different times resulting in what is known as *time walk*; the time that the trigger occurs relative to the start of the pulse is amplitude dependent. If the timing of the radiation event is not critical or the amplitude range of pulses is small then time walk may be perfectly tolerable. However if the arrival time of events is of importance (for example in time-of-flight measurements or for certain pulse shape discrimination schemes) then the LED trigger method is inadequate and a more accurate trigger scheme is required.

Constant fraction discrimination [4, 5] is a more sophisticated trigger scheme that is designed to deal with the issue of time walk and generate a more time-accurate trigger time than the LED method. The CFD is designed to trigger at the optimum fraction of the pulse amplitude for any given pulse amplitude. In the CFD method, an attenuation-summation operation is performed on the current pulse from the detector to produce a bipolar with a zero-crossing point. The attenuation-summation operation (with a fraction setting of 0.2) is illustrated in Fig. 7.11.

The current pulse is first attenuated to a fraction of its initial amplitude. The current pulse is also inverted and delayed. The delay is chosen to make the fraction point on the leading edge of the delayed pulse line up with the peak amplitude of the attenuated pulse. The two altered pulses are summed to produce a bipolar pulse.

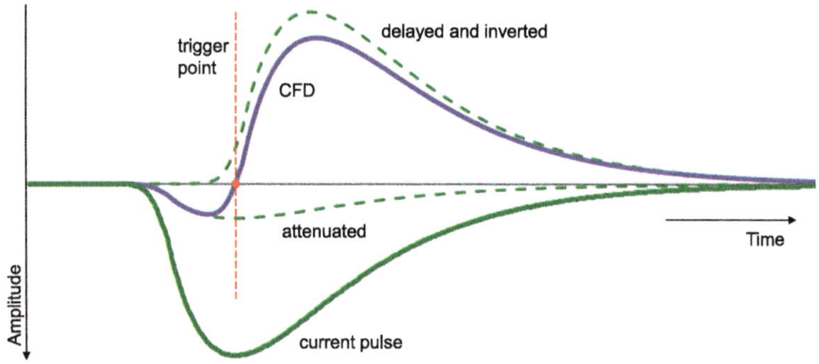

Fig. 7.11 Illustration of the constant fraction discrimination trigger method and how the trigger time is amplitude invariant

The zero-crossing point of the bipolar pulse is used to generate a trigger signal that is amplitude invariant and has negligible time walk.

LED and CFD stages are typically available as separate DAQ modules but may also be built into other analog pulse processing DAQ modules.

7.2.10 Dead Time

The dead time of a detector system is defined as the minimum amount of time that must separate two events in order that they are recorded as two separate pulses [1]. For the vast majority of radiation detection systems used for radiological and nuclear security applications there will be a dead time associated with the pulse processing electronics (where the duration of the shaped pulse sets the minimum pulse separation before pile-up occurs). There is also a dead time associated with signal conversion for storage, as the detection system will require a finite amount of time to perform analog-to-digital conversion and to store data to memory, during which time no further pulses can be accepted.

There are two commonly used models to describe the ideal behavior of a counting system known as paralyzable and non-paralyzable response. The behavior of the two response types are virtually the same at low count rates. In a non-paralyzable system, an event that occurs during the dead time period is simply lost; so with an increasing event rate the detection system will reach a saturation count rate equal to the inverse of the dead time. In a paralyzable system, an event that occurs during the dead time period will not only be lost but will restart the dead time; so with an increasing event rate the detection system will reach a saturation point where it will be totally paralyzed and unable to record any events at all [6].

Dead time in the pulse processing electronics, as well as pileup, can lead to a loss of events; it follows that the magnitude of these losses increases with increasing count rate.

In summary, one should avoid measurement conditions under which dead time losses are high in order to minimize the errors that occur in making dead time corrections. When losses are greater than 30–40%, the calculated true count rate becomes very sensitive to small changes in the measured rate and the assumed system behavior [1].

7.2.11 Analog Pulse Shape Discrimination

It is fair to say that all of the widely used neutron detection materials have some sensitivity to γ rays. In some cases the neutron response is the wanted signal and so the γ-ray response is considered a nuisance. In some cases both the neutron and γ-ray responses are important in measurements. In either case, it is necessary to separate the neutron events from the γ-ray events.

For neutron detectors that employ one of the three main nuclear absorption reaction materials (^{3}He, ^{10}B and ^{6}Li), neutron interactions and γ-ray interactions deposit different amounts of energy [2]. So, discrimination of neutron and γ-ray events by energy for these detector types is both feasible and adequate. The DAQ chains described previously in Sects. 7.2.3–7.2.8 are all suitable for neutron capture-based detectors and examples of such detectors include ^{3}He tubes, boron-lined tubes and some LiF/ZnS scintillators. Neutron detectors based on elastic scattering (such as PSD liquids and PSD plastics) do not enjoy a separation of neutron and γ-ray events by energy deposition and so some other method must be used to discriminate the two radiation types. There are some scintillator materials, such as certain types of liquid hydro-carbons, where neutrons and γ rays give up their energy in different ways as they traverse the detection medium and this manifests as a difference in the shape of the light pulse emitted by the scintillator; in these materials, neutrons produce light pulses with a longer tail (slower decay constant) when compared with γ rays [7]. The cartoon of Fig. 7.12 illustrates how the PMT current pulse might look for a γ ray and neutron when viewed on a oscilloscope. In practice the observable difference in the γ-ray and neutron pulse shapes will be quite small but this difference has been exaggerated in Fig. 7.12 for the purposes of illustration.

This pulse shape difference can be used as a way of separating neutron and γ-ray events using a technique commonly referred to as *pulse shape discrimination* (PSD). Indeed, with a suitable choice of scintillator material, PSD can be used to separate many different radiation types (for example, beta particle and γ-ray separation in a CsI(Tl) crystal [8] and much of what follows in this section can be equally applied to discriminating radiation types other that neutrons and γ rays. For AI the key focus is usually on neutrons and γ rays so this will be the focus here.

There are several well established methods for performing PSD with analog electronics, the most common of which are listed below:

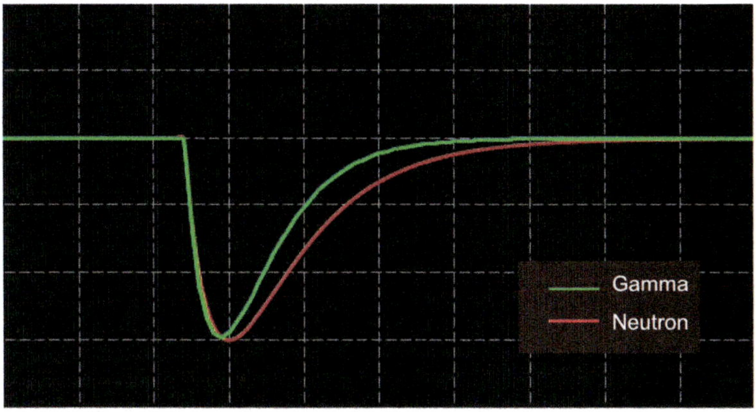

Fig. 7.12 Illustration of the pulse shapes resulting from a γ ray and neutron as might be viewed on an oscilloscope

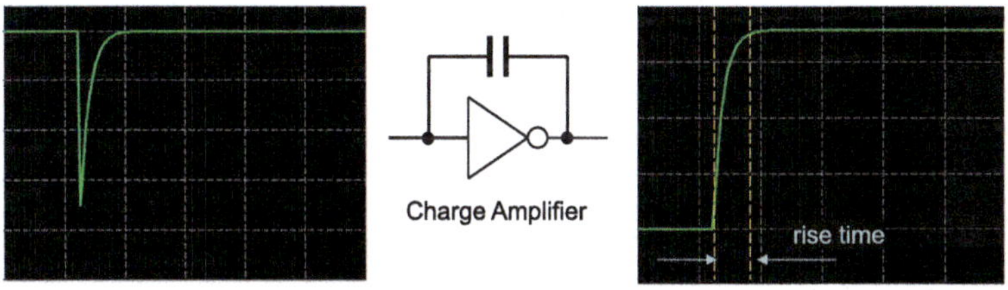

Fig. 7.13 Illustration of the rise time discrimination method for pulse shape discrimination, as might be viewed on an oscilloscope

- Rise time discrimination
- Zero cross-over/constant-fraction discriminators
- Charge comparison
- Constant time discrimination

In the *rise time discrimination* (RTD) PSD method, timing measurements are made on the rising edge of a charge integrated detector pulse [9]. The rising edge of the charge integrated pulse is a function of the entire time development of the current pulse from the detector, which will be different for neutrons and γ rays. The RTD concept is illustrated in Fig. 7.13.

The amount of time taken for the charge integrated detector pulse to rise from zero to full amplitude (in practice the time interval is measured between 10% and 90% of full amplitude, or some other suitable percentage) will be different for neutrons and γ rays and thus this time duration measurement can be used to discriminate between the two radiation types.

One implementation of a DAQ chain to perform RTD PSD is shown in Fig. 7.14. Referring to Fig. 7.14, a *delay-line amplifier* (DLA) is used to shape the preamplifier pulse to return the signal to the baseline ready to accept the next pulse. By

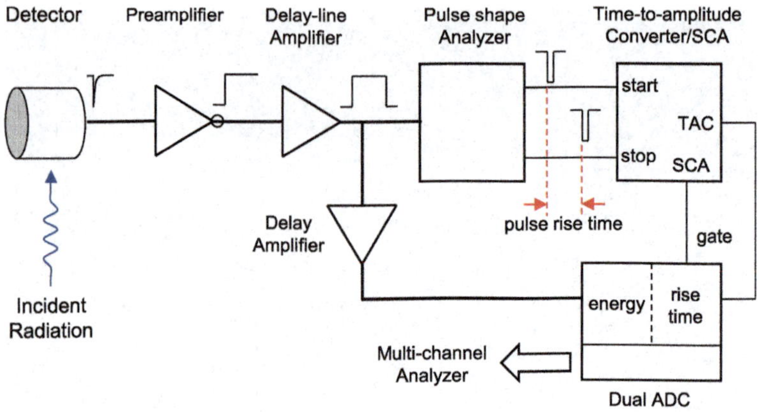

Fig. 7.14 Rise time discrimination method pulse shape discrimination data acquisition chain

definition, the rise and fall time of the DLA is symmetrical and in principle the rise time can be measured on either edge of the DLA output signal. In practice the rise time is measured on the falling edge of the DLA output signal because the timing measurements are taken at fractional levels of the maximum amplitude, and the measurement is much simpler to make if the maximum amplitude is known beforehand. The *pulse shape analyzer* (PSA) generates a *start* signal for the *time-to-amplitude converter* (TAC) when the falling edge of the DLA output signal falls below a certain fraction of the maximum amplitude (usually 90%). The PSA generates a *stop* signal when the DLA output signal falls below a smaller fraction of the maximum amplitude (usually 10%). The TAC produces an output voltage level that is proportional the time interval defined by the start and stop signals, which in this case is a measure of the rise time of the charge integrated pulse from the detector. The TAC output can be fed to an ADC and multichannel analyzer (see Sect. 7.2.7) to produce a histogram of the rise time, which will have a form as shown in Fig. 7.15. With properly configured hardware, the γ rays and neutrons will fall into two distributions determined by their rise time differences.

In practice, there is a wide statistical variation in the pulse shapes for both γ rays and neutrons and separation can only ever be performed to some value of statistical significance. Moreover, the amount of separation varies with energy; separation is better at higher energy due to better statistics resulting from the higher number of scintillation photons generated by the radiation interaction. It follows that the amount of separation is also dependent upon the characteristics of the scintillator material and in particular the number of photons produced per keV of deposited energy. The example of Fig. 7.15 is the rise time histogram that can be expected for a xylene-based liquid scintillator detector at an energy deposition in the region of 100 keVee (electron equivalent energy). The width of the γ distribution is in the region of a few nanoseconds and the neutron distribution roughly an order of magnitude greater.

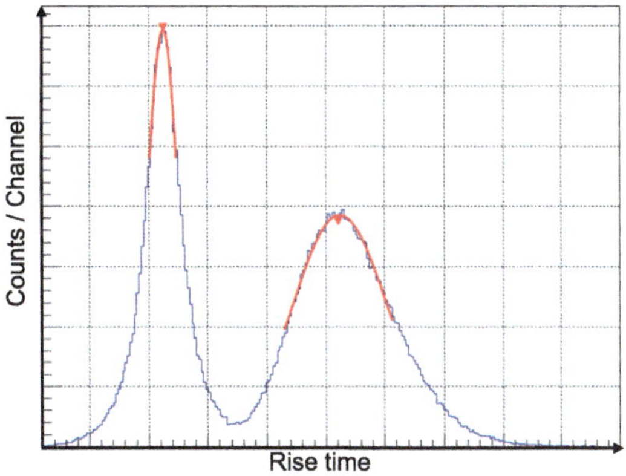

Fig. 7.15 Example of a pulse shape discrimination rise time histogram (typical of a liquid or PSD plastic scintillator)

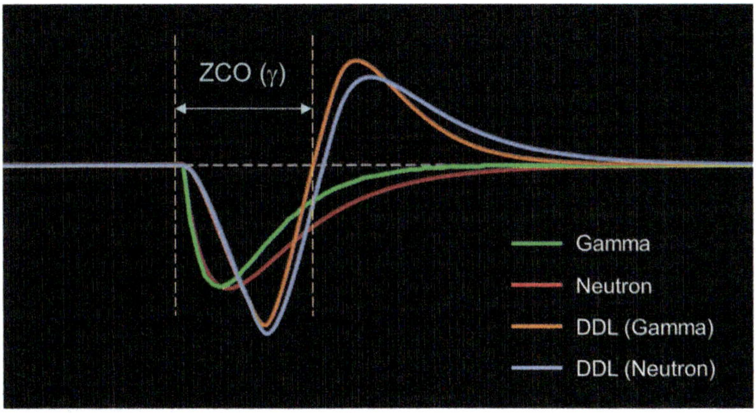

Fig. 7.16 Illustration of the zero cross-over method for pulse shape discrimination

The delay amplifier output can be fed to a second ADC and MCA to produce an energy histogram. It is also possible to combine the energy value and rise-time value for each event to produce a two-dimensional histogram of energy versus rise-time.

In the *zero cross-over* (ZCO) method the individual pulses from the detector are passed through an appropriate shaping network (such as a CR-RC-CR network or a double delay line shaper) to create a bipolar pulse [10, 11]. The time at which the bipolar pulse crosses zero is an amplitude invariant function of the detector pulse shape and rise time [12]. The ZCO concept is illustrated in Fig. 7.16. The time interval between the start of the detector pulse and the zero cross-over point of the bipolar pulse will be different for neutrons and γ rays and thus this time interval measurement can be used to discriminate between the two radiation types. The two Double Delay-Line (DDL) traces in Fig. 7.16 are the results of the two traces shown in Fig. 7.12 after they have each undergone charge integration and fed through two stages of a delay-line circuit; the zero cross-over time for the γ-ray pulse is labelled in Fig. 7.16 and it can be seen that the zero cross-over time for a neutron will be longer.

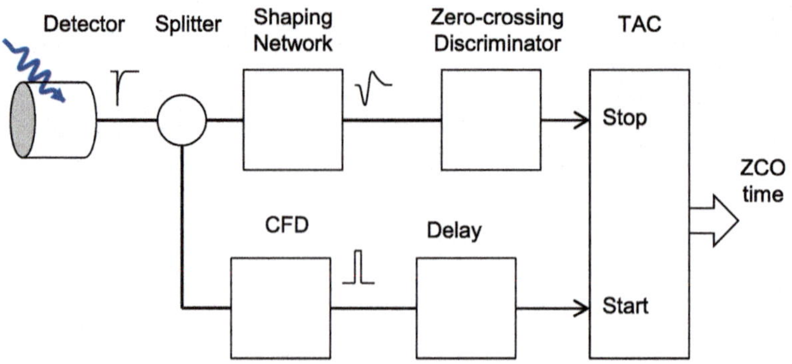

Fig. 7.17 Illustration of the zero cross-over method for pulse shape discrimination

One implementation of a DAQ chain to perform ZCO PSD is shown in Fig. 7.17. The constant-fraction discriminator (CFD) generates an amplitude invariant trigger, which serves as a start signal for the Time-to-Amplitude Converter (TAC). The shaping network contains the circuitry to create the bipolar signal, which would typically be a charge integration stage followed by a double delay line (but there are many variations on this circuitry). The shaping network is followed by a zero-crossing discriminator, which generates the stop signal for the TAC. The TAC produces an output voltage level that is proportional to the time interval defined by the start and stop signals. A delay is often needed after the CFD to bring the start signal into the same time range as the stop signal.

The TAC output can be fed to an ADC and multichannel analyzer (see Sect. 7.2.7) to produce a histogram of the zero cross-over time, which will look similar to the form shown in Fig. 7.15 (generated by the rise time discrimination PSD method). The addition of a QDC stage (see Sect. 7.2.5) would allow one to generate an energy value and a ZCO value for each event making it possible to produce a two-dimensional plot of energy versus ZCO.

In the charge comparison method every detector pulse is integrated over a short time window (short gate) and a long time window (long gate). The relative amounts of the charge in these integrated gate periods determine whether the event is a neutron or a γ ray [13]. The charge integration periods of the charge comparison PSD method are illustrated in Fig. 7.18. The prompt and delayed time periods relate to the prompt and delayed fluorescence from the scintillator [1].

A single PSD parameter that can be used as a measure of neutron and γ-ray separation can be calculated as the ratio of the short gate charge to the long gate charge; or more commonly the ratio of the delayed charge component (see Fig. 7.18) to the overall charge (the long gate charge).

In the constant time discrimination (CTD) method the amplitude of every detector pulse is taken at some constant time relative to the start (or trigger point)

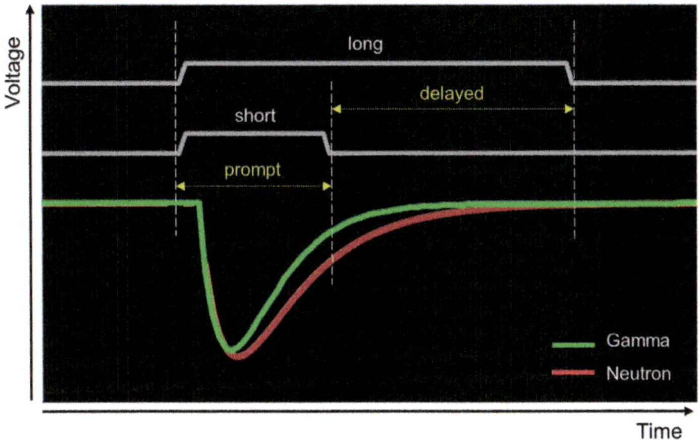

Fig. 7.18 Illustration of the time gates associated with the charge comparison pulse shape discrimination method. The prompt and delayed time periods relate to the prompt and delayed fluorescence from the scintillator

of the pulse. The amplitude of the pulse at this constant time value will be different for neutrons and γ rays since they have different pulse shapes. Although the CTD method is relatively simple to implement in analog electronics this PSD method is particularly susceptible to time jitter and the pulse triggering scheme used (see Sect. 7.2.9) so should be used with caution.

Similar DAQ chain implementations exist for the charge comparison method and the constant time discrimination method each of which can be used to generate a histogram plot similar to that shown in Fig. 7.15 to quantify separation between neutrons and γ rays. The amount of separation that can be achieved will depend on a number of factors including the detection material itself, the PSD method used as well as the particular DAQ components chosen and the way in which they are configured.

PSD in the analog domain can be quite complex in terms of the hardware required and the expertise needed in setting it up. All of the PSD methods described here typically require a fair number of individual DAQ modules, which can rapidly become quite cumbersome especially when one moves beyond a handful of detection channels. The hardware complexity and high module count associated with analog PSD is a serious limitation when building the multichannel detector arrays associated with many AI systems; one is forced down the route of designing custom DAQ units just to keep the electronics to a reasonable size. The hardware complexity problem has been addressed quite elegantly with the introduction of digital DAQ and, in general, PSD can be much simpler to implement in digital hardware; see Sect. 7.3.7.

7.3 Digital Data Acquisition

Analog data acquisition systems, as described in Sect. 7.2, are perfectly suitable and adequate for simple radiation detection systems that have a small number of detectors but the DAQ system becomes rapidly cumbersome and costly as the number of detector channels is increased.

The detector systems that are commonly used for AI will typically be made up of large area detector arrays of which each element will require its own data acquisition channel. Digital DAQ systems are particularly suited to these multi-channel implementations and significantly outperform their analogue counterparts. Another significant advantage of digital processing is that the detected radiation events can be analyzed in far greater detail opening up the possibility of sophisticated detection algorithms thereby increasing overall detection performance.

7.3.1 Advantages of Digital DAQ

Digital DAQ offers a number of significant advantages over analog DAQ and can now be considered the DAQ of choice when designing a new radiation detection system. A summary of some of the advantages of digital DAQ is given below and the sections of this chapter that follow demonstrate how these advantages have been realized in modern digital DAQ systems:

- A significant reduction in hardware and compact solutions. A single digital hardware module is capable of performing multiple functions such as energy calculations, event timing, counting and pulse shape analysis
- More cost-effective and reliable compared with analog
- Good linearity and stability leads to good reproducibility
- Wider dynamic range and uniformity
- Digital techniques allow better correction of pile-up and noise effects due to baseline fluctuations and ballistic deficit [1],
- Easy synchronization and correlation over several DAQ channels
- Low dead time in the acquisition leading to high count rates
- Flexibility—algorithms can be tailored (changed and adapted) to better fit the application
- Tuning and calibration: software programming instead of manual control; faster and automatic

Despite the huge range of benefits that a digital DAQ system can offer there are also some disadvantages that one should always be aware of when considering whether analog or digital DAQ is most appropriate for a given application. Properly setting up a digital DAQ system requires good knowledge of digital signal processing algorithms and the relevant control parameters; more so than with an equivalent analog DAQ system. It can take more time for beginners to understand and configure

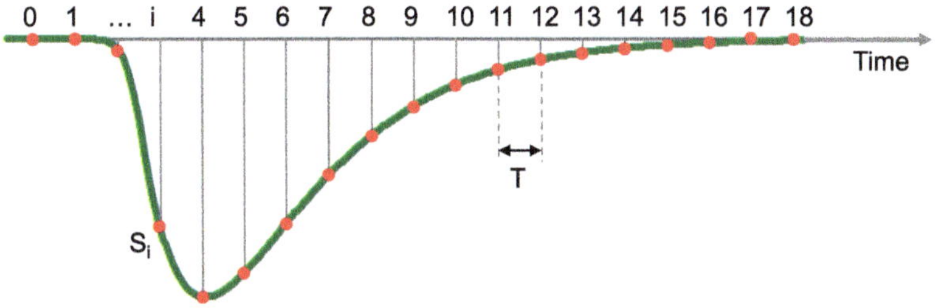

Fig. 7.19 Illustration of the sampling process

a digital DAQ system because the user interface generally has more parameters to vary, the layout is not necessarily standardized and the configuration process is not always obvious.

7.3.2 Digital Sampling

Digitization is the process by which a continuous time signal is converted to a discrete time signal. Sampling is performed by measuring the value of the continuous signal every T units of time, where T is referred to as the sampling interval. The digitization process results in a sequence of samples (that can be stored in memory on a digital device) that represent the original signal. The sampling of a detector current pulse is illustrated in Fig. 7.19, where S_i is the sampled value of the pulse amplitude at sample number i.

For the sampled signal to be an adequate representation of the original analog signal, there are certain rules that must be obeyed. These rules are fundamental in ensuring the preservation of information when moving from the analog domain to the digital domain and result from a subject are known as *information theory*. Linked to information theory is the *sampling theorem* that dictates the minimum sampling period required to faithfully represent a continuous time signal in the digital domain.

Any continuous time signal can be represented by a summation of sine waves at different frequencies, each of which represents a frequency component. The continuous time signal can be represented as a *spectrum* of frequency components in what is known as the frequency domain. The continuous time signal is said to reside in the time domain and it can be converted to the frequency domain via the Fourier transform [14]. An example of this transform is illustrated in Fig. 7.20, which shows a time domain signal on the left and its frequency domain representation on the right, which in this case is composed of four sinusoidal components of varying amplitude and frequency. These frequency components span a bandwidth, B.

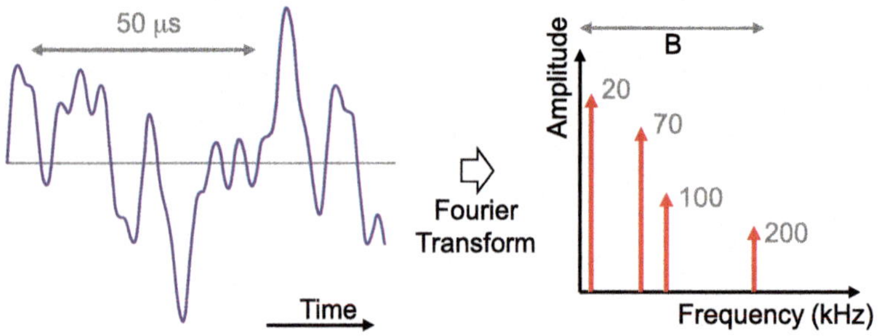

Fig. 7.20 Time domain and frequency domain representations of a continuous time signal

A signal is said to be band-limited if it contains no energy at frequencies higher than some bandwidth B. A signal that is band-limited is constrained in how rapidly it changes in time, and therefore how much detail it can convey in a certain time interval.

Referring to Fig. 7.19, the Sampling Theorem states that the uniformly spaced discrete samples are a complete representation of the signal if its bandwidth is less than half the sampling rate. Put another way, for an exact digital representation of the continuous time signal, the sampling rate must be at least twice the highest frequency in that signal; known as the Nyquist rate [14]. So in determining the required sampling rate for a detector output signal, the main consideration is the highest frequency component of that signal. The consequence of sampling below the Nyquist rate is an effect known as aliasing, where different frequency components of the signal become indistinguishable when sampled. To avoid this problem, a digital sampling system (such as that described in Sect. 7.3.3) will have an anti-aliasing filter as its first stage to remove any frequency components that violate the Nyquist criterion. When choosing or designing a digital sampling system, one should take care that the anti-aliasing filter is not removing wanted high frequency components.

In practice the sampling rate is chosen to be somewhat higher than the Nyquist rate to allow for imperfections in the design of the anti-aliasing filter.

7.3.3 The Digitizer

Recent developments in high-speed digitizers have paved the way for a more elegant hardware solution to the analog DAQ chains presented in Sect. 7.2 particularly so for DAQ systems with a high number of channels. Digitizers offer a one-box solution that can be dynamically configured to perform many different DAQ functions. The analog DAQ chain is replaced by a digital DAQ chain. The essential difference between the analog and digital systems is that the latter digitizes the detector output

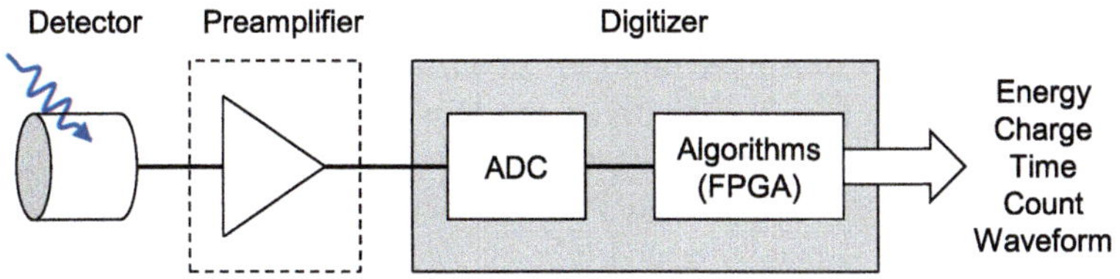

Fig. 7.21 Digital data acquisition chain

signals very early in the chain (usually at the detector output or just after the charge integrating preamplifier if one is used in the DAQ chain) and then performs the pulse processing in the digital domain.

The digitizer is today the core component of most digital DAQ systems. The analog DAQ chains made up of a series of hardware modules is replaced by a single hardware digitizer module that performs analog-to-digital conversion as well as applying digital algorithms. These algorithms mimic the functions of the modules one would have in an analog DAQ chain such as shaping amplifiers, discriminators, peak sensing, scalers and charge-to digital converters. A typical block diagram for a digitizer-based DAQ chain is shown in Fig. 7.21.

The key components of the digitizer are the analog-to-digital (ADC) conversion stage and a processing stage where different pulse processing algorithms can be applied to the digitized detector pulses.

The specification of a digitizer is largely defined by the specification of the ADC and the amount of processing power available to run algorithms. The ADC is largely defined by the rate at which it samples the input signal from the detector and the precision to which it produces those samples. The required specification of ADC is dictated by the frequency content of the input pulse and the energy resolution of the detector. The faster the time development of the detector pulse, the higher the sampling rate needed. The required sampling rate can be chosen in accordance with the sampling theorem described in Sect. 7.3.2. As a guideline, charge integrated pulses from detectors such as sodium iodide and high-purity germanium have a time development in the μs range and 100 MSa/s (million samples per second) is sufficient to digitally capture these signals. The current pulse from a liquid scintillator or PVT (polyvinyl toluene) plastic has a time development in the range of 10s to 100s of ns; a sampling rate in the region of 250 MSa/s might be needed for energy measurements and perhaps 500 MSa/s for Pulse shape discrimination measurements. The higher the energy resolution of the detector, the more digital bits needed to faithfully represent the measured energy values. Examples of the number of ADC bits needed for different detector types can be found in Sect. 7.2.4. The output of the ADC is a sequence of digital samples (representing the analog pulse presented at its input) which is fed to the digital pulse processing stage.

In its simplest form, there is no pulse processing carried out on the digitizer hardware and the digitized detector pulses are passed directly to a computer, where the pulse processing can be performed in software. This mode of operation is commonly referred to as *waveform* or *raw* capture mode. The obvious disadvantage of raw mode is in the sheer amount of data that must be passed between the hardware and the host computer and this can often preclude real-time operation of the DAQ system; this issue is discussed in detail in Sect. 7.4. The ability to carry out pulse processing on the digital hardware itself opens up the possibility of real-time operation of complex algorithms. Pulse processing is performed on a hardware chip (integrated circuit), such as a field-programmable gate array (FPGA), and just the results of those pulse processing algorithm operations are passed to the host computer rather than the whole digital pulse waveform. Not only can the algorithms be performed faster in hardware but the amount of data passed to the host can be reduced substantially, and in the process easing the data transfer specification on the host interface. The hardware pulse processing device of choice is the FPGA (but there are alternatives such as microcontrollers, Digital Signal Processors or Complex Programmable Logic Devices) which is an array of reconfigurable gates (groups of transistors) that can be programmed (and re-programmed) to form processing blocks such as logic functions, comparators, latches and memory [15]. These elementary processing blocks are used to build up more complex functions such as a shaping network, constant fraction discriminator or charge-to-digital converter.

7.3.4 Triggering

The concepts of triggering are explained in Sect. 7.3.4 and apply equally in the digital domain. Both the leading edge discriminator and constant fraction discriminator can be implemented digitally. But digital DAQ systems have the advantage of being able to store data and "look backwards in time", and so the trigger schemes can be more sophisticated if desired. For example, a system can be set to produce a trigger only if the input signal remains above a fixed threshold for a certain period of time; or only trigger if a certain amount of time has elapsed since the previous trigger. With digital systems, it is also relatively easy to generate a trigger based on the time coincidence of two or more events either on the same DAQ channel or across different channels of a multi-channel DAQ system. This can be extended further to generate triggers based on a sequence of events that fit a certain pattern, which could be a pattern based on, say, the energy of the event or the particle type. Smart triggering of this type and the benefits that it can bring are discussed further in Sect. 7.4.

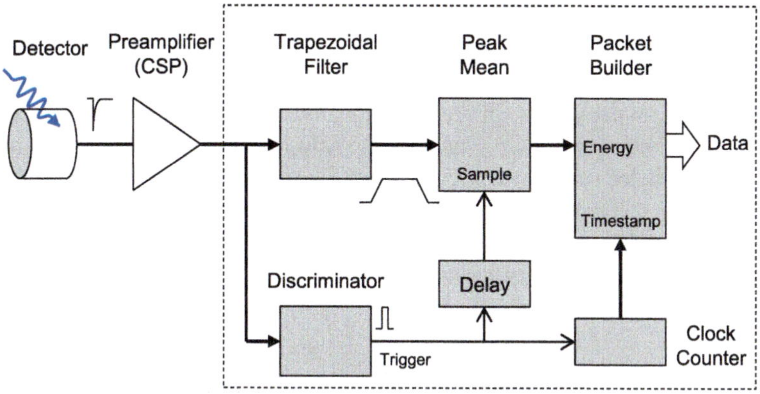

Fig. 7.22 Digital pulse height analysis data acquisition chain. The processing blocks within the dotted outline are performed in the digital domain

7.3.5 Pulse Height Analysis

Pulse height analysis (PHA) is the digital equivalent of the analog peak sensing ADC data acquisition chain described in Sect. 7.2.4. The PHA digital DAQ chain takes the charge integrated preamplifier signal as its input. The preamplifier signal is appropriately sampled and the remainder of the processing is performed in the digital domain. One possible DAQ chain implementation for digital PHA is illustrated in Fig. 7.22. The processing blocks within the dotted outline are performed in the digital domain; for real-time operation these blocks would typically reside in programmable hardware (e.g. a field-programmable gate array) on the digitizer module.

The trapezoidal filter creates a trapezoid or *flat-top* output signal, whose maximum amplitude is proportional to the peak height of the preamplifier pulse. The trapezoidal filter is the digital equivalent of the analog shaping amplifier shown in Fig. 7.5. The trapezoidal filter is typically composed of the digital equivalent of a delay line or similar filter network [15], to produce the flat-top output signal. The width of the flat-top can be controlled, in the same way that the shaping time can be varied on an analog shaping amplifier. The flat-top should be sufficiently wide so that its peak amplitude can be sampled accurately; in practice, the peak amplitude will be taken as the average over a pre-defined number of samples. The discriminator (which can be either a leading edge triggered or a constant-fraction discriminator; see Sect. 7.3.4) detects the start of the preamplifier signal. The discriminator provides a trigger to the clock counter, which counts the number of events and registers the event time (the time stamp). The trigger is also used as the timing signal to determine when the Peak Mean stage is to sample the flat-top amplitude; the delay can be tuned to make the sample time line up with the centre of the flat-top. The sampled value is proportional to the deposited charge and hence the energy of the radiation event.

Depending on the specific implementation, the time stamp and energy value (and any other relevant information such as the DAQ channel number, event number or DAQ parameter settings) can be built into a data packet to be sent to the host computer. The relevant information from these data packets can be taken to form a list-mode data file (see Sect. 7.3.8), which can be further processed to generate an energy or time spectrum (histogram).

7.3.6 *Charge Integration*

Charge integration (CI) is the digital equivalent of the Charge-to-Digital analogue data acquisition chain described in Sect. 7.2.5. The CI digital DAQ chain takes the detector current pulse as its input signal. The current pulse is appropriately sampled and the remainder of the processing is performed in the digital domain. A typical data acquisition chain for digital charge integration is shown in Fig. 7.23. The processing blocks within the dotted outline are performed in the digital domain; for real-time operation these blocks would typically reside in programmable hardware (e.g. a field-programmable gate array) on the digitizer module.

The *charge accumulator* essentially sums the digital sample values of the detector current pulse, which is equivalent to integrating the signal. The Charge Accumulator requires a charge integration window (a time period over which to integrate the detector current pulse), which is provided by the timer. The start signal for the timer is provided by the discriminator (which can be either a leading edge

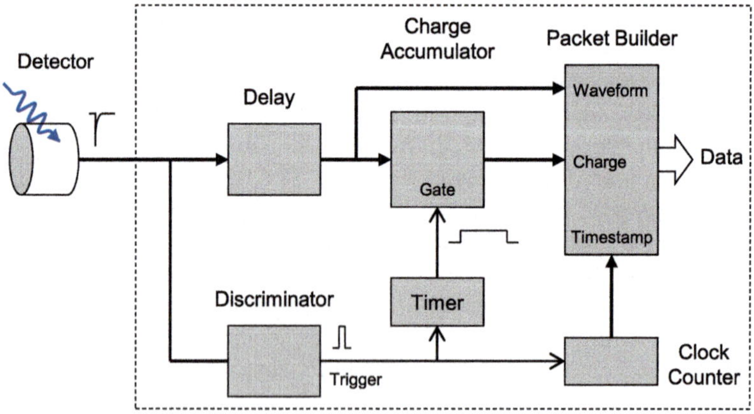

Fig. 7.23 Digital charge integration data acquisition chain. The processing blocks within the dotted outline are performed in the digital domain

triggered or a constant-fraction discriminator; see Sect. 7.3.4), which triggers when an input pulse is detected. The delay unit (prior to the Charge Accumulator) is necessary to align the current pulse with the charge integration window, as there is an inherent delay introduced by the discriminator relative to the current pulse. The discriminator output also triggers the clock counter, which counts the number of events and registers the event time (the time stamp). The "raw" digitized current pulse (waveform) is usually available as an output on such a digital DAQ system and may be recorded to file for later processing.

Depending on the specific implementation, the time stamp, charge value and waveform samples (and any other relevant information such as the DAQ channel number, event number or DAQ parameter settings) can be built into a data packet to be sent to the host computer. The relevant information from these data packets can be taken to form a list-mode data file (see Sect. 7.3.8), which can be further processed to generate an energy or time spectrum (histogram).

7.3.7 Digital Pulse Shape Discrimination

The use of PSD to separate neutrons from γ rays was described in some detail in Sect. 7.2.11 and a number of analog PSD methods were presented, namely rise time discrimination, zero cross-over, charge comparison and constant time discrimination. Any one of these methods may be implemented in the digital domain because all of the building blocks that make up the analog DAQ chains for each of these methods can be implemented digitally. Charge integration is the method that has been most widely adopted for digital PSD and is the method that will be described in detail here.

The DAQ chain for the charge integration PSD method takes the detector current pulse as its input signal. The current pulse is appropriately sampled and the remainder of the processing is performed in the digital domain. A typical data acquisition chain for charge integration PSD is shown in Fig. 7.24, which is essentially a modified version of the CI DAQ chain shown in Fig. 7.23 to include a second charge gate. One gate is used to measure the short duration (or prompt) charge of the detector current pulse and a second gate used to measure the long duration (or total) charge; as is required for the charge integration PSD method (see Fig. 7.18).

The processing blocks within the dotted outline are performed in the digital domain; for real-time operation these blocks would typically reside in programmable hardware (e.g. a field-programmable gate array) on the digitizer module.

The charge accumulator sums the digital sample values of the detector current pulse (which is equivalent to integrating the signal) over the short and long time windows (charge gates). The two charge integration intervals are provided by two timers. The start signal for both timers is provided by the discriminator (which can be either a leading edge triggered or a constant-fraction discriminator; see Sect. 7.3.4), which triggers when an input pulse is detected. The delay unit (prior

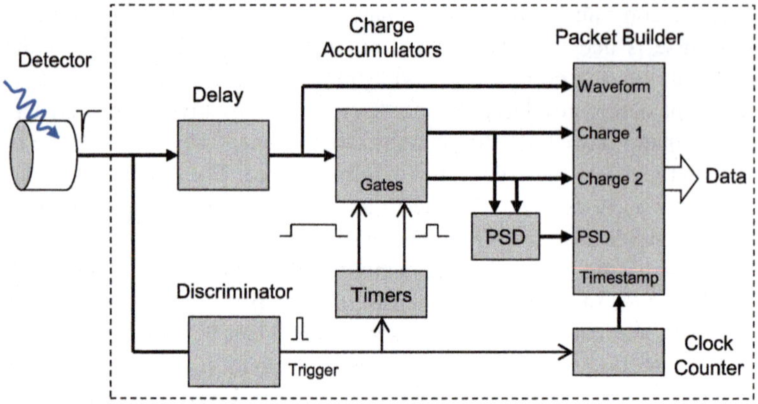

Fig. 7.24 Digital PSD data acquisition chain based on charge integration. The processing blocks within the dotted outline are performed in the digital domain

to the Charge Accumulator) is necessary to align the current pulse with the charge integration window, as there is an inherent delay introduced by the discriminator relative to the current pulse. The discriminator output also triggers the clock counter, which counts the number of events and registers the event time (the time stamp). The PSD stage in Fig. 7.24 calculates a PSD value (a measure of neutron and γ-ray separation) from the two charge integration values, which is typically calculated as the ratio of the delayed charge component (see Fig. 7.18) to the overall charge (the long gate charge).

The "raw" digitized current pulse (waveform) is usually available as an output on such a digital DAQ system and may be recorded to file for later processing.

Depending on the specific implementation, the time stamp, two charge values, a PSD value and waveform samples (and any other relevant information such as the DAQ channel number, event number or DAQ parameter settings) can be built into a data packet to be sent to the host computer. The relevant information from these data packets can be taken to form a list-mode data file (see Sect. 7.3.8), which can be further processed to generate an energy, time or PSD histogram. Further, one can combine and of these histograms to produce two dimensional (2D) or even three dimensional (3D) histograms to illustrate the radiation behavior. A useful and commonly used 2D histogram is energy versus PSD value, which shows neutron and γ-ray separation as a function of energy. An example of such a 2D histogram is shown in Fig. 7.25, which is for a hydrocarbon-based liquid scintillator.

7.3.8 *Time Stamp List Mode*

Time stamp list mode (or more commonly just list mode) is the capability to record and retain the time and energy of each radiation event (as opposed to traditional

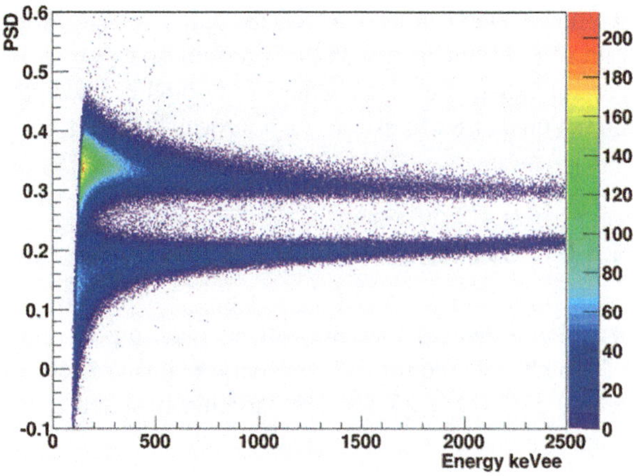

Fig. 7.25 Two-dimensional histogram of PSD value versus energy for a hydrocarbon-based liquid scintillator. The histogram was created using a digital PSD DAQ chain based on charge integration

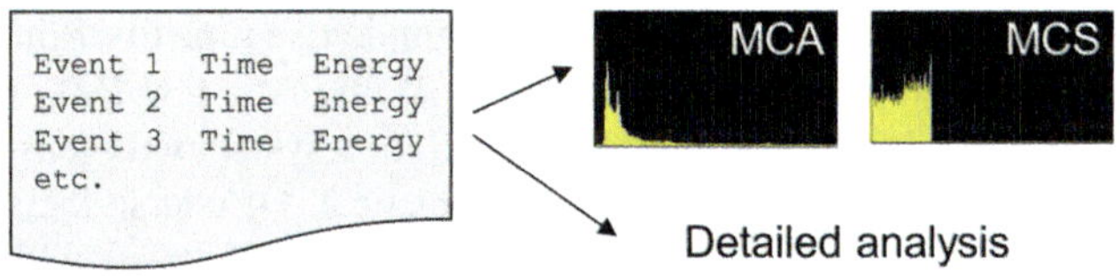

Fig. 7.26 Time stamp list mode data acquisition

multichannel analyzers that throw away timing information and multichannel scalers that throw away energy information). List mode is most certainly the future direction of DAQ systems; each radiation event is tagged (as illustrated in Fig. 7.26) and maintained in a list, which can just be a simple text file on the host computer. In its simplest form, the list might contain just a time stamp and an energy value for each radiation event. From the list mode data it is then possible, either in real-time or offline, to create either the energy spectrum of an MCA (see Sect. 7.2.7) or the time spectrum of an MCS (see Sect. 7.2.8).

For PSD analysis the list would include additional parameters per event that describe the shape of the pulse (in the case of pulse shape discrimination based on the charge integration method the list might include two additional parameters with values for the prompt and delayed charge; see Sect. 7.3.7).

The main advantage of list mode (compared with DAQ systems that just produce histograms as their output) is that a much more sophisticated analysis can be performed on the measurement or experiment data. With list mode it is possible to produce energy histograms that span a certain time interval or produce time histograms that cover a given energy range. List mode provides a means of replaying the entire acquisition and displaying the same information in different ways. For example, it is possible to post-process the list mode file to identify coincident events of, say, events of a given energy within a certain time window; the same list mode file can be parsed multiple times to look at different energy events or different time windows. One can also search for patterns such as a neutron event followed by γ rays of a certain energy within a certain time period. In AI, the time development of certain radiation events is often of importance and appropriate algorithms can be run on list mode files to extract such information; this is discussed further in Sect. 7.5.4.

7.3.9 Multichannel DAQ Systems

Any large detector system will be made up of multiple detector elements each of which will require its own data acquisition channel. There are many reasons why digital DAQ systems are particularly suited to multi-channel implementations when compared with their analog counterparts. One of the biggest advantages is the physical size of the electronics modules needed. A simple analog DAQ chain (for example Peak Sensing ADC; Sect. 7.2.4) requires at least three different modules and a more complex analog DAQ chain (for example rise time discrimination PSD; Sect. 7.2.11) may require six modules or more. With just a few DAQ channels the total number of modules becomes excessive and for a system with tens or hundreds of channels, the implementation becomes impractical. Of course there are always analog modules that combine, say, multiple shaping amplifiers, which can help to reduce the module count but this only goes part-way to solving the issue. In comparison, it is already possible to get 8 or 16 channels of digital DAQ on a single module roughly the same dimensions as a single analog DAQ module. One can envisage that in future, with advancements in integrated circuit technology, it will be possible to get more digital channels per module.

Another advantage that digital DAQ has is in the flexibility of the hardware. The pulse processing functions of a real-time digital DAQ chain will be held in a programmable device such as a field-programmable gate array (FPGA). The pulse processing functions can be updated and reconfigured in a matter of seconds by downloading new functions to the FPGA. This introduces the notion of a multipurpose multichannel DAQ system, where the DAQ chain implementation for individual DAQ channels can be tailored to different detector hardware or experimental configurations.

Other advantages of digital DAQ for multichannel implementations include the relative ease in which cross-channel processing can be done (for example coincidence triggering) and being able to update DAQ parameters on a channel-by-channel basis or on a group of channel simultaneously.

7.3.9.1 Channel-to-Channel Synchronization

A single digitizer hardware module will have a number of identical acquisition channels (typically 4, 8, 16 or 32 channels on modern digitisers for radiation detection). The sampling clock and the clock signal required by internal logic and memory units will typically be generated by a clock tree, which is itself driven by

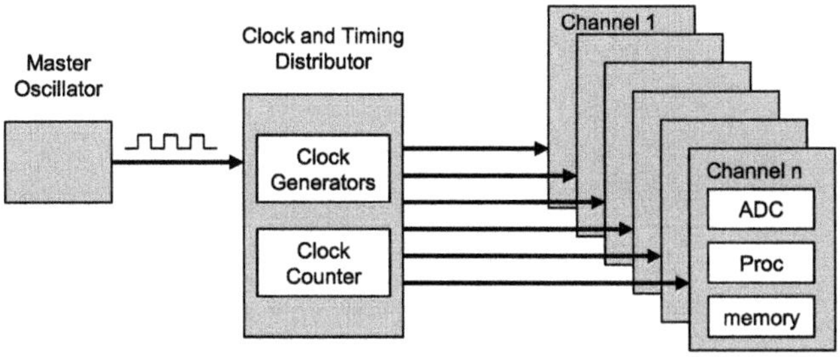

Fig. 7.27 Typical clock generation and distribution in a hardware digitizer module

a master oscillator. An example of this is shown in Fig. 7.27. The clock generators derive the various clock signals needed by the system (typically through cascaded multiplication and division stages) and importantly these different clock signals are phase and time aligned. Each channel also has access to the same system clock counter (which might be centrally located or reside as a separate synchronized unit on each channel). Each channel will have its own analog to digital converter (ADC), pulse processor (Proc) and working memory.

Such a clock generation scheme means that signal sampling and data processing can be handled synchronously. By the very nature of the design of the digitizer, individual channels will be synchronized to each other and so a time stamp on one channel can be compared relative to the timestamp on another channel. Channel-to-channel synchronization is particularly important where, say, signals from a multi-element detector are being summed or if channel-to-channel event timing is required (such as for coincidence or time-of-flight measurements).

An issue only arises when channel-to-channel synchronization is required between two channels on two separate digitizer modules. Each digitizer will have its own master oscillator so in general both the frequency and phase of the clock signals on the two digitizers will be different. Furthermore, the notion of "time zero" on one digitizer will be different for the two digitizers. In summary, the timestamp of a given channel on one digitizer has no relationship to the timestamp of another channel on a different digitizer. If more DAQ channels are required than are available on a digitizer module then some means of synchronizing digitizer modules is needed. Fortunately there are digitizer solutions available from certain manufacturers that offer synchronization functionality.

7.3.9.2 Synchronization of Two or More Digitizer Boards

Figure 7.28 details the clock distribution topology between two digitizer modules in a Master-Slave configuration. A clock input and clock output is provided on each digitizer module so that the clock signal can be distributed across multiple digitizers. A multiplexer (MUX) is used to select whether or not a digitizer uses its own internal oscillator or takes its clock from a master digitizer. The Clock Distributor on the

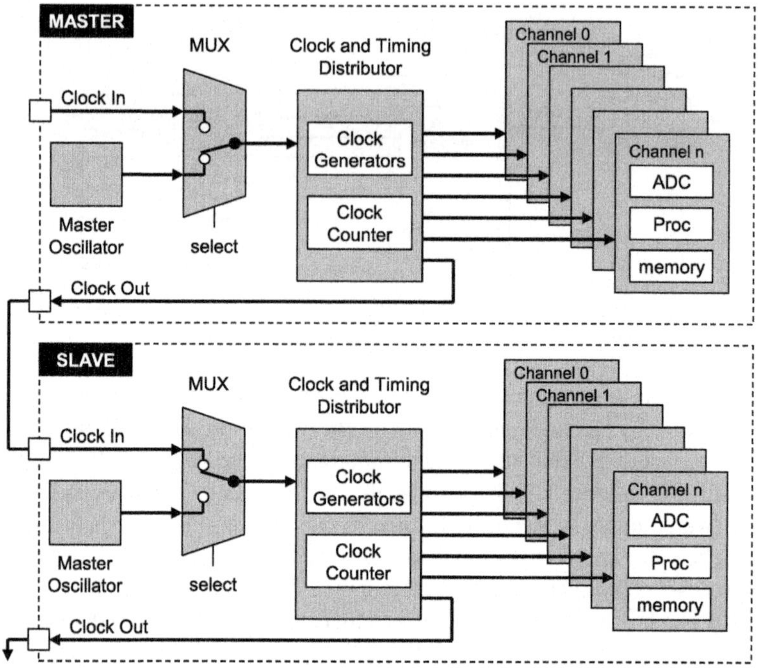

Fig. 7.28 Clock distribution topology between two digitizer modules in a Master-Slave configuration

Master board generates the sampling clocks for each of its DAQ channels, which are inherently synchronized. In addition, the Clock Distributor on the Master provides an output to feed the Slave board. The Slave uses this input clock rather than its own internal clock. Further slave digitizers can be added to the external clock chain.

The Master-Slave topology provides a means of synchronising the time-base of multiple digitizers but the local clock signal on each digitizer will not necessarily be phase aligned (the rising and falling edges of the clocks will not necessarily line up); in general, clocks on different digitizers will be out of phase. Depending on the timing accuracy required, the phase misalignment may or may not be an issue. If clock phase alignment is required then a phase adjustment mechanism is required; this can normally be achieved by applying a small delay to the output clock on the master digitizer. In practice, phase alignment can be a delicate operation and the exact procedure will be specific to the design of the digitizer.

It has been described how the clocks on two different digitizers can be frequency and phase aligned but there is the remaining issue of aligning "time zero" across all digitizers in a Master-Slave configuration. Alignment of "time zero" is necessary so that the time stamp registers on different digitizers are synchronized. It is usually insufficient to use a software signal to start the acquisition across multiple digitizers due to latencies in software and on the digitizer-host interface. The usual method for aligning "time zero" is though a cable connection at a hardware level so that all digitizers see the start signal at the same time.

Clock and timing synchronization across digitizers is very complex and great care must be taken in matching cable lengths and setting phase delays. At the high clock rates of 10s and 100s of MHz, signal propagation times along cables become comparable with clock periods and compensating for cable delays can be difficult. For these reasons, the external clocks that go from digitizer to digitizer are often made to run at a lower frequency than the sampling clock of the digitizer. Phase-locked loops are then used to multiply the external clock frequency to the higher sampling frequency required by the ADC and signal processing logic.

7.4 Data Processing and Storage

The way in which data is processed, stored and transferred through different stages is an important consideration in the design of a data acquisition system. This is particularly so for digital DAQ systems, where vast amounts of data can be moving through the DAQ chain.

7.4.1 Analog Systems

Figure 7.29 illustrates the flow of data in a generic analog data acquisition system. Analog processing (such as charge integration and pulse shaping; as described in Sect. 7.2.3) is performed on the current pulse from the radiation detector producing an analog value representing a quantity of interest. That quantity could be, for example, the energy of the radiation event or pulse rise-time for a pulse shape discrimination measurement. That analog value is converted to a digital value using

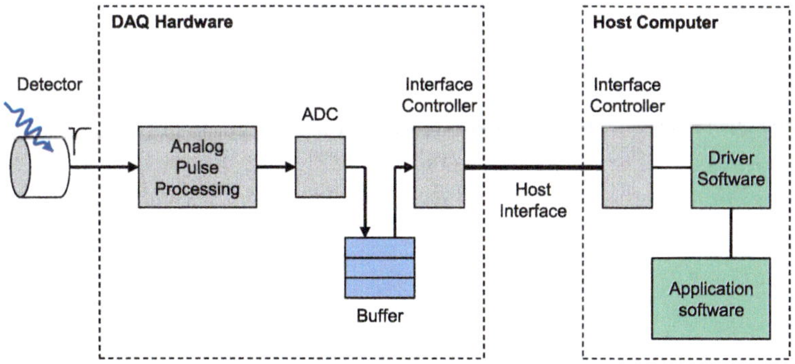

Fig. 7.29 Data flow in a generic analog data acquisition system

an analog-to-digital converter (ADC) and stored in a local buffer on the DAQ hardware. This buffer will have a certain depth, meaning that it can hold a certain number of events (ADC values) ready for transfer to the host computer. Event data from this local buffer is transferred to the host computer via the host interface (managed by the interface controllers on the DAQ hardware and host computer).

There is a notion here of data being produced by the DAQ hardware and being consumed by the host computer. The maximum rate (count rate) at which the DAQ hardware can produce data is limited by the dead time of the DAQ hardware. The dead time is made up of a settling time for the analog signal, ADC conversion time and time to clear the circuits ready for the next event. The DAQ hardware dead time is typically of the order of a few microseconds. The ultimate count rate that can be achieved is limited by the total dead time of the detection system with contributions from the detector itself (for example the decay constants of scintillator light pulses), the front end electronics (for example the time constant of the shaping amplifier) and the aforementioned digital conversion dead time. Depending on the value of these dead time contributions, the theoretical count rate could be many tens of kHz or even hundreds of kHz (however, in practice, the random nature of radioactivity and the resulting effects of pulse pile-up, will be a further limitation on the achievable event rate; also, other system design considerations such as maximizing energy resolution performance will be at the expense of count rate).

The maximum rate at which the digitized event data can be transferred to the host depends on the maximum data transfer rate of the host interface. The most favored interface for modern systems is USB (Universal Serial Bus) running in high speed mode, which has a theoretical data transfer rate of 480 Mb/s (megabits per second) or 60 MB/s (megabytes per second); the transfer rate that can be achieved in practice (useable data rate) is about half this value. If it is assumed that each event (count) produces a digital value of no more than 16 bits (2 bytes) then a USB interface operating at high speed could sustain an event rate of around 15 ME/s (million events per second), which is significantly higher than the rate that the DAQ hardware is able to produce. In practice the data transferred to the host may contain supplementary information or metadata but even taking this into account, the USB host interface is more than good enough to support a single channel of analog DAQ hardware. Indeed, many analog DAQ channels can be supported over a single USB connection.

As mentioned earlier, the DAQ hardware will have a data buffer, which should be sufficiently deep to allow for the fact that the host computer is invariably not a real-time system. The buffer is a "holding bay" for the digital event data for those periods of time where the host interface controller at the host end is either preparing for a data transfer or busy due to another host system operation.

7.4.2 Digital Systems

Figure 7.30 illustrates the flow of data in a generic digital data acquisition system. Unlike analog DAQ systems, the current pulse from the radiation detector (or the charge integrated pulse following a charge sensitive preamplifier) is converted to a digital waveform and the remainder of the DAQ processing is carried out in the digital domain. It is possible that those digital waveforms could be passed directly to the pulse processor stage but it is more usual that the digital pulse waveforms will be held in a local waveform buffer. This buffer will have a certain depth, meaning that it can hold a certain number of pulse waveforms ready for the pulse processor stage. This buffer will also have a certain width, which is equal to the number of samples in the pulse waveform. Each sample will have a number of bits (binary digits). This buffer arrangement is illustrated in Fig. 7.31, where waveform S_0 has samples S_{00} through S_{0n} (where n is the number of samples per waveform) and each sample is a 16-bit (2-byte) number. The buffer can hold a maximum of m waveforms.

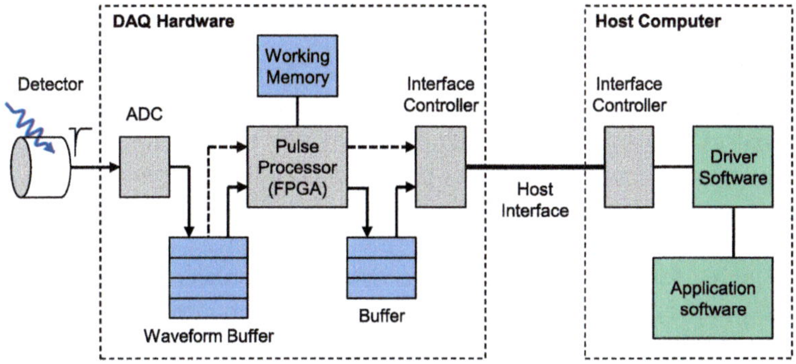

Fig. 7.30 Data flow in a generic digital data acquisition system

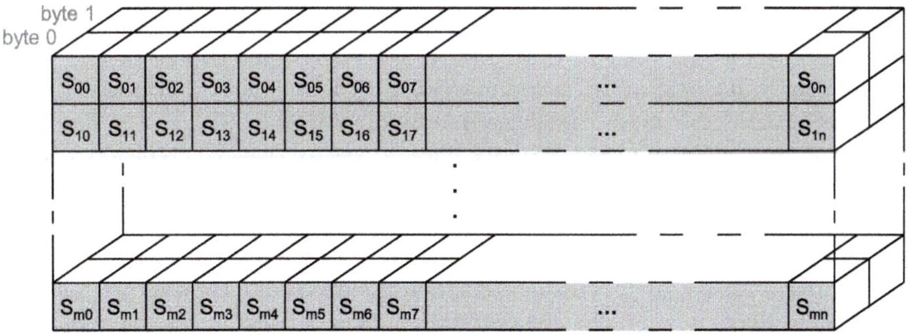

Fig. 7.31 Example of the arrangement of sample data in the local waveform buffer

The width of the buffer is equal to the pulse duration divided by the ADC sampling period. The number of bits in each sample depends on the bit resolution of the ADC. As an example, consider a detector current pulse of 1 μs being sampled at 500 MSa/s by a 14-bit ADC. The sampling period is 2 ns (reciprocal of the sampling rate) giving a buffer width of 500 samples, which will be taken as 512 samples since memory arrangements work better in powers of 2. The depth of each sample is traditionally stored in units of bytes (8 bits) so each 14-bit sample will occupy 2 bytes of memory. If we wanted to store, say, up to 256 waveforms in the local buffer then this would require a memory storage block of 256 kB.

The local waveform buffer is made sufficiently deep to allow the pulse processor stage time to perform whatever functions are necessary on the raw waveforms. If the pulse processor can operate in real-time or near real-time then the buffer can be quite shallow because the waveforms can be processed at the same rate (or similar rate) at which they arrive from the ADC. However, there is a mode of operation where the raw waveforms are passed directly to the host without any (or very little) processing on the DAQ hardware and in this case the buffer needs to be sufficiently deep to allow for the host interface data transfer speed and latency; and to allow for the fact that the host computer is invariably not a real-time system. It is worth exploring the raw waveform mode a little further to understand the data transfer rates involved. Expanding on the example given earlier of a 1 μs being sampled at 500 MSa/s by a 14-bit ADC, the theoretical pulse waveform storage rate is 1 million pulses per second (reciprocal of 1 μs). At 1 kB per waveform, the theoretical incoming data rate from the ADC is approximately 1 GB/s (gigabytes per second), which is approximately 30 times greater than the 30 MB/s or so that can be sustained over a USB interface running in high speed mode. Indeed, 1 GB/s would even be challenging for USB running in super speed mode (theoretical transfer rate of 640 MB/s; commonly referred to as USB 3.0). Furthermore, the DAQ system is likely to have more than one channel. So it can be seen that sustained throughput of waveform data is challenging at best. Fortunately in AI, advantage can be taken of the burst nature of radiation (e.g. pulsed interrogation from a linear accelerator and the resulting bursts of induced radiation) to avoid overflow of the waveform local buffer. The buffer needs to have sufficient depth to acquire the detector pulses resulting from a single interrogating pulse; the waveform data can be processed and/or sent to the host in the intervening time before the next interrogating pulse arrives. Pulsed interrogation is discussed further in Sect. 7.5.5.

Referring once again to Fig. 7.30, the pulse processor (which is typically a field-programmable gate array) operates on the data in the local waveform buffer to perform functions such as PHA, CI and PSD; as described in Sects. 7.3.5–7.3.7. The pulse processor will have some working memory to hold parameters and calculation results from the signal processing operations performed. The working memory would most likely be internal to the FPGA or might be a combination of internal and external memory. The result of the pulse processor will be a handful of values for each processed waveform; for example, the energy and time of arrival of the event or the PSD value. These values are stored in another local buffer, similar to that at the final stage of the analog data acquisition data flow diagram of Fig. 7.29. This buffer will have a certain depth, ready for transfer to the host computer. Event data from this buffer is transferred to the host computer via the host interface (managed by the interface controllers on the DAQ hardware and host computer). Similar to the description in Sect. 7.4.1, there is a notion here of data being produced by the DAQ hardware and being consumed by the host computer. The buffer is a "holding bay" for the post processing results data (timestamps, energy values etc.) and should be sufficiently deep to allow for the fact that the host computer is invariably not a real-time system.

Returning once again to the example given earlier, the pulse processing results for, say, a charge integration digital DAQ chain is a timestamp and a charge value. The timestamp might be a 32-bit (4 byte) number and the charge value a 16-bit number. Comparing these 6 bytes with the 1 kB needed to represent the whole pulse waveform, that's a ratio of more than two orders of magnitude. Looking at it a different way, the data transfer rate to the host for our theoretical event rate of 1 million pulses per second, reduces from 1 GB/s (for the raw pulse waveform) to 6 MB/s (for the timestamp and charge value). In practice the data transferred to the host may contain supplementary information or metadata, which will modify these data rate numbers to a certain extent but the basic argument still holds true. The relative merits of waveform output versus pulse processed values are discussed further in Sect. 7.4.4.

It can be seen that data flow management is a major consideration in the design or choice of a digital DAQ system. The size of the various buffers and data transfer rates of the busses and interfaces that run between them must be carefully specified or chosen to achieve a balanced system. Invariably there will be a limiting factor (or bottle-neck) that will affect the overall performance of the DAQ system; depending on the particular measurement or experimental configuration, it could perhaps be a buffer size, a bus speed or the processing speed of the pulse processor that is this limiting factor. The data processing chain is only as strong as its weakest link and the result of any link not performing to the required level is loss of data (usually the overflow of a data buffer).

Figure 7.30 is a simplified representation of a digital data processing chain but in practice there are various techniques that can be employed to alleviate potential bottlenecks in the chain. For example, depending on the pulse processor architecture, it may be more efficient to process multiple samples or waveforms in parallel or use pipe-lining techniques to make best use of processor clock cycles. There may be DAQ hardware user configurable parameters that can be set to make the most efficient use of available memory or control data transfer modes (e.g. data block size and frequency of transfer). Individual DAQ systems vary greatly in their implementation and so a detailed description is not possible here, but one should be aware that the default settings of such configuration parameters may not be optimal for a particular measurement or experiment and should be adjusted accordingly.

7.4.3 Data Formats and Data Storage

Choices can be made on the format of digital data and the manner in which that data is stored and manipulated. There are advantages and disadvantages of different data formats. It was described in Sect. 7.4.2 that pulse waveform samples are a digital value, typically ranging from 0 to 65,536 (a 16-bit value). When stored in DAQ hardware memory (for example in the local waveform buffer) this 16-bit value will occupy 2 bytes. But if that sample value is transferred to the host computer then most likely we will want that value to be stored in a file on the computer data storage drive. The two most common formats for storing that data is ASCII or binary. ASCII stands for American Standard Code for Information Interchange and is the most common format for storing text files on a computer. Each sample value is converted to a decimal number and stored in a text file typically with a space character or carriage return after each sample; that file is human-readable and can be opened with any standard text reader or editor. In binary format each sample is stored as a native binary number and stored in a file as a contiguous sequence of binary values. The binary file is not human readable and one needs to know the specific format for that file in order to read or process it.

The advantage of the ASCII format is that it can be read like a normal text file and the sample values just read as a sequence of numbers. Of course, one needs to know such information as how many samples there are in each waveform but this information if often included as ASCII metadata at the start of the file. The disadvantage is that ASCII characters on average take up more file space compared with a binary file. The ASCII file is also of variable length whereas the binary file is of fixed size for a given number of samples. As a simple example, if the sample values were to be stored as 16-bit values then in binary format each possible sample value from 0 to 65,536 will occupy just two bytes each; the equivalent ASCII file would have sample values of either one, two, three, four or five characters depending on the sample value. The size of an ASCII file can vary depending on the data being stored but in practice ASCII files tend to be two or three times larger than a binary file containing the same information. If waveform data files are several megabytes or gigabytes in size then there are clear advantages for using a binary file format. But, as mentioned earlier, the disadvantage of using binary files is that one must keep a separate record of the file format because it will not ordinarily be possible to read or process the file without knowledge of the file format. The same principles of ASCII versus binary apply to the storage of other data such as charge values or timestamps. The binary format can be taken several steps further to minimize file sizes and this is described in Sect. 7.4.5.

7.4.4 Real-Time Versus Offline Processing

Analog systems of the type described in Sect. 7.2 are, by their nature, real-time systems. Real-time means that detector pulses are being captured, processed and stored on the host computer at the same rate that the pulses are occurring. It was explained in Sect. 7.4.1 that there is at least one local buffer in the hardware of an analog DAQ system to help with the flow of data to a non-real-time host computer. But the DAQ hardware itself runs in real time and on average, the whole DAQ chain from the arrival of pulses from the detector to writing processed data (e.g. energy value, event time or pulse rise time) to file can be taken as running in real-time. Digital DAQ systems based on digitizer hardware (as described in Sect. 7.3.3) produce vast amounts of data (in particular the sampled waveform data) and real-time operation is more of a challenge. Where it is neither possible nor necessary to process the data in real-time, one can transfer the raw waveforms to the host computer and the pulse processing algorithms can be run there. Offline processing gives the option of running more sophisticated algorithms (perhaps taking much longer to run but producing more precise results) in non-real-time and it is also possible to "replay" the same data using different algorithms.

Historically (less than a decade ago), the normal operational mode for a digital DAQ system would be to collect sampled waveform data using digitizer hardware and passing this data directly to the host computer to perform the pulse processing, which might be PHA, CI or PSD. It was explained in Sect. 7.4.3 how there are challenges associated with the interface bandwidth required to transfer raw waveform data to the host. As DAQ hardware waveform sampling rates and sample resolution continue to increase (as digitizer hardware technology advances), greater demands are placed on the host interface bandwidth. So the host interface is one limitation in the ability to process data in real-time when running in raw waveform (offline) mode. Another limitation is the speed at which the host computer can run the required pulse-processing algorithms, meaning that the host computer must be well specified and configured to maintain sensible processing times.

The way to avoid these limitations associated with the host interface bandwidth and host computer processing power is to perform the pulse processing on the DAQ hardware itself; this is described in Sect. 7.4.2. It is only in the last 5 years or so that DAQ hardware pulse processing has become widely available in everyday off-the-shelf digital DAQ systems. In particular, digitizer hardware technology has evolved to include programmable hardware integrated circuits (typically field-programmable gate arrays) that are specialized in digital pulse processing and can run in real-time. With an FPGA-based digitizer, the pulse processing is off-loaded from the host computer to the DAQ hardware with only basic control data (very low bandwidth) flowing between the host computer and DAQ hardware. There are however challenges in developing efficient pulse processing algorithms that are capable of running in real-time. There are also challenges in creating firmware (software programmed into read-only memory) code that can fit in the limited space and use the limited resources available on a typical cost-effective FPGA. But, as always, there are advantages and disadvantages of operating a digital DAQ system in real-time versus offline mode. The advantages have already been discussed in Sect. 7.4.2, but the main disadvantages are a loss of data fidelity and an accompanying loss of flexibility in how data is processed, stored and (potentially) post-processed.

To explain the relative merits in terms of data storage and processing for real-time versus offline operation, we have to look at how and where data is stored through a DAQ processing chain. We start by looking at how data might be stored with an analog DAQ system compared to a digital one. With an analog system the data values transferred to the host (e.g. energy, event time or pulse rise time) are typically built into a histogram and then that data value is thrown away. The histogram file that is eventually written to file will be of a (preconfigured) fixed number of histogram channels (or "bins"). With a 14-bit analog to digital converter this would be 16,384 bins (2^{14}) and in an ASCII file format the histogram would just be a list of 16,384 numbers representing the number of counts (radiation events) in each bin. This histogram file would perhaps be just a few tens of kilobytes in size, which by modern standards is a tiny file. Moving on to digital DAQ systems, the detector data could be stored as raw waveform samples (which takes up a significant amount of file space; see Sect. 7.4.2) or as list mode data (which takes up perhaps two or more orders of magnitude less file space than raw waveforms); or the same data can be reduced to a histogram file that is many more orders of magnitude smaller again. But the reduction in file space in moving from waveforms to list-mode to histograms is accompanied by a commensurate loss of information; if one is only interested in, say, a time spectrum or an energy spectrum then the histogram format would be the best solution for long-term storage of measurement data. But if there is a need to post-process the data at a later date (for example to generate an energy spectrum for just a particular time interval) then one would want to keep data in list mode format but at the expense of using more file space. In the case of histogram mode or list-mode, operation in real-time, with the pulse processing performed in hardware would be the preferred solution because it is not a problem that the waveform data gets thrown away.

However, if there is a need to post process the data to, say, apply an alternative pulse shape discrimination algorithm then one must transfer the raw pulse waveforms to the host computer to store to file (which invariably means that real-time operation is challenging or not possible at all).

7.4.5 Data Reduction

Section 7.4.2 highlighted the challenges that can be encountered storing and transferring the large amounts of data often associated with digital DAQ systems, particularly those systems with high sampling rate and high sample resolution.

It can be advantageous to the data throughput and data management of such systems to reduce the amount of data to be processed, transferred or stored. Data reduction can be achieved in many ways ranging from more efficient methods of arranging a given set of data through to eliminating superfluous data altogether.

There are several techniques that can be used to achieve data reduction in digital DAQ systems, the most noteworthy are listed below but many other techniques exist:

- Smart triggering
- Data cuts
- Data packing
- Data compression
- Partial waveform capture
- Waveform parameters

Smart triggering is a general term applied to triggering schemes that are more sophisticated than the standard trigger mechanism where each pulse that meets the trigger criteria (either leading edge or constant fraction discrimination; see Sect. 7.3.4) is duly processed by the rest of the data acquisition chain. One example of smart triggering is hardware coincidence logic where a pulse is only processed if it occurs within a given time window of another pulse (either on the same DAQ channel or a different DAQ channel). For analog DAQ systems, additional hardware modules are required to implement this coincidence logic, which ensures that data is only recorded for those pulses that meet the coincidence criteria and thereby reduces the amount of data being passed to the host computer. In a digital DAQ system the coincidence logic can be implemented in re-configurable logic (for example, a field-programmable gate array) at an early stage of the digital DAQ chain; pulses that do not meet the coincidence criteria are discarded as early as possible to reduce both the load on the hardware pulse processing and the transfer of data to the host computer. Hardware coincidence logic can be particularly effective in a detector system where the radiation rate is very high but the wanted events are just a small fraction of those radiation events that initially trigger the DAQ system.

Other examples of smart triggering are where a trigger is produced only if the input signal remains above a fixed threshold for a certain period of time, or only trigger if a certain amount of time has elapsed since the previous trigger. Another example is that the DAQ system be configured to only trigger if it detects a sequence of events that fit a certain pattern, which could be a pattern based on, say, the energy of the event or the particle type (for example a neutron followed by a gamma ray above a certain energy). These types of smart triggering schemes are generally reserved for digital DAQ systems, which have great flexibility in implementing the logic and memory allocations required. But these sophisticated schemes require careful implementation in the DAQ hardware if they are to be effective in reducing the data at the appropriate stages in the DAQ chain and thereby easing the processing load on the DAQ system overall.

Data cuts generally operate much further along the DAQ chain than smart triggering and the main aim with data cuts is to reduce the load on the host interface. As described in Sect. 7.4.2, host interface bandwidth limitations are more of an issue with digital DAQ rather than analog DAQ. Referring to Fig. 7.30, with data cuts, pulse-processed data (e.g. charge value, timestamp or PSD value) is only transferred to the host if it meets certain pre-specified criteria. For example, the system can be configured to only transfer events that are classified as neutrons based on the PSD value. In a radiation environment that is heavily polluted with (potentially unwanted) gamma rays, this can bring about a significant reduction in data transfers. As another example, the system can be configured to only transfer events that are above a certain charge threshold; this can be particularly beneficial in a pulse shape discrimination system where a standard voltage threshold (as used for leading edge and constant fraction triggering schemes) is not suitable for masking out unwanted low energy events.

More effective than cuts to the pulse-processed data are cuts to the waveform data associated with that pulse-processed data. That is, a waveform is not transferred to the host unless the pulse-processed data resulting from that waveform meets certain pre-specified criteria. Since waveform data can be many orders of magnitude greater that pulse-processed data (as described in Sects. 7.4.2 and 7.4.4), the load on the host interface can be reduced significantly.

Data packing is a technique that can be employed to make best use of the space available for data storage. This space could be memory blocks on the DAQ hardware, random access memory (RAM) on the host computer or file storage space. Commercially available storage memory is almost exclusively configured as storage elements that are multiples of a byte (8 bits); it was mentioned in Sect. 7.4.2 that a 14-bit value from an analog-to-digital (ADC) conversion would most likely be stored in 2 bytes, which means that 2 bits (12.5% of the available memory) are effectively wasted for each 14-bit value that is stored. A way to avoid this wastage is to pack data to fill all of the available bits; this is depicted in Fig. 7.32, where eight 14-bit samples (S0 through S7) are packed to fit the space that would normally hold seven 14-bit samples.

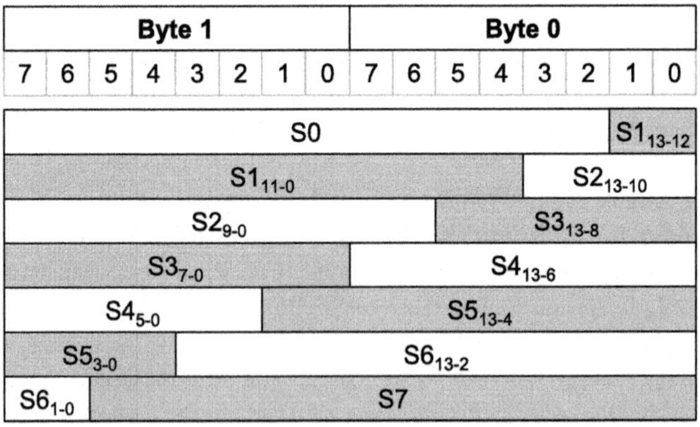

Fig. 7.32 An example of data packing for 14-bit data values. Eight 14-bit samples (S0 through S7) are packed into the space of seven (2-byte) words. The subscript notation indicates the range of bits for a particular sample

Data packing is most effective for data values that just exceed a byte boundary. For example, in a DAQ system with a 10-bit ADC, 6-bits (37.5% of the available memory) are wasted for every stored event, which is a significant amount of storage space; but that wasted space can be clawed back utilizing data packing. Data packing not only makes more efficient use of data storage space but also reduces the data transfer bandwidth requirements by the same factor as the data space savings. Data packing can yield big savings in data storage and data transfer bandwidth requirements but the disadvantage is that the data value boundaries have to be carefully managed and there can be circumstances where the management overhead can out-weigh the savings.

Data compression is a general term used to describe techniques that can be applied to data to reduce the amount of space it occupies in storage memory and reduce data transfer bandwidth required. Examples of data compression that might be suitable for DAQ systems include run-length encoding (RLE), Huffman coding and zero suppression, which are all lossless compression techniques (meaning that there is no loss of information in the compression process) [16]. Data compression offers savings in data storage and data transfer bandwidth requirements but the disadvantage is that compression and decompression algorithms take up processing time themselves and there can be circumstances where the compression overhead out-weighs the savings.

Partial waveform capture is a compromise to full waveform mode in digital DAQ systems (described in Sects. 7.4.2 and 7.4.4). Instead of transferring all samples of the complete pulse waveform to the host computer, only the relevant parts of the waveform are transferred. For example, if the rising part of the detector pulse is the only part of interest then the host interface bandwidth requirements can be alleviated somewhat by not having to transfer all of the samples that make up the complete pulse. As the rising part of the waveform is generally much shorter than the falling part (the pulse decay), the data transfer bandwidth requirement can be reduced by a substantial factor. In a more sophisticated DAQ implementation, it is also possible to capture more than one section of the detector pulse waveform (for example, to capture the baseline prior to the pulse rise, or the region around the pulse peak, or maybe sections in the pulse tail for the identification of pulse pile-up). Another partial waveform capture technique is to decimate (only retain every nth sample, where n is the decimation factor) some parts of the waveform (typically those parts that are slow moving and so do not require a high sampling rate). In summary, with partial waveform capture it is possible to capture enough of the pulse waveform to allow some offline post-processing (see Sect. 7.4.4) but without suffering the problem of excessive data transfer to the host computer.

Lastly, the detector pulse can be represented by a number of values that describe the pulse waveform. In comparison with partial waveform capture, a value is calculated to represent a section of the pulse waveform rather than capturing all of the samples that make up that section. A range of waveform samples is replaced with a single value giving a reduction in data. As an example, consider the pulse waveform shown in Fig. 7.33.

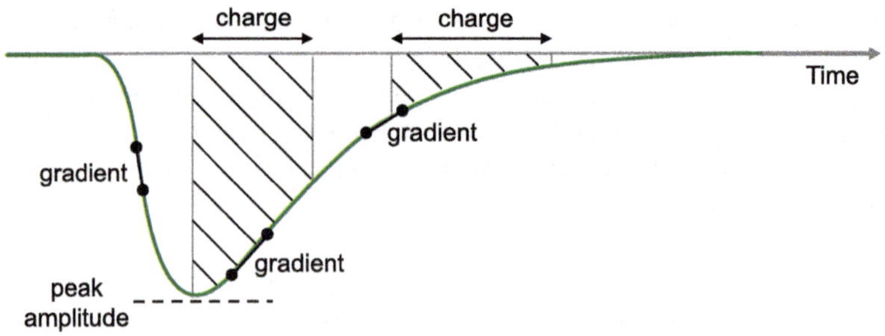

Fig. 7.33 An example of pulse waveform representation as a series of values that capture features such as pulse gradient, charge integration and peak amplitude

This example shows multiple values based on the pulse gradient, charge integrated periods and also the peak amplitude (but there are many other features that could be represented by a value). The idea here is that the calculations for these values are performed in real-time on the DAQ hardware resulting in a reduced set of data to transfer to the host (compared with transferring all of the samples of the pulse waveform). Calculating values in this way is just an extension to the time stamp list mode (TSLM) representation described in Sect. 7.3.8; the difference here is that the list of values are more extensive and can be customized to suit the application. With the use of multiple values to represent certain features of a pulse waveform, it is possible to capture enough of the waveform features to allow some reasonably detailed offline post-processing (see Sect. 7.4.4) whilst at the same time maintaining a very low data transfer rate to the host computer.

Some of the data reduction techniques described in this section (namely, basic smart triggering, data cuts, basic partial waveform capture, some compression algorithms) are readily catered for on some commercial digital DAQ systems. But other techniques require modification to the hardware pulse processor firmware (typically code running on a field-programmable gate array) of commercial DAQ systems or would have to be realized through a custom digital DAQ implementation.

7.5 AI Data Acquisition Challenges

This section covers some aspects of DAQ system design and configuration that are of particular importance and relevance to AI.

7.5.1 Event Rate

The single most challenging issue with DAQ systems for AI is the event rate both from the interrogating source and the radiation induced in the material under inspection. The radiation particle rates that might be encountered in an AI system (such as the nuclear car wash [17, 18]) very much depends on the particular configuration and position of the source and detectors. But it is not unusual to see both neutron and photon rates of the order of 10^3–10^5 $s^{-1}cm^{-2}$ at several meters stand-off, particularly in the immediate microseconds and milliseconds following an interrogating pulse. Such high count rates induce unwanted effects in any DAQ system, analog or digital.

The obvious effect of a high count rate is pulse pile-up, which is discussed in Sect. 7.5.2. Another less obvious effect is in how the baseline estimation calculations for the incoming detector pulse can be upset; this is covered in Sect. 7.5.2.1.

Another obvious adverse effect of a high event rate is in the high resulting data rate that must be processed, transferred and stored throughout the DAQ chain (see Sect. 7.4).

7.5.2 *Pulse Pile-Up*

The interfering effect between successive pulses from a detector is commonly referred to as pulse pile-up. Since radiation events are random, pile-up can in theory occur even at low count rates, but in practice pile-up only becomes significant at higher counting rates.

7.5.2.1 Pile-Up in Analog DAQ Systems

In analog DAQ systems, pile-up phenomena are of two general types: The first type is known as tail pile-up and is a result of (charge integrated and shaped) detector pulses being superimposed on a voltage or current offset from a preceding pulse. This superposition results in a change in the pulse height, which manifests as a shift in its measured energy. Figure 7.34 illustrates the case of pulse pile-up due to

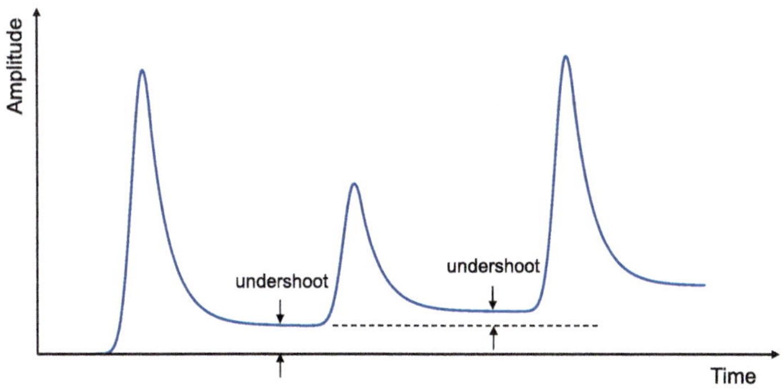

Fig. 7.34 Tail pile-up resulting from undershoot of a preceding pulse in an analog DAQ system

undershoot of a preceding pulse. In the figure, the second pulse is superimposed on the tail of the first pulse and so the peak amplitude (as measured relative to zero) is higher than it should be, giving a false energy value. The pile-up effect can be compounded for successive pulses.

One remedy for tail pile-up is pole-zero compensation. As described in Sect. 7.2.3, the output signal from the charge sensitive preamplifier is shaped with a multistage CR-RC^n network into a near-Gaussian pulse waveform. The assumption in the design of the shaping network is that the input signal is a step voltage, but in reality the preamplifier has an appreciable decay (typically a 50 μs decay constant) that has a measurable effect on the response of the shaping network [1]. The result is that the tail of the shaped pulse will show an undershoot (zero crossover) before returning to the baseline. The way to remedy this undershoot is to introduce a pole-zero compensation resistor across the capacitor of the first stage (differentiator) of the shaping network (see Fig. 7.4), which modifies the transfer function accordingly and compensates for the undershoot. Overcompensation however, will result in overshoot (the shaped pulse not returning to the baseline quickly enough; as depicted in Fig. 7.34). Both overshoot and undershoot can lead to the undesirable effect of tail pile-up.

The second type of pulse pile-up is known as peak pile-up and occurs when two pulses are sufficiently close together to be counted as a single pulse. The apparent pulse amplitude will be equivalent to the sum of the two individual amplitudes giving an energy value that is the sum of the energy of the two individual events. In an analog DAQ system there is little that can be done to detect and/or remedy peak pile-up, but one can limit the effects of peak pile-up by suitable design and configuration of the setup to keep pulse durations short and event rates low at the DAQ input.

7.5.2.2 Pile-Up in Digital DAQ Systems

In a digital DAQ system, the current pulse (or charge integrated pulse) from the detector is digitized and the pulse processing is carried out on that digitized waveform. Following a trigger, an acquisition window is opened and the analog waveform is sampled for a predetermined number of samples (known as the record length). Pile-up is said to have occurred if another pulse arrives within the acquisition of the preceding pulse. Some examples are illustrated in Fig. 7.35.

For both PHA and CI digital DAQ topologies, the presence of an unwanted pulse within the acquisition window can result in an erroneous calculation of the charge value or energy value. For the PSD digital DAQ topology, the erroneous calculation of a charge value directly affects the calculated PSD value and results in particle discrimination performance degradation.

The solution for pulse-pile up in digital DAQ systems is to employ algorithms that operate on the contents of the acquisition window prior to the pulse data being passed to the pulse processing stage. These algorithms typically fall into two categories: Pile-up rejection and pile-up correction. A pile-up rejection algorithm

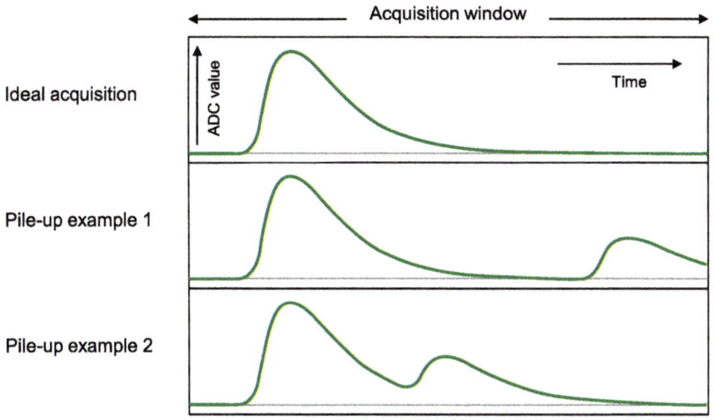

Fig. 7.35 Pulse pile-up in the acquisition window of a digital DAQ system

will search for the presence of additional pulses within the acquisition window and rejects the entire record (data packet) if it fails to meet the rejection criteria (the rejection criteria is set according to the level of DAQ performance degradation due to pile-up that can be tolerated for a particular measurement). A pile-up correction algorithm is more sophisticated and will attempt to not only identify the presence of additional pulses within the acquisition window but also attempt to either remove the errant pulse(s) or separate the superimposed pulses into individual pulses that can be passed to the pulse processor in turn. There are many well established techniques and algorithms for combatting pulse pile-up, with new algorithms being developed all the time [19–21].

7.5.3 Baseline Estimation

In any DAQ system, the measurements on the detector current pulse (or the charge-integrated pulse) are made relative to a baseline (that is taken as the zero reference for any amplitude measurements). In an analog DAQ system, this baseline is generally fixed with perhaps some rudimentary baseline compensation control. In a digital system, the baseline can be fixed, but more usually the DAQ attempts to make baseline estimation measurements when there are no incoming pulses (i.e. the baseline is continually estimated and adjusted between the pulses). In a digital system the baseline will be estimated by taking the average value of a set number of samples during a period where there are no pulses. The longer the time duration (or the more samples used in the calculation) the more accurate the baseline estimation will be.

At a low event rate there is plenty of opportunity for the DAQ system to make estimations of the baseline and there is the luxury of being able to configure the baseline estimation period to be relatively long. But as the event rate increases, the time period between pulses decreases and there is less time available to make baseline estimation measurements. The net effect of an increasing event rate is that the baseline will be estimated less often and the most recent baseline estimation may not necessarily be appropriate for the pulse that is about to be processed; and so the amplitude measurements made on that pulse are relative to a false baseline.

The false baseline problem can be minimized to some extent by reducing the baseline estimation period (albeit at the expense of a less accurate baseline estimation measurement, which may or may not be an issue). But as the event rate increases even further, there will come a point where baseline estimation measurements become too infrequent or cannot be made at all because the event rate is too high. Some digital DAQ systems have various baseline estimation modes and settings that can be optimized to suit the application and this can help to improve high event rate performance and accuracy but there will come a point where normal DAQ functionality breaks down.

7.5.4 Time Development of Active Signatures

AI involves stimulating some material of interest with a radiation source and then making measurements on the induced radiation. The induced radiation will have certain characteristics (commonly referred to as a signature) that help to identify the type and composition of that material. These signatures are typically distinct by radiation type (neutron, gamma ray etc.), energy and time development (for example, the radiation decay profile of induced fission). So, in AI, it is desirable to have a DAQ output that shows an energy spectrum as a function of time, which in practice would be a two dimensional histogram of energy and time. In the analog domain this amounts to combining the functionality of a MCA with that of a MCS to produce a two-dimensional spectrum of energy versus time; see Sects. 7.2.7 and 7.2.8. With AI, the MCS dwell time is likely to be quite short (in the order of microseconds or milliseconds). These short timescales mean that in practice, this 2D histogram functionality is difficult to achieve using standard commercially available analog MCA and MCS units and a custom DAQ solution will most likely be needed. In the digital domain however, this 2D histogram functionality can be achieved relatively easily by creating a sequence of energy histograms from the list mode data (see Sect. 7.3.8) at set time intervals.

Figure 7.36 is an illustration of a 2D histogram of data acquired from a ^{3}He proportional counter. In this case, the histogram is showing the decay profile of the energy spectrum of the radiation being measured. Calculations can be made on the decay constant(s) to give information on the material under inspection in an AI detection system. The example of Fig. 7.36 is composed of 100 histogram bins each for energy and time but these bin sizes would be adjusted to fit the required energy resolution and dwell time.

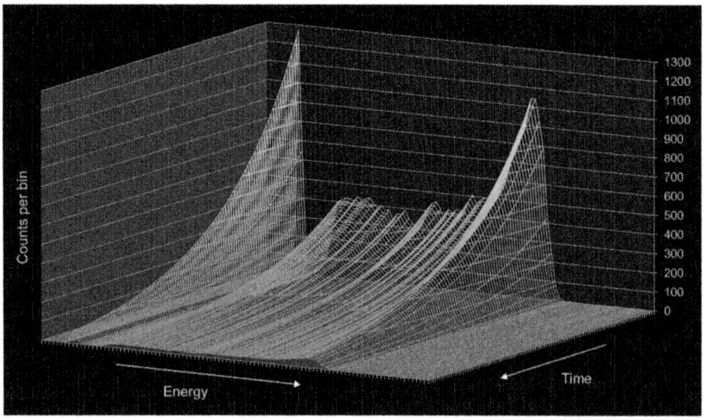

Fig. 7.36 Two dimensional histogram showing the energy spectrum as a function of time

Depending on the particular AI system, a measurement of the type shown in Fig. 7.36 would most likely be made following a pulse of interrogating radiation. This might be a one-shot pulse from, say, a flash x-ray machine or a sequence of pulses from a linear accelerator or neutron generator. In the latter cases the DAQ measurements would be synchronized with the interrogating pulse stream.

7.5.5 Pulsed Interrogation

In an AI system (such as the Nuclear Carwash [17, 18]), the interrogation source is typically a high intensity beam of radiation that is pulsed for a given time duration at a given repetition rate. AI sources of this type include linear accelerators, continuous wave x-ray generators and neutron DD or DT generators. For clarity, the pulse of radiation from these accelerators and generators is hereafter referred to as a radiation "burst" to be distinct from detector pulses in a DAQ chain.

The radiation bursts from these accelerators and generators will be short in duration (typically tens or hundreds of microseconds) but very high intensity (for example, 10^7–10^8 neutrons/s into 4π for a DT generator); the intensity is sufficiently high to overwhelm the DAQ system. So it is usual to inhibit the DAQ system (and often the detector systems too) during the interrogating radiation burst and just enable the DAQ during the periods between the bursts. The inhibit control to the DAQ is derived from a logic level synchronization (sync) signal provided as an output on the accelerator or generator. In an analog DAQ system, the sync signal can be used as a gate to inhibit or enable the acquisition of data. In a digital system, the sync signal can also be recorded a separate channel and can be used (either in real-time or offline) to identify the time regions of interest relative to the interrogating pulse.

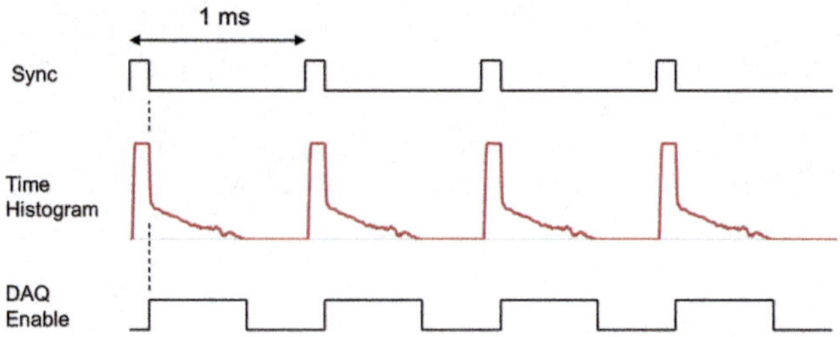

Fig. 7.37 Synchronization of data acquisition to the pulsed interrogating source

An example of how the DAQ might be inhibited and enabled in a pulsed AI system is shown in Fig. 7.37, which is for a particular AI technique known as Differential Die-Away Analysis (DDAA) [22, 23].

In this example, the source emits high intensity radiation in 100 s bursts at a burst repetition rate of 1 kHz (period of 1 ms). Referring to Fig. 7.37, the sync pulse from the source (in this case a DT neutron generator) is at a logic level "high" when the radiation is being emitted and logic level "low" when the radiation beam is off. Figure 7.37 shows the die-away signal (the AI signature of interest) of the induced radiation following every pulse of source radiation. In this example, the die-away signal is represented as a time histogram of neutron events (in this case neutron events detected by a hydro-carbon based liquid scintillator). Simple logic and a timer can be used to create the DAQ enable signal so that the DAQ is only acquiring data during the radiation periods of interest and, importantly, is not overloaded by the intense source radiation.

It is also interesting to note that the pulsed nature of the interrogating source (and the resulting induce radiation signatures) can help with data flow management throughout the DAQ chain. Data buffers on the DAQ hardware can be used to manage the flow of data to the host computer (see Figs. 7.29 and 7.30). With knowledge of the interrogating pulse structure, it is possible to configure the size of those data buffers to be deep enough to hold all of the data resulting from a single interrogating pulse. There is then a short period of time (depending on the burst repetition rate) before the next burst, which can be used to transfer the data to the host computer, thereby avoiding excessive data bandwidth demands on the host interface.

References

1. G. Knoll, *Radiation Detection and Measurement* (Wiley, New York, 2010). https://books.google.com/books?id=4vTJ7UDel5IC
2. K. Krane, *Introductory Nuclear Physics* (Wiley, New York, 1987). https://books.google.com/books?id=ConwAAAAMAAJ

3. R. Evans, *The Atomic Nucleus* (Krieger Publishing Company, Malabar, 2003). https://books.google.com/books?id=mWTFPQAACAAJ
4. D. Gedcke, W. McDonald, Nucl. Inst. Methods **55**, 377 (1967). http://dx.doi.org/10.1016/0029-554X(67)90145-0. http://www.sciencedirect.com/science/article/pii/0029554X67901450
5. J. Gál, G. Bibok, Nucl. Inst. Methods **163**(2), 535 (1979). http://dx.doi.org/10.1016/0029-554X(79)90145-9. http://www.sciencedirect.com/science/article/pii/0029554X79901459
6. W. Leo, *Techniques for Nuclear and Particle Physics Experiments: A How-to Approach* (Springer, Berlin, 2012). https://books.google.com/books?id=yc4qBAAAQBAJ
7. L.M. Bollinger, G.E. Thomas, Rev. Sci. Instrum. **32**(9), 1044 (1961). http://dx.doi.org/10.1063/1.1717610
8. S. Wu, Q. Yue, W. Lai, H. Li, J. Li, S. Lin, Y. Liu, V. Singh, M. Wang, H. Wong, B. Xin, Z. Zhou, Nucl. Instrum. Methods Phys. Res. Sect. A Accel. Spectrom. Detect. Assoc. Equip. **523**(1), 116 (2004). http://dx.doi.org/10.1016/j.nima.2003.12.024. http://www.sciencedirect.com/science/article/pii/S0168900204000336
9. C. Cialella, J. Devanney, Nucl. Inst. Methods **60**(3), 269 (1968). http://dx.doi.org/10.1016/0029-554X(68)90131-6. http://www.sciencedirect.com/science/article/pii/0029554X68901316
10. G. McBeth, J. Lutkin, R. Winyard, Nucl. Inst. Methods **93**(1), 99 (1971). http://dx.doi.org/10.1016/0029-554X(71)90144-3. http://www.sciencedirect.com/science/article/pii/0029554X71901443
11. T. Miller, Nucl. Inst. Methods **63**(1), 121 (1968). http://dx.doi.org/10.1016/0029-554X(68)90314-5. http://www.sciencedirect.com/science/article/pii/0029554X68903145
12. D.W. Glasgow, D.E. Velkley, J.D. Brandenberger, M.T. McEllistrem, Nucl. Inst. Methods **114**(3), 535 (1974). http://dx.doi.org/10.1016/0029-554X(74)90179-7. http://www.sciencedirect.com/science/article/pii/0029554X74901797
13. T. Alexander, F. Goulding, Nucl. Inst. Methods **13**, 244 (1961). http://dx.doi.org/10.1016/0029-554X(61)90198-7. http://www.sciencedirect.com/science/article/pii/0029554X61901987
14. E. Ifeachor, B. Jervis, *Digital Signal Processing: A Practical Approach*. Electronic Engineering Series (Addison-Wesley, Wokingham, 1999). https://books.google.com/books?id=ZxPwygAACAAJ
15. P. Horowitz, W. Hill, *The Art of Electronics* (Cambridge University Press, Cambridge, 2015). https://books.google.com/books?id=LAiWPwAACAAJ
16. S. Smith, *The Scientist and Engineer's Guide to Digital Signal Processing* (California Technical Publishing, San Diego, 1997). https://books.google.com/books?id=rp2VQgAACAAJ
17. J. Hall, S. Asztalos, P. Biltoft, J. Church, M.A. Descalle, T. Luu, D. Manatt, G. Mauger, E. Norman, D. Petersen, J. Pruet, S. Prussin, D. Slaughter, Nucl. Instrum. Methods Phys. Res. Sect. B Beam Interact. Mater. Atoms **261**(1), 337 (2007). http://dx.doi.org/10.1016/j.nimb.2007.04.263. http://www.sciencedirect.com/science/article/pii/S0168583X07007355. The Application of Accelerators in Research and Industry
18. D. Slaughter, M. Accatino, A. Bernstein, P. Biltoft, J. Church, M. Descalle, J. Hall, D. Manatt, G. Mauger, T. Moore, E. Norman, D. Petersen, J. Pruet, S. Prussin, Nucl. Instrum. Methods Phys. Res. Sect. A Accel. Spectrom. Detect. Assoc. Equip. **579**(1), 349 (2007). http://dx.doi.org/10.1016/j.nima.2007.04.058. http://www.sciencedirect.com/science/article/pii/S0168900207006456. Proceedings of the 11th Symposium on Radiation Measurements and Applications
19. M. Nakhostin, Z. Podolyak, P.H. Regan, P.M. Walker, Rev. Sci. Instrum. **81**(10), 103507 (2010). http://dx.doi.org/10.1063/1.3499241
20. J. Liu, H. Li, Y. Wang, S. Kim, Y. Zhang, S. Liu, H. Baghaei, R. Ramirez, W.H. Wong, in *2007 IEEE Nuclear Science Symposium Conference Record*, vol. 6 (2007), pp. 4230–4232. https://doi.org/10.1109/NSSMIC.2007.4437051

21. M.D. Haselman, J. Pasko, S. Hauck, T.K. Lewellen, R.S. Miyaoka, IEEE Trans. Nucl. Sci. **59**(5), 1823 (2012). https://doi.org/10.1109/TNS.2012.2207403
22. K.A. Jordan, T. Gozani, Nucl. Inst. Methods Phys. Res. B **261**, 365 (2007). https://doi.org/10.1016/j.nimb.2007.04.294
23. T.J. Shaw, D.A. Strellis, J. Stevenson, D. Keeley, T. Gozani, AIP Conf. Proc. **1099**(1), 633 (2009). http://aip.scitation.org/doi/abs/10.1063/1.3120116

Chapter 8
Data Interpretation and Algorithms

Miltiadis Alamaniotis

Abstract Detection of nuclear threats such as the presence and movement of materials that could be used to construct nuclear weapons is essential to nuclear security and may be accomplished by advanced active interrogation sensing systems. A key to performance of these sensing systems is the ability to analyze and interpret the acquired data. This chapter discusses the analysis and interpretation algorithms applied to data collected by sensing systems in the context of nuclear security. The structure of the chapter is as follows: Sect. 8.1 briefly discusses data analysis in nuclear security. Section 8.2 presents planar and tomographic imaging systems, and the following Sect. 8.3 presents data reduction techniques with the main focus on principal components. Section 8.4 introduces data unfolding methods, while Sect. 8.5 introduces sensor networks and distributed detection algorithms. The main conclusions are summarized at the end of the chapter.

8.1 Introduction

The viability of the transportation system, upon which globalization rests, strongly depends upon national and international security. In this context, significant effort is devoted to nuclear security and nonproliferation. The impact of nuclear security on globalization has been evident for an extended period. In addition to the apocalyptic specter of nuclear war, failure to control the use and storage of nuclear material may destabilize the global system through nuclear terrorism. Lack of effective nuclear security may unleash forces of disintegration on the modern global system in the form of nuclear and radiological attacks on urban communities [1]. Therefore, advancements of nuclear security technologies is of paramount importance.

Digital connectivity is often seen as the emblematic unifying force of the global system [2]. Myriads of services enabled through connectivity continuously draw the

M. Alamaniotis (✉)
Purdue University, West Lafayette, IN, USA
e-mail: malamani@purdue.edu

I. Jovanovic, A. S. Erickson (eds.), *Active Interrogation in Nuclear Security*,
Advanced Sciences and Technologies for Security Applications,
https://doi.org/10.1007/978-3-319-74467-4_8

world closer through a wide-ranging array of applications pertaining, but not limited to, security, trade, education, transportation, entertainment, weather prediction, logistics, manufacturing, and the rapid dissemination of knowledge from life-saving medical and other scientific research [3]. At first glance, digital connectivity may appear antithetical to nuclear security and nonproliferation (obviously, sensitive data must not be shared over the commercial Internet and sensor networks [4]). In actuality, nuclear security is not only about securing information but also infrastructure, and connectivity provides rich context for integrating nuclear research, sensing and industrial capacities to protect, advance and enhance nuclear security and nonproliferation goals on a global scale [5]. Fundamentally, digital connectivity ensures that the higher the density of sensor population the better the detection resolution since every participating individual and device may be turned into a sensor—their collective behavior providing a statistical basis for robust and accurate decision-making.

Hence, the increasing demands on nuclear security also increase the challenges in data analysis, representation of obtained data, and drawing inferences from the analysis [6]. From geographic information systems (GIS) to localized representation of information and data, all can be fruitfully utilized to improve verification and resolution, and enhance the chances of threat aversion. To that end, significant research has already been conducted, and further will focus on innovative sensors concepts, as well on data collection, representation and analysis methods pertained to nuclear security [7].

The time-critical nature of nuclear security problems forms the fundamental underlying constraint in the development of new data analysis methods. During time-critical operations many tasks must be performed in close synchronization. These include accessing appropriate information, interpreting the state of the operation, and making decisions. Frequently, extensive knowledge resources must be accessed, including consulting experts and searching knowledge bases in a time dictated by the needs and particular dynamics of the processes involved. For example, in data acquisition and prediction for remote sensing, a plethora of sensing systems can provide voluminous data-sets potentially relevant to nuclear security [8]. What the data means, however, needs to be decided in a timely manner, and it has to be useful in predicting developments or patterns of interest to nuclear security. On one hand, to miss uninteresting piece of data may be an acceptable omission provided that a relevant pattern or trends are not missed (ignored or misinterpreted) within a relevant decision time. On the other hand, in nuclear forensics, in-situ characterization and attribution tasks must be achieved during the timeframe of inspection, especially as additional measurements and/or audit activities may be in order before the inspection team leaves the premises [9]. The goal of data analysis and interpretation algorithms is to evolve integrated methods in nuclear security and nonproliferation systems to facilitate decision making in these crucial time-critical operations [10].

The second fundamental characteristic of nuclear security and nonproliferation is the uncertainty of evidence available for identification (or *characterization*) and decision-making (or *attribution*) [11]. This evidence may consist of data, information, and knowledge, and may exist at different levels of abstraction, both at the level of individual instruments and sensors and at the level of a system as a whole. There are many different sources of uncertainty. At the system level uncertainty arises from the dynamics and complexity of a task itself, from incomplete and/or conflicting information, and/or from multiple and possibly contradictory task goals. At a lower physical level uncertainty arises from ambient radiation background, environmental conditions, measurement errors, drifting instruments, non-ideal detector characteristics, faulty sensors and/or improperly processed and missing data [12].

Uncertainty and time-critical functions take on multiple and varying meanings due to the inherent complexity of nuclear security systems. Data complexity arises from the multiplicity of ways the sensors, components, systems, interact within the various systems. Managing complexity is the special promise of data interpretation and analysis algorithms [13]. Therefore, advanced data analytics tools are needed, which can facilitate and validate the procurement of information from different modalities and multiple interacting processes, i.e., heterogeneous sensor networks [14].

Overall, data analysis and interpretation algorithms comprise an integral part of sensing system development in nuclear security applications. Their performance significantly affects the overall system performance pertaining to advanced prediction, verification, and resolution in nuclear security.

8.2 Planar and Tomographic Imaging Systems

Imaging systems utilizing various physical processes offer many advantages in applications involving non-destructive interrogation, security screening, detection and verification of nuclear material processing. The general block diagram of an imaging system is depicted in Fig. 8.1. Imaging systems utilize the penetration property of physical particles or waves in materials [15]. They consist of a transmitter that sends a beam toward the targeted object and an array of receivers that measure the beam that passes through that object. The receivers turn the detected

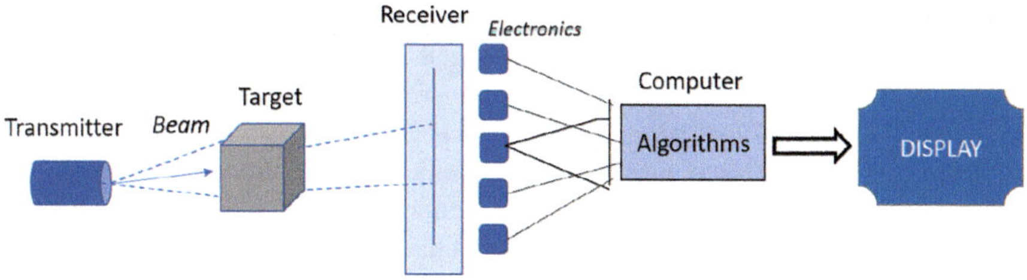

Fig. 8.1 Principle of operation for a generic imaging system

beams into electronic signals using specialized hardware; signals are subsequently sent to a computer that uses tailored algorithms to display the interior of the targeted object.

It should be noted that the imaging system in Fig. 8.1 is representative for an active interrogation system. In passive imaging systems, there is no transmitter and a spontaneous emission from the target is characterized. In general, imaging systems may be considered as information systems that (1) form information by probing the object, (2) record information with detectors, and (3) present information in the form of images.The following subsections present imaging systems that have been used and/or proposed in active and passive interrogation [16].

8.2.1 X-ray Radiography/Computed Tomography

The physical principle upon which X-ray imaging systems operate is the degree of absorption of X-rays in the penetrated material. A beam of X rays is transmitted through the target material, while an array of X-ray detectors behind the target measures the spatially-varying intensity of outcoming X-rays [17]. Such a radiographic system is consistent with the general principle upon which imaging systems operate (Fig. 8.1).

The X-ray beam is produced by a source, which is usually a linear accelerator, and contains highly energetic photons over a continuous spectrum of energies (up to ∼10 MeV). The X rays penetrate the targeted material, and their attenuation depends on the effective atomic number (Z_{eff}) of the penetrated material [18]. The higher the Z_{eff}, the higher the specific attenuation of the beam. In the simplest description, the photon attenuation follows the Beer-Lambert law:

$$I = I_0 \exp(-\mu \Delta x), \tag{8.1}$$

where I_0 is the intensity of the incident beam, I is the transmitted intensity, μ is the attenuation coefficient of the target, and Δx is the thickness of the material along the beam path. Effective radiography requires that there is adequate penetration and the X-ray detectors provide a signal of adequate contrast for items of interest. In other words, adequate penetration, sensitivity, and resolution are fundamental for efficient radiography systems. An example of an X-ray image of a scanned cargo container simulated with Geant4 [19] is depicted in Fig. 8.2. The signal contrast reveals the existence of five different objects in the container, with two of them (the ones on the left) being less dense than the remaining objects [20].

X-ray radiography is a planar imaging technique since it provides a 2D representation of the radiographed object. A 3D representation of the object may be obtained with an X-ray CT (*computed tomography*) technique. CT images provide a cross-section of the object from various views and use these views to reconstruct a 3D

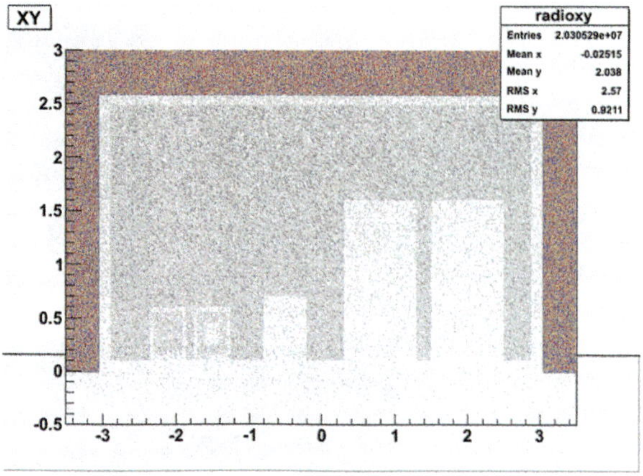

Fig. 8.2 X-ray image of a cargo container simulated in the Geant4 framework

representation of an object [21]. CT utilizes the projections of an object obtained by radiographic measurements at range of angles θ (in other words, multiple images are taken from various views and angles) to reconstruct the object in 3D. Various algorithms have been developed and applied to CT reconstruction and they are largely considered as special cases of the *Radon transform* [22].

One of the widest used reconstruction algorithms is the *filtered backprojection* algorithm, which is essentially a stabilized and discretized form of the inverse Radon Transform [23]. Assuming a set of N samples, the discretized inverse Radon Transform is

$$f(x, y) = \frac{1}{2\pi} \sum_{n=0}^{N} \Delta\theta_n g_n (x_n \cos\theta_n + y \sin\theta_n) \tag{8.2}$$

with

$$g_\theta(t) = p_\theta(t) k(t), \tag{8.3}$$

where $k(t)$ is the Radon kernel, $p_\theta(t)$ is the *Hilbert transform*, and $\Delta\theta_n$ is the angular spacing between the projections. In this algorithm, the 1D projections are filtered by the Radon kernel in order to obtain a 2D signal; therefore, a back projection from 1D to 2D is achieved. Other reconstruction algorithms include the *Fourier-domain reconstruction, iterative reconstruction*, and *fan-beam reconstruction* [22]. To illustrate the difference between planar imaging and CT scan, an example of an X-ray image and the corresponding CT-based 3D reconstruction is provided in Fig. 8.3.

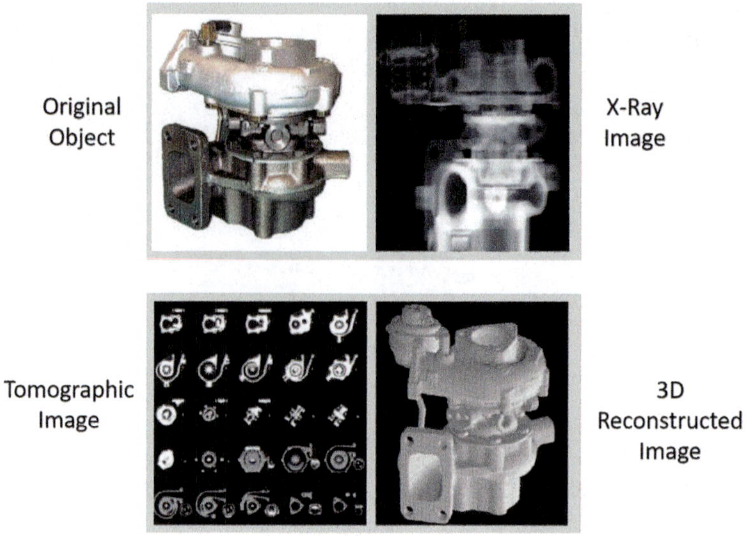

Fig. 8.3 Example of X-ray image of an object and its 3D reconstruction. Reproduced from Ref. [24]

8.2.2 *Gamma-Ray Radiography*

The basic operational principle of γ-ray imaging systems is similar to X-ray imaging. The difference lies in the probing source. Gamma-ray radiography may utilize radioactive sources such as ^{60}Co and ^{137}Cs, along with arrays of γ-ray detectors [25]. The obtained images can be of high quality and used to identify high-density regions of the object, where the presence of nuclear material is more likely [26]. Gamma-ray imaging is also considered as a non-intrusive inspection method [27]. The main disadvantages of γ-ray radiography (as well of X-ray radiography) include the limitations in material penetration and the existence of scattered background that may impose difficulties in obtaining accurate images [28].

8.2.3 *Neutron Radiography*

The fundamental approach of neutron radiography is the same as in X-ray radiography, with neutrons being the probing source [29]. Neutron imaging systems utilize a neutron source such as AmBe or ^{252}Cf and obtain an image based on neutron attenuation. Unlike X rays that get absorbed by high-Z materials, neutron attenuation does not depend directly on the atomic number Z (albeit it is significant for some low-Z materials). Some nuclides such as ^{10}B strongly absorb neutron of lower energies, while hydrogen efficiently scatters them at higher energies. In addition, various metals allow neutrons to pass through them without significant interaction. Therefore, the difference in attenuation properties of neutrons and X rays may allows objects that are not visible with X rays to be recognized in neutron images, and vice versa [30].

In thermal neutron radiography, a neutron source must be first adopted for neutron generation. The generated fast neutrons are slowed down in a moderator such as water. Because of the fact that neutrons travel within the moderator in all directions, a collimator is adopted. The collimator passes the neutrons in a specific direction, which allows a high-quality image to be generated. Behind the probe object, a conversion screen is placed that captures neutrons and produces visible light, which is measured more easily [31]. At last, the measured optical image is used to create an image using a film or a camera device.

8.2.4 Muon Tomography

Muon tomography (MT) is an innovative technique that has found various applications in detection and imaging. For instance, one interesting application of MT has been the imaging and detection of hidden chambers in pyramids [32]. However, it was not until recently that muon tomography was applied for imaging in nuclear security, and more particularly in cargo scanning [33].

Muons are highly energetic particles generated in the upper layers of atmosphere, and were first discovered in 1936. They are generated by the interaction of cosmic rays with atmospheric atoms and they have 207 times greater mass than that of the electron. Because of their mass, muons are not abruptly slowed down, and hence, emit only a very small amount of bremsstrahlung. In addition, the muon average kinetic energy on the surface of the Earth is approximately 4 GeV. Because of these properties, muons have very high penetration capabilities compared to photons and electrons; for instance, they can penetrate and reach great depths [34]. Furthermore, the flux of muons reaching thc Earth surface is high: about $10^4\,\mathrm{m}^{-2}$ muons per minute.

Exploiting the flux of muons reaching the Earth's surface eliminates the need for utilization of active sources. In other words, information probing in MT is driven by the natural source of muons, and hence, a dedicated transmitter depicted in Fig. 8.1 is eliminated. The high penetration ability of muons makes it almost impossible to shield even large and dense objects from muon interrogation. This property makes the muon tomography an attractive potential imaging technique for applications such as scanning containers [35].

8.2.4.1 Muon Scattering

MT utilizes the Coulomb interaction of muons to provide the information on the mean contents of the path traversed by muons. Thus, the signal of interest is the

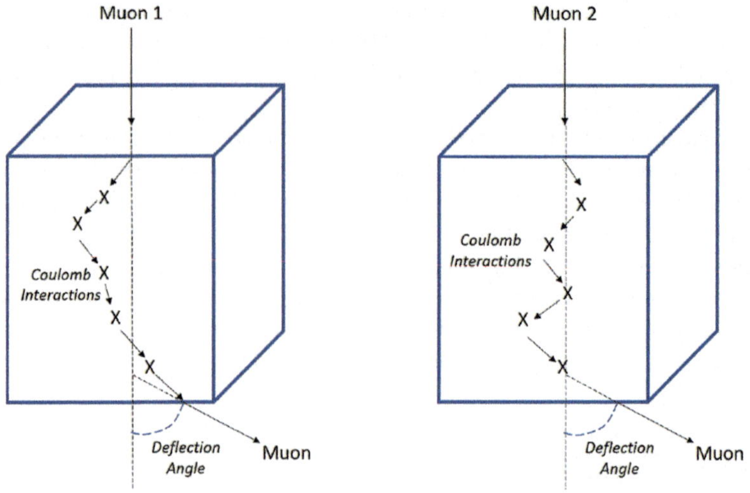

Fig. 8.4 Visualization of two different muons having different paths resulting in the same deflection angle

Coulomb scattering; while a muon passes through a material, it interacts multiple times through Coulomb scattering [36]. The muon traversing an object may be scattered multiple times over its path, but the observed deflection angle may be exactly the same with that of muon following a different path, as shown in Fig. 8.4. Multiple interactions build up to provide a single observed angular deviation of the muon once it emerges from the object. The deviation follows a Gaussian distribution given by

$$f(\theta_x) = \frac{1}{\sqrt{2\pi}\sigma_\theta} \exp \frac{\theta_x^2}{2\sigma_\theta^2}, \tag{8.4}$$

where the width of the distribution (σ_θ) is

$$\sigma_\theta = \frac{13.5}{p\beta}\sqrt{\frac{T}{X}}, \tag{8.5}$$

with X being the radiation length in the material, p the muon momentum, β the particle speed divided by the speed of light, and T the material thickness [37].

Figure 8.4 clearly exhibits the possibility that the deflection angle of two different muons may be exactly the same, despite the fact that the number of Coulomb interactions and trajectories may be very different. Equations (8.4) and (8.5) imply that the radiation length decreases with increasing Z (atomic number). Thus, the mean scattering increases with increasing Z, allowing the distinction of low-Z from high-Z material.

This material distinction property of muons, together with their high penetration ability, make the muon scattering a method that has high potential for applications in nuclear security. In particular, measuring the scattering deflection angle of muons allows reconstruction of 3D images of an object, and subsequent designation of high-Z areas in it.

8.2.4.2 Point of Closest Approach (POCA) Reconstruction

Tomographic image reconstruction using muon scattering is conducted by adopting statistical algorithms that take into consideration the muon deflection angle, while identifying the difficulty in accurately knowing the trajectory of a muon. The simplest and widely used algorithm in the field of muon imaging is the *point-of-closest approach algorithm* (POCA) [38].

POCA is a pure geometric-based reconstruction algorithm that does not take into consideration the underlying physics. The fundamental assumption behind the POCA algorithm is that a muon interacts only at a single point within the traversed mean. Therefore, it identifies the closest geometrical point between the input and output muon extrapolated directions as the point of interaction, as sketched in Fig. 8.5.

Muon tomography requires a system of sensors that is capable of detecting the track of muons entering the object and the track of the same muon leaving the object. The directions are used for linearly extrapolating the incoming and outgoing tracks of the muon. In the vast majority of the cases, those two lines do not meet at the same point because of multiple scattering and uncertainty in the measurement. In

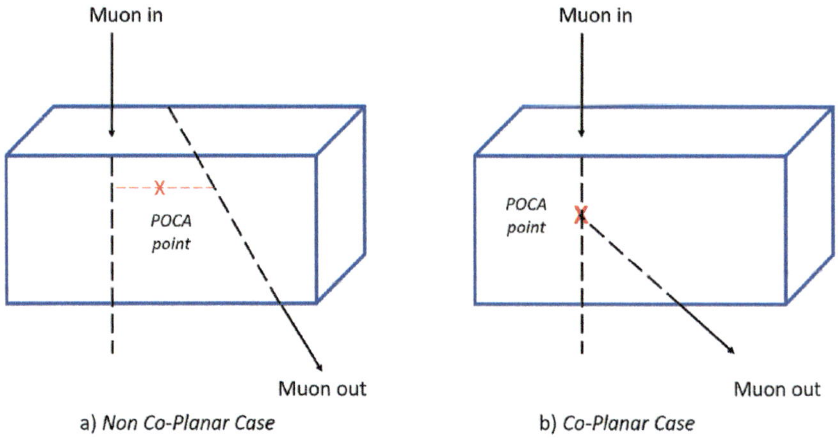

Fig. 8.5 Illustration of the POCA algorithm in muon imaging for (**a**) non co-planar and (**b**) co-planar input and output muon directions

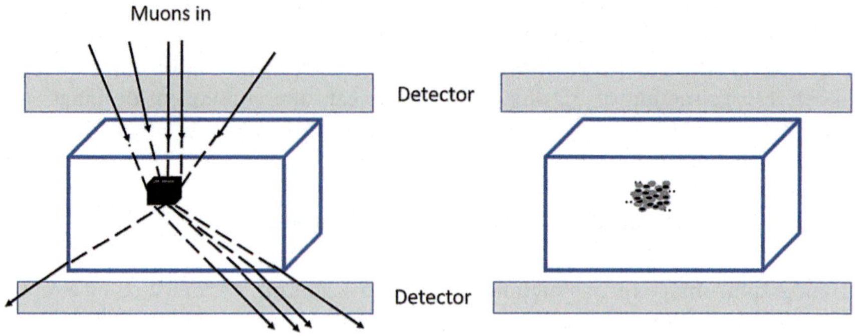

Fig. 8.6 Sketch of muon tomography using the POCA algorithm

that case, where tracks do not belong to the same geometric plane, for each line the closest point of the two extrapolated track lines is computed using a geometric formula [39]

$$T_{POCA} = \frac{1}{2}(T_{in} + T_{out}), \tag{8.6}$$

where T_{in} and T_{out} is the vector of incoming and outgoing muon track. The geometric approach identifies the midpoint along the shortest distance between the two extrapolated tracks as the point of interaction, as depicted in Fig. 8.5a. In the case when two tracks are in the same plane (i.e., they are coplanar), there is a single point where the two tracks meet that coincides with the POCA interaction point, as presented in Fig. 8.5b. In addition, the deflection angle of the muon is also computed; if it is significantly different from zero, it is recorded. Zero or close to zero angles (parallel input and output tracks) are of no use in reconstruction; parallel tracks geometrically cannot provide a POCA point. Therefore, POCA reconstruction utilizes the POCA points that result from non-zero deflection angles [40]. A diagram of POCA reconstruction is given in Fig. 8.6. Overall, in muon tomography, high deflection angles indicate the presence of high-Z material, while POCA points designate their area in the probed volume.

8.3 Principal Component Analysis and Related Methods

Raw sensor data are seldom utilized for making direct inferences. In most cases, an additional layer of analysis is required before any inference is made. In this section, a set of analytic tools is presented that find wide use in processing of data such as energy spectra in nuclear security applications.

8.3.1 Principal Component Analysis

Principal component analysis (PCA), is concerned with reducing data dimensionality, extracting features, visualizing data, and explaining the variance-covariance structure of a set [41]. The objectives of PCA are to achieve: (1) data reduction and (2) data interpretation. In several fields, PCA is also known as the Karhunen-Loeve transform [42].

Assuming that the variability of a dataset comprised of N measurements is reproduced using a set of D variables, often much of the variability can be represented by a smaller number M of variables called principal components. In other words, with little loss of information we may project the initial N measurements defined on D variables to a set of N measurements defined on M variables. PCA often reveals relationships that were not observed, and thus, allows interpretations that would not initially be made.

The definition of PCA starts by considering a dataset of observations $\mathbf{x}_n$, $n = 1, \ldots N$ with N denoting the population of observations. The observation $\mathbf{x}_n$ is variable of dimension D. The goal of PCA is to project the initial set of observations to a lower space, i.e., $M < D$, while maximizing the amount of initial variability explained by the projected data [41].

The value of M may be determined either by the modeler or by using various existing techniques (for instance, scree plots). In order to derive the PCA framework, the case of projecting to $M = 1$ is considered here. Then, the direction of the D space may be defined as a D-dimensional vector $\mathbf{u}_1$. It is a convenient choice to consider $\mathbf{u}_1$ as a unit vector, thus, giving $\mathbf{u}_1^T \mathbf{u}_1 = 1$. Selection of a unit vector allows projection of each datapoint on a scalar value, and more particularly on the value $\mathbf{u}_1^T \mathbf{x}_n$.

Therefore, the mean of the projected data is $\mathbf{u}_1^T \mathbf{x}_m$, with $\mathbf{x}_m$ being the sample mean computed by

$$\mathbf{x}_m = \frac{1}{N} \sum_{n=1}^{N} \mathbf{x}_n \tag{8.7}$$

and the variance of the projected data by

$$\mathbf{u}_1^T \mathbf{S} \mathbf{u}_1 = \frac{1}{N} \sum_{n=1}^{N} \{\mathbf{u}_1^T \mathbf{x}_n - \mathbf{u}_1^T \mathbf{x}_m\}^2, \tag{8.8}$$

with $\mathbf{S}$ being the data covariance matrix given by

$$\mathbf{S} = \frac{1}{N} \sum_{n=1}^{N} (\mathbf{x}_n - \mathbf{x}_m)(\mathbf{x}_n - \mathbf{x}_m)^T. \tag{8.9}$$

Next, maximization of the variance shall occur. However, maximization should be conducted with respect to constraint $\mathbf{u}_1^T\mathbf{u}_1 = 1$ that prevents $|\mathbf{u}_1| \rightarrow \infty$ to become true. To apply this constraint, a Lagrange multiplier is introduced, i.e., l_1, and therefore the following is obtained [41]:

$$\mathbf{u}_1^T\mathbf{S}\mathbf{u}_1 + l_1(1 - \mathbf{u}_1^T\mathbf{u}_1). \tag{8.10}$$

Differentiating Eq. (8.10) with respect to $\mathbf{u}_1$ and setting the derivative to zero results in

$$\mathbf{S}\mathbf{u}_1 + l_1\mathbf{u}_1. \tag{8.11}$$

which reveals the $\mathbf{u}_1$ is an eigenvector and l_1 is the respective eigenvalue of $\mathbf{S}$. Furthermore, multiplying both sides of Eq. (8.11) with the $\mathbf{u}_1^T$, the variance of projected data results is obtained as:

$$\mathbf{u}_1^T\mathbf{S}\mathbf{u}_1 + l_1, \tag{8.12}$$

implying that the variance is maximized if the eigenvector $\mathbf{u}_1$ coincides with the eigenvector of the largest eigenvalue. This eigenvector is known as the *principal component* [41].

Once a principal component has been identified, additional ones may be defined by choosing a direction that maximizes the projected data variance and is also orthogonal to directions already considered. This procedure is iterated until M principal components are found, resulting in an M-dimensional space defined by the M eigenvectors of the respective largest M largest eigenvalues. A simple illustration of how PCA interprets variance in data is depicted in Fig. 8.7, where two principal components interpret the variance in a 2D space.

To sum up, PCA is a powerful method that projects down a dataset of D dimensions by computing the eigenvectors of the data covariance matrix and keeping those eigenvectors that correspond to M largest eigenvalues. Integration of

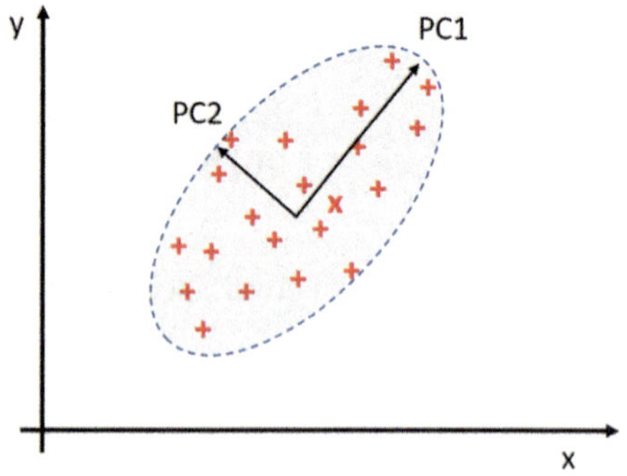

Fig. 8.7 Illustration of PCA by interpreting the maximal variance using two principal components

PCA with a fast and powerful method of computing eigenvectors may significantly decrease the computational cost, especially in big data applications. Lastly, it should be noted that within the context of machine learning, PCA may be represented with the aid of a kernel function leading to a more generalized nonlinear form of PCA, known as *kernel PCA* [43].

8.3.2 Data Smoothing

In the context of nuclear security, removing data fluctuation with the process of data smoothing is a common first step in enhancing isotope detectability potential in spectroscopy applications [44]. From a theoretical point of view, data smoothing has no substantial foundation; it represents a technique that is empirically proven to be beneficial. In practice, it is a data manipulation technique aiming at filtering statistical fluctuation while retaining the features of the data.

From a mathematical point of view, data smoothing is a process that replaces the value at hand with a new value taken as the arithmetic average of a window of values centered at the value at hand. Averaging is iterated for every single value in the dataset by shifting the center of window by one value. In general, the window length is predetermined by the modeler and remains constant throughout the smoothing process.

Analytically, the general formulation of data smoothing is given by

$$y_j = \frac{1}{N} \sum_{i=-n}^{n} w_i y_{j+i}, \tag{8.13}$$

where y_i is the smoothed value, N is the length of averaging window, n is an index determining the number of points used in the smoothing according to $2n + 1 = M$, and w_i is the weight coefficients of smoothing. It should be noted that the weight coefficients depend on the type of smoothing that has been selected for the problem at hand. In general, coefficients are evaluated by a least squared polynomial fitting process (see Sect. 8.3.3) with the modeler defining the degree of polynomial [12].

In practice, the most commonly used data smoothing process is the N-point zeroth-order smoothing, where

$$w_1 = w_2 = \ldots = w_N = 1 \tag{8.14}$$

with the smoothing equation in the form

$$y_j = \frac{1}{N} \sum_{i=-n}^{n} y_{j+i}, \tag{8.15}$$

also known as *moving average*. In other words, the N-point zeroth-order smoothing simply drops down to the sample mean value of the window centered at the smoothed value. Smoothing may be iterated on the same data multiple times until a desired level of smoothness is attained. Figure 8.8 illustrates the smoothing of a γ-ray spectrum with the 3-, 5-, and 7-point zeroth-order smoothing process resulted after multiple iterations of smoothing [44].

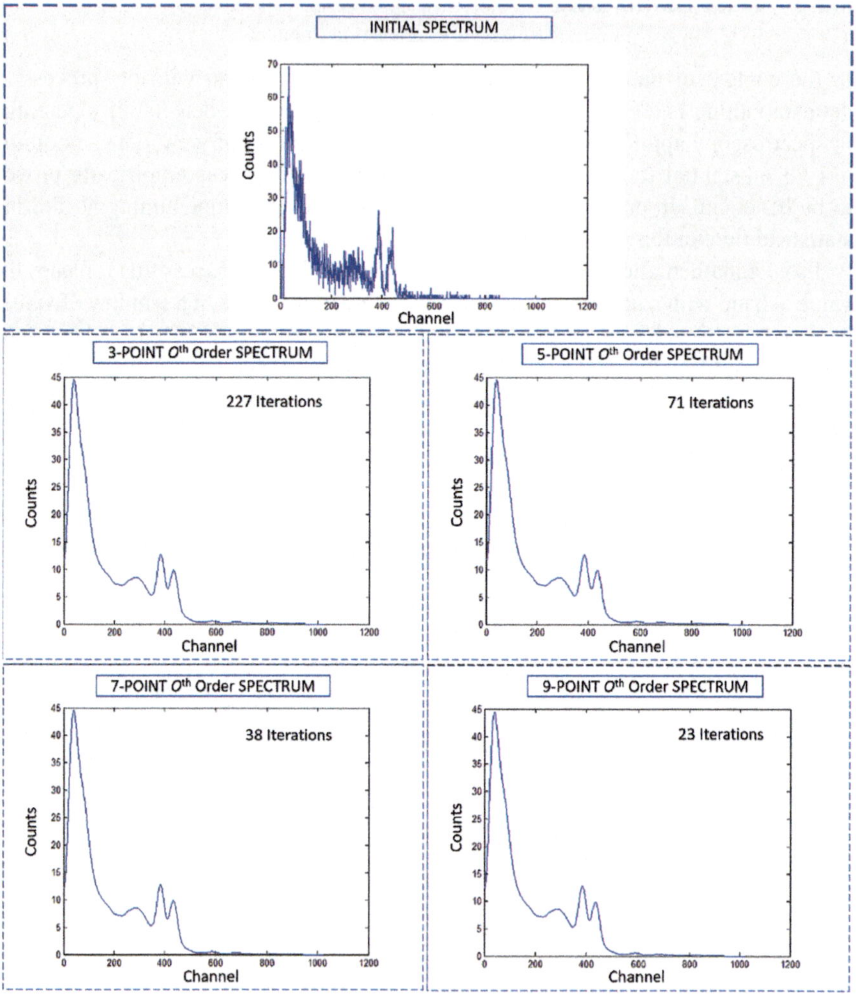

Fig. 8.8 Gamma-ray spectrum taken with a NaI(Tl) low-resolution detector and respective smoothed spectra for iterative (# of iterations displayed in the inset), 3-, 5-, 7-, and 9-point zeroth-order smoothing. Reproduced from Ref. [45]

Overall, the difficulty in data smoothing is selecting a window of length M (and implicitly of n in Eq. (8.13)). With regard to spectral smoothing as applied to nuclear security, smoothing may have undesirable effects; it tends to flatten the isotopic peaks and fill the valleys. For instance, in Fig. 8.8, smoothing has filled up the valley between the two peaks with non-existing counts. Therefore, selection of M strongly depends on the analyst's experience and the effect of smoothing method on data.

8.3.3 Least Squares Curve Fitting

The usual experimental data come in the form of pairs (x, y), where x is the variable controlled by the experimenter and y is the observed value. Repetition of an experiment N times provides the experimenter with a set of pairs $(x_n, y_n), n = 1, \ldots N$. The experimenter aims at finding the underlying pattern that associates x to y values. However, in most cases the underlying physical processes are not known or are too complex, imposing significant difficulties to find the exact pattern that observations follow. Therefore, it is often desirable to find an analytic function that represents the pairs $(x_n, y_n), n = 1, \ldots N$ to a satisfactory degree of accuracy [12].

A simple but widely used approach in representing observed data is based on curve fitting. In particular, the observed data are fit by a polynomial function as given below [46]:

$$y(x, \mathbf{w}) = w_D x^D + \ldots + w_2 x^2 + w_0 = \sum_{d=0}^{D} w_d x^d, \tag{8.16}$$

where D is the order of polynomial function, w_d are the polynomial weights (with $\mathbf{w}$ being the vector that contains all w_d), and x_d denotes the dth power of x. It is observed in Eq. (8.16) that the polynomial is a linear function of the weights w while being a nonlinear function of the variable x. Furthermore, in Eq. (8.16), the unknown parameters are only the weight coefficients, given that (x, y) pairs are experimentally known [46].

Evaluation of coefficients is performed with respect to an error function. The error function quantifies the degree of misfit between the polynomial function and the observed data. In the context of least squares, the error function is given by

$$E(\mathbf{w}) = \sum_{n=1}^{N} \{y(x_n, \mathbf{w}) - y_n^{ob}\}^2, \tag{8.17}$$

with y_n^{ob} being the observed (i.e., target) value, and N the population of observations. The goal of the least squares method is to find those w_n that minimize the error function in Eq. (8.17). For convenience, visualization of the geometrical interpretation of least squares fitting is given in Fig. 8.9. The degree of polynomial

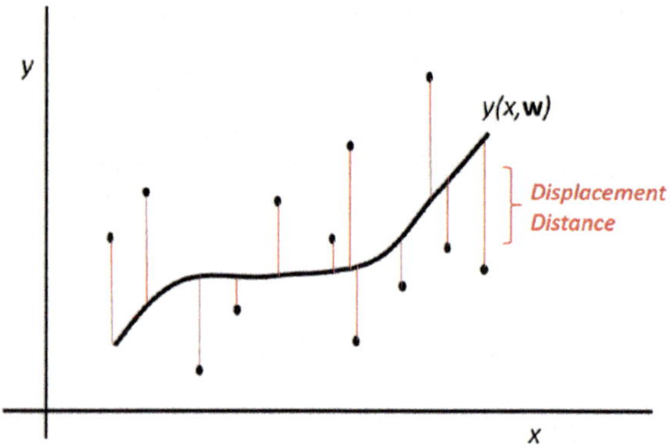

Fig. 8.9 Geometrical visualization of least squares curve fitting. The curve provides the minimum of squared displacement distances. Reproduced from Ref. [46]

curve determines the form of fitting curve. In the case of $d = 1$, the polynomial takes the form of a line, while in the case of $d = 2$ it takes the form of a quadratic curve.

Determination of the degree of polynomial is a difficult task. A low degree may give a high misfit, while a very high degree may give zero misfit. The former case is not desirable because it represents poorly the observations, while the latter is also not desirable because it gives poor generalization. In particular, the function represents perfectly the current set of data but fails to accurately predict new observations, a phenomenon known as *overfitting*. Therefore, the degree of polynomial should be selected as a tradeoff between misfit and overfitting [46].

8.3.4 Maximum Likelihood

Least squares fitting has been proved an effective tool for fitting experimental data. In principle, it is a special case of the maximum likelihood (ML) method [46]. Maximum likelihood is also applied to curve fitting problems, and is more flexible compared to least squares. However, ML is slower than least squares and the difference in their speed increases for larger datasets.

The fundamental framework of ML is rather simple. Assuming a set of N experimental data with dependent variables $x_i, i = 1, \ldots N$ and the respective independent variables $y_i, i = 1, \ldots, N$, each pair (x_i, y_i) may be considered as a single event, thus, providing a set of N events. Next, a function $y(x, \mathbf{w})$ is utilized for fitting the experimental data. Fitting is performed by evaluating the M weight parameters $\mathbf{w}$ of the function $y(x, \mathbf{w})$ [47]. Each event is modeled as a normalized probability density function given by

$$P_i = P(x_i; w_i, \ldots, w_M) = P(x_i; \mathbf{w}) \tag{8.18}$$

and is evaluated at value x_i. Next, the likelihood function is taken as the product of independent probability density functions [48]. In particular, by denoting the likelihood as L then the likelihood function is given by

$$L(w_i, \ldots, w_M) = \prod_{m=1}^{M} P(x_i; \mathbf{w}) = \prod_{m=1}^{M} P_i. \tag{8.19}$$

Maximization of the likelihood function in Eq. (8.19) with respect to weights $w_i, n = 1, \ldots N$ provides a set of parameter values, defined as the maximum-likelihood values [46].

In nuclear security problems, in the majority of individual events follow either a Gaussian (normal) or a Poisson density functions. In those cases, it is more convenient to maximize the logarithm of the likelihood

$$\log L = \log\{\prod_{m=1}^{M} P_i\} \tag{8.20}$$

than the likelihood function directly. For example, assuming the data follow a Gaussian density distribution with two unknown parameters, namely, mean μ and variance σ^2, then the log-likelihood function of (8.20) is given by [49]:

$$\ln L = -\frac{1}{2\sigma^2}\sum_{m=1}^{M}(x_m - \mu)^2 - \frac{M}{2}\ln\sigma^2 - \frac{M}{2}\ln(2\pi). \tag{8.21}$$

Maximization of Eq. (8.21) with respect to μ provides the ML solution of the mean:

$$\mu_{ML} = \frac{1}{M}\sum_{m=1}^{M} x_m, \tag{8.22}$$

while maximization of Eq. (8.21) with respect to σ^2 provides the respective ML solution of the variance:

$$\sigma^2_{ML} = \frac{1}{M}\sum_{m=1}^{M}(x_m - \mu_{ML})^2. \tag{8.23}$$

It should be noted that in the case of Gaussian distribution given above, Eq. (8.22) should be evaluated prior to Eq. (8.23).

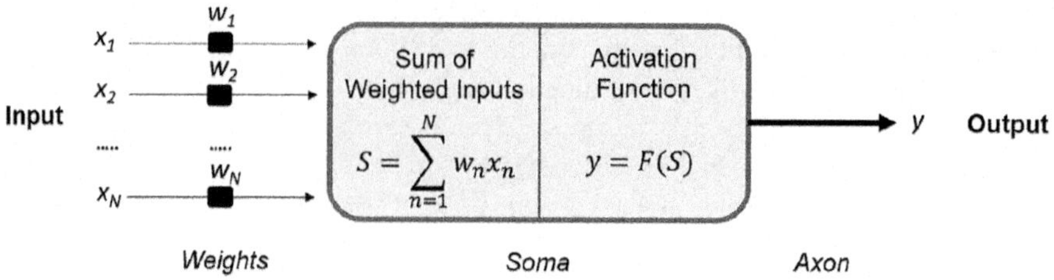

Fig. 8.10 Architecture of an artificial neuron

8.3.5 *Artificial Neural Networks*

One of the preeminent set of tools in machine learning library are the *artificial neural networks* (ANN) that have found wide application in various domains [50]. Development of ANN has been inspired by the biological operation of the human brain. The fundamental cellular unit of the human brain is the neuron consisted of the *soma*, *synapses*, and *dendrites*, while it is connected to other neurons through the *axon*.

Similar to biological neurons, the fundamental operational unit in ANN is the artificial neuron [50]. The architecture of an artificial neuron is depicted in Fig. 8.10, where we observe that it is comprised of three parts, namely, the *weights*, *soma*, and *axon*. The input signal to the neuron denoted by $x_1, \ldots, x_N$ is multiplied by the respective weights $w_1, \ldots, w_N$ providing a set of N factors. The computed factors are forwarded to the *soma* of the neuron. As shown in Fig. 8.10, the *soma* is comprised of two parts. In the first part, the incoming factors are added as

$$S = \sum_{n=1}^{N} w_n x_n, \tag{8.24}$$

where S denotes the sum of the factors. In the second part, the computed sum is used as an input to a predetermined function $F(x)$, known as the *activation function*. The activation function provides a value y that is the final output of the neuron [50].

The activation function may take the form of any valid mathematical function [46]. The most common form of activation function is the *logistic function*, given by

$$F(S) = \frac{1}{1 + e^{-\alpha S}}, \tag{8.25}$$

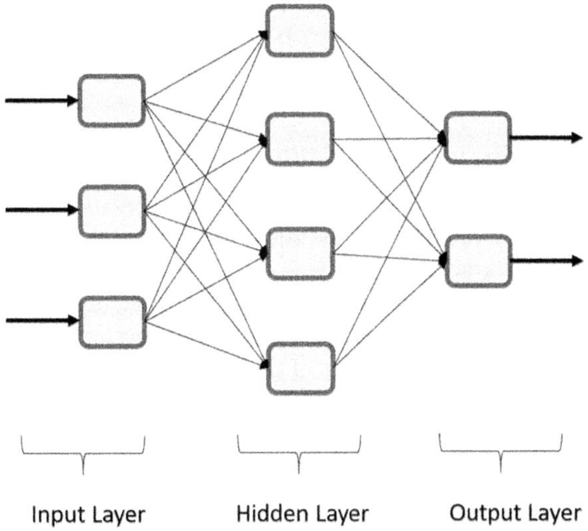

Fig. 8.11 Architecture of an artificial neural network with three layers: input, hidden and output

with α being a parameter that controls the abruptness of the logistic function. Another popular activation function is the *threshold function*, whose analytical form is

$$F(S) = \begin{cases} 1, & S > T \\ 0, & S \leq T, \end{cases} \tag{8.26}$$

where T is a threshold value determined by the system modeler.

Several neurons may be connected to each other and form a more complex system. Therefore, the output of a neuron may consist of the output of the next neuron. By connecting neurons in an architecture similar to the structure of the human brain, we obtain an artificial neural network. From a computational point of view, ANN are information processing elements that model complex systems [50]. The neurons of an ANN are grouped into layers, namely, the input, hidden and output layer, as presented in Fig. 8.11. There is no restriction in the number of neurons per layer, as well as in the number of hidden layers. Selection of number of neurons and hidden layers depends on the specifics of the application. Every neuron may be connected to any other neuron in the ANN; if neurons are connected only to neurons of the next layer the ANN is called *feedforward* [50].

The weights of an ANN are evaluated through a process called *training*. In the training, the neural network is exposed to output values for which the input is known. Adjustment of weight values for every pair of input-output is performed by training algorithms such as the *backpropagation algorithm* [50]. Training of ANN is continued until an error function drops below a prespecified threshold.

ANN may be used in a variety of data processing problems including, but not limited to, prediction, classification, compression, projection, pattern recognition, and data reduction. The strength of ANN is their universality; for any valid function, there is a neural network that can do the same job [51].

8.4 Signature Unfolding Techniques

Radiation measurements consist of the aggregation of individual contributions coming from various sources. With regard to energy spectra, a measurement is comprised of the aggregation of signature spectra of the ambient sources. Identification of constituents of a spectrum is performed by unfolding the measurement into individual signatures. This section will examine unfolding techniques that lead to subsequent source identification.

8.4.1 Elements of Signature Unfolding

In the ideal case, an unfolding technique analyzes a measurement into a set of signatures that, if aggregated, give the initial measurement. Unfolding assumes that spectra are comprised of a set of bins, with every bin representing an energy interval $\Delta E_i = E_{i+1} - E_i$.

Assuming that a measured spectrum is represented as $M(E)$, the general formulation of spectrum unfolding known as *Fredholm equation* is the following [12]:

$$M(E) = \int_0^{\infty} R(E, E')S(E)dE', \tag{8.27}$$

where $R(E, E')$ is a detector response function and $S(E')$ is the source signature spectrum. Taking into consideration that a spectrum is comprised of N energy channels (bins), Eq. (8.27) may be cast as follows:

$$M(E) = \sum_i^N \int_{E_i}^{E_{i+1}} R(E, E')S(E)dE', \tag{8.28}$$

where the initial integral over all energies has been rewritten as a sum of integrals over N energy channels (bins). It should be noted that Eq. (8.28) is continuous, but spectrometers do not measure a continuous function $M(E)$ but rather a histogram comprised of N energy bins. Hence, spectrometers measure a discrete function comprised of the following quantities:

$$M_i = \int_{E_i}^{E_{i+1}} M(E)dE, \tag{8.29}$$

where $i = 1, \ldots, N$. By applying Eq. (8.29) to Eq. (8.28), the measurement function is discretized:

$$M_i = \sum_{i}^{N} \int_{E_i}^{E_{i+1}} \int_{E_i}^{E_{i+1}} R(E, E')S(E)dE'dE, \tag{8.30}$$

with $i = 1, \ldots, N$ representing the measured quantities per energy bin.

In Eq. (8.31), the source spectrum $S(E)$ is the unknown parameter, while the detector response function $R(E, E')$ is known either by measurement or by simulation. Consolidating Eq. (8.31) for all energy channels provides a matrix equation:

$$\mathbf{M} = \mathbf{R} \cdot \mathbf{S}, \tag{8.31}$$

whose solution gives the source spectra. However, a matrix solution requires the use of matrix inversion, i.e., $\mathbf{R}^{-1}$, that is not always feasible or may provide an erroneous solution.

Possible contribution from multiple sources adds complexity to the source spectrum, and imposes difficulties in unfolding it. Additionally, inherent uncertainties due to non-ideal detector characteristics, or systematic errors in measurements enhance complexity of the measured spectrum. In a highly complex measured spectrum, low contribution sources may be masked by high contribution sources, hence leading unfolding algorithms to wrong inferences. There is a variety of unfolding algorithms in the literature following different approaches in order to determine $S(E)$. In this section, we present full spectrum (template based) and peak based unfolding [52].

8.4.2 Least Squares Full Spectrum Unfolding

A common approach in spectrum unfolding is to utilize the full spectrum curvature. In particular, unfolding is performed using the whole set of energy bins, where each bin contains contributions of one or more sources. Essential in this type of unfolding is the existence of a library that contains the full spectrum of nuclide signatures (or, in other words, a set of templates). Theoretically, the library should contain and utilize the whole set of nuclides, but in practice, only a subset of them is used due to computational resource constraints. In the remainder of this section, it is assumed that a library contains a population of N signature spectra.

In the full spectrum unfolding, Eq. (8.31) is approached by a linear combination of the signature spectra:

$$E_i = \alpha_1 S_{1i} + \ldots + \alpha_N S_{Ni} = \sum_{n=1}^{N} \alpha_n S_{ni}, \tag{8.32}$$

where E_i is the estimated source spectrum, S_N are the signature spectra, and α_i are the linear coefficients that denote the relative abundances of the respective signatures. Equation (8.32) that denotes contribution for a single energy bin can be written in a matrix form to contain the whole spectrum:

$$\mathbf{E} = \boldsymbol{\alpha} \cdot \mathbf{S}, \tag{8.33}$$

where $\boldsymbol{\alpha}$ is a $1 \times N$ vector containing the linear coefficients and $\mathbf{S}$ is a $b \times N$ matrix, with b being the number of energy bins. A more accurate representation is taken by decomposing the matrix $\mathbf{S}$ into its columns, where each column stands for a signature spectrum:

$$\mathbf{E} = \alpha_1 \cdot \mathbf{S}_1 + \ldots + \alpha_N \cdot \mathbf{S}_N = \sum_{n=1}^{N} \alpha_n \mathbf{S}_n. \tag{8.34}$$

The goal of the above approach is to set the estimated spectrum $\mathbf{E}$ equal to the measured spectrum $\mathbf{M}$:

$$\mathbf{M} = \mathbf{E} = \sum_{n=1}^{N} \alpha_n \mathbf{S}_n \tag{8.35}$$

and subsequently evaluate the coefficients α_n. Coefficient evaluation is expected to drive coefficients of respective signatures that are not present in the measured spectrum to zero, and the rest coefficients to non-zero values. The latter values denote the true relative abundance of the signatures in the spectrum.

Evaluation of coefficients is performed by formulating an optimization problem. The objective function takes the form of the least squares, which minimizes the mean squared differences between the measured and the estimated spectrum. The analytic formula of the least squares L_s is given by:

$$L_s = \frac{1}{N}(\mathbf{M} - \mathbf{E})^2 = \frac{1}{N}(\mathbf{M} - \sum_{n=1}^{N} \alpha_n \mathbf{S}_n)^2, \tag{8.36}$$

which is minimized with respect to coefficients a_n. A solution to Eq. (8.36) is obtained by setting

$$\frac{\partial L_s}{\partial \alpha_n} = 0, \quad n = 1, \ldots, N \tag{8.37}$$

that provides a set of N values. Signatures associated with non-zero values are considered as the truly identified source signatures in the measured spectrum.

8.4.3 Chi-Square Full Spectrum Unfolding

Similar to the least squares technique, the *chi-square unfolding* is also driven by minimizing the distance between the measured spectrum and a synthesized spectrum [53]. A synthesized spectrum may be taken as the linear combination of nuclide templates—computed or simulated by respective software codes. In this technique, the metric of the distance between measured and synthesized spectrum is quantified as the chi-squared error metric given by

$$\chi^2 = \sum_{n=1}^{N} \chi_n^2 = \sum_{n=1}^{N} \frac{[M(n) - S(n)]^2}{S(n)}, \tag{8.38}$$

where N is the number of energy bins, $S(n)$ the synthesized spectrum, and $M(n)$ is the measured spectrum.

Minimization of the above error metric is performed by an optimization algorithm—for instance genetic algorithms [54]—that is adopted to minimize the value of χ^2 with respect to synthesized spectrum parameters.

8.4.4 Peak Driven Spectrum Unfolding

In contrast to full spectrum unfolding, there are techniques that utilize only a part of the spectrum. Such techniques are the peak driven ones, that seek and extract specific features in the measured spectrum. In particular, they locate the peaks in the spectrum and match them with known nuclide peaks [55].

Peak driven techniques share a common framework independent of the mechanisms behind each individual technique. Figure 8.12 depicts the general framework of peak driven unfolding, where we observe the individual steps. Initially, the measured spectrum is forwarded to a spectral peak localization module. There, the peak in the measured spectrum are localized and their features are extracted. Next, the extracted peak features are compared to stored features in a library. The library contains known spectral peak features of radionuclides. Matching the

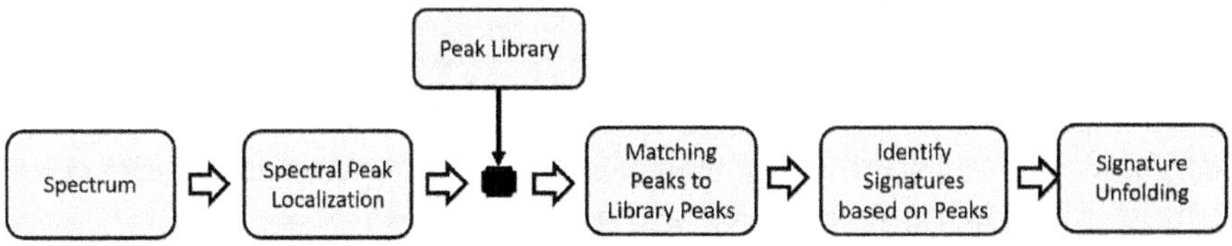

Fig. 8.12 Block diagram of the general peak driven signature unfolding technique

extracted features with stored features allows identification of source signatures in the measured spectrum. The set of identified signatures is then forwarded as the final output set of signatures of the unfolding technique.

There are several techniques that have been proposed that implement the above framework. The majority of them adopt tools from statistical pattern recognition, and machine learning. For instance, wavelets [56], fuzzy logic [57], and neural networks [58] are tools that have been employed in peak driven unfolding.

8.5 Advanced Algorithms for Distributed Detection Systems

8.5.1 Elements of Sensor Networks

Although sensors and computer based instrumentation have existed for long time, it was only until recently their integration to a single monitoring system made possible. In particular, advances in sensing technology and pervasive networking accommodated the development of sensor networks that can be applied to monitoring spaces, things, and/or specific actions [59].

A sensor is an artificial element that monitors and measures the values of a variable of interest. In the most usual case it is comprised of a single hardware element, i.e., the sensing element, that acquires measurements over a specific variable, e.g., temperature. Beyond the conventional sensors, there exists the class of intelligent sensors that are equipped with a software module that implements data processing and storage [60].

Sensor networks are comprised of several interconnected sensors known as the network nodes [59]. If the nodes of a network are of the same type, then the sensor network is characterized as homogeneous. In case, the network nodes are of various types, then the network is characterized as heterogeneous. Notably, the vast majority of sensor networks are comprised of wireless nodes that transmit their data via wireless links employing powerful communication protocols, and therefore are called *wireless sensor networks* (WSN). Development of WSN has significantly facilitated the use of mobile sensors that allow nodes to change locations and dynamically reconfigure the topology of the network [14].

Detection of radioactive material with multiple sensors is a topic of great interest because of its potentiality to increase detection probability. Deployment of multiple detectors and their interconnection forms a *radiation sensor network*. Acquisition of measurements (information) coming from various sources and their subsequent fusion enhances detection capabilities pertained to nuclear security. Detection enhancement is enabled by algorithms that process data from various sources and are known as distributed detection algorithms.

8.5.2 Bayesian Methods in Distributed Detection

Detection and localization of a source with mobile detectors in urban environments is an important challenge in nuclear security. Background radiation is the major challenge to overcome in identifying threats. In the sensing parlance, this challenge is quantified by the signal-to-noise ratio (SNR).

In Ref. [61] a Bayesian method for estimating the radioactive source parameters using SNR characterization in a sensor network has been presented. The main assumption of the method is that the rad detectors are placed on a straight line, while the source moves parallel to the detectors.

In particular, the detectors are placed uniformly across the line with distance between detectors set to d. The population of detectors is equal to Ψ and they acquire measurements over time intervals of length τ, while the background rate is constant and equal to β. The source moves with constant speed v at a distance $h > 0$ from the detector lineup. The source has an activity α and dimensions T^{-1}.

The detection method is based upon estimation of parameters α, v, and h and utilization of the counts recorded by the sensors s_n, $n = 1, \ldots, \Psi$ in a time interval $t, t = 1, \ldots, \Theta$ of length τ; the counts measured by sensor s in interval t are denoted as c_{st}. The key is to express the probability of observing the counts c_{st} in terms of β, δ, and τ, while assuming that they follow Poisson distribution. Therefore, the following probability expression is obtained:

$$P(c_{11}, \ldots, c_{\Psi\Theta}|\alpha, h, v; \beta, d, \tau) = \prod_{s=1}^{\Psi}\prod_{t=1}^{\Theta} \exp\{-\mu_{st}\}\frac{\mu_{st}^{c_{st}}}{c_{st}!}, \tag{8.39}$$

with μ_{st} being the expected counts measured by sensor s at time interval t. Further, the sensor measurement rate is approximated by $\chi\alpha r_s^{-2}$, where α is the activity, and χ the cross-section coefficient of the source, while r_s is the source-detector distance. Thus, the number of non-background counts is given by [61]:

$$\frac{\chi\alpha}{hv}\int_{x_{st}}^{x_{st}+v\tau} \frac{1}{r_s^2(x)}dx, \tag{8.40}$$

where x_{st} is the coordination of the source at the beginning of the time interval t and r_s is the distance of the source from sensor s. Hence, the expected counts in sensor s are taken by

$$\mu_{st} = \frac{\chi\alpha}{hv}\tan^{-1}\left(\frac{x_{st} + v\tau}{h}\right) - \tan^{-1}\left(\frac{x_{st}}{h}\right) + \beta\tau \tag{8.41}$$

with the $\tan^{-1}$ functions yielding values constrained to $[-\pi/2, \pi/2]$.

Assuming β, d, and τ are a priori known constants, along with a uniform prior distribution, the Eq. (8.39) provides *a posterior* distribution of α, h, and v given the data. This posterior distribution is characterized through its moments denoted as $\langle \alpha^i \rangle$, $\langle h^j \rangle$, and $\langle v^k \rangle$ and are defined as [61]

$$\langle \alpha^i, h^j, v^k \rangle = \frac{\iiint_0^{\alpha_m, h_m, v_m} \alpha^i h^j v^k P(c_{11}, \ldots, c_{\Psi\Theta} | \alpha, h, v; \beta, d, \tau) d\alpha dh dv}{\iiint_0^{\alpha_m, h_m, v_m} P(c_{11}, \ldots, c_{\Psi\Theta} | \alpha, h, v; \beta, d, \tau) d\alpha dh dv}, \tag{8.42}$$

where α_m, h_m, v_m stand for upper bounds.

The first moments $\langle \alpha_m \rangle$, $\langle h_m \rangle$, $\langle v_m \rangle$ computed by Eq. (8.42) consist of the estimates of the source parameters. Additionally, the second moments may also be obtained and subsequently utilized to obtain an ellipsoidal confidence domain [61].

Overall, computation of the source parameters utilizing information of multiple detectors enable the detection and localization of a source. In addition, the presented method also provides a confidence level over the computed parameters.

8.5.3 Hierarchical Source Model Detection

In Ref. [62], an algorithm that proposes a hierarchical source model detection. The algorithm assumes a sensor network architecture that acquires measurements continuously. By assuming that there are L sensors in the network, p_i is the probability of detection and q_i is the probability of false alarm for the local sensor i. Thus, a hypothesis test may be formulated as

$$H_0 : \phi = \phi_0 \tag{8.43}$$

$$H_1 : \phi = \phi_1 > \phi_0, \tag{8.44}$$

where $\phi_1 = [p_1 \ldots . p_L]$ and $\phi_0 = [q_1 \ldots .]$. By denoting the decision taken from a single sensor over N time intervals as u_i, the respective decision vector that takes into consideration all the sensors is $\mathbf{u} = [u_i \ldots . u_L]$.

Next, using the above equations, the *generalized likelihood ratio* (GLR) is formed as

$$\Lambda_G(\mathbf{u}) = \frac{P(\mathbf{u}|H_1, \phi_1^{ML})}{P(\mathbf{u}|H_0)}, \tag{8.45}$$

where P denotes the Bernoulli distribution probability and ϕ_1^{ML} is the maximum likelihood estimation of ϕ_1. The key idea is that the individual sensor decisions, i.e., $\mathbf{u}_i$, is a sufficient statistic for estimating p_i and eliminating the spatiotemporal dependence of the sensor [62]. Therefore, the GLR becomes:

$$\Lambda_G(\mathbf{u}) = \frac{\prod_{i=1}^{L} \prod_{n=1}^{N} P(u_{in}|H_1, \phi_1^{ML})}{P(u_{in}|H_0)} \tag{8.46}$$

and the maximum-likelihood of p_i can be taken as the argmax $P(\mathbf{u}_i|p_i)$, where

$$P(\mathbf{u}_i|p_i) = \prod_{n=1}^{N} p_i^{u_{in}}(1-p_i)^{1-u_{in}}. \tag{8.47}$$

However, a low quality of data will subsequently provide low quality parameter estimates, resulting in bringing detection probability at a lower level than the false alarm probability. In that case, the overall performance of the GLR test is significantly degraded [62].

To avoid any degradation, the *restricted range ML estimation* (RMLE) has been developed while evaluating the GLR test statistic. The RMLE adopts prior knowledge in order to constrain the range of the parameter to be estimated. In the radioactive source detection, such constraint is imposed by setting the detection probability higher than that of the false detection [62].

The RMLE as opposed to MLE prevents the global degradation of the sensor network performance. In particular, sensors that consistently provide false alarms (*bad* sensors) are censored by RMLE and ignored. Isolation of bad sensors prevents degradation of the detection performance.

8.5.4 Other Distributed Detection Methods

The above presented distributed detection methods exhibit the dependence of distributed detection via sensor networks on use and fusion of data using statistical tools [63]. Except for the aforementioned algorithms, there is a variety of methods that have been proposed and are of interest in nuclear security. In Ref. [64], algorithms that use statistical thresholds and model γ-ray intensities in terms of model states were proposed. In Ref. [65], a statistical method that models the correlation between sensors that observe the same event was presented. In Ref. [66], a statistical approach that takes into consideration the temporal periodicity of signal observed by the sensors in the network was introduced and tested. Lastly, in Ref. [67], a weighted graph based algorithm is introduced that maps with respect to space the risk of a radioactive source detection by utilizing mobile detector measurements.

8.6 Conclusion

Utilization of imaging and data processing algorithms is essential toward developing and deploying efficient systems in nuclear security. Automated algorithms promote interpreting various physical processes, making predictions, and facilitating human operator decisions.

In this chapter, an overview of the basic principles of algorithms and methods pertaining to imaging, data processing, spectrum signature unfolding, and distributed detection in nuclear security is given. With the advent of the era of big data, fueled by new technologies in sensing and data acquisition and storage technologies, more sophisticated algorithms will be needed to satisfy the nuclear security needs. However, the underlying principles behind any algorithm will remain unchanged. This chapter aspired to present methods in data processing by emphasizing the principles that drive the methods rather than focusing on the specifics of their implementation.

References

1. C. Wirz, E. Egger, Int. Rev. Red Cross **87**(859), 497 (2005)
2. M.D. Chinn, R.W. Fairlie, Oxf. Econ. Pap. **59**(1), 16 (2007)
3. L. Devriendt, B. Derudder, F. Witlox, Telecommun. Policy **34**(8), 417 (2010)
4. L. Atzori, A. Iera, G. Morabito, Comput. Netw. **54**(15), 2787 (2010)
5. C.S. Cho, W.H. Chung, S.Y. Kuo, IEEE Trans. Syst. Man Cybern. Syst. **46**(3), 356 (2016)
6. P.L. Lagari, V. Sobes, M. Alamaniotis, L.H. Tsoukalas, Int. J. Monit. Surveill. Technol. Res. **4**(4), 54 (2016)
7. M. Alamaniotis, L.H. Tsoukalas, in *2015 IEEE 27th International Conference on Tools with Artificial Intelligence (ICTAI)* (IEEE, Piscataway, 2015), pp. 1114–1121
8. A. Burger, M. Groza, Y. Cui, U.N. Roy, D. Hillman, M. Guo, L. Li, G.W. Wright, R.B. James, Phys. Status Solidi C **2**(5), 1586 (2005)
9. M.J. Kristo, S.J. Tumey, Nucl. Instrum. Methods Phys. Res. Sect. B Beam Interact. Mater. Atoms **294**, 656 (2013)
10. M. Alamaniotis, S. Terrill, J. Perry, R. Gao, L. Tsoukalas, T. Jevremović, Nucl. Technol. Radiat. Prot. **24**(1), 46 (2009)
11. J.A. Favorite, Z. Perkó, B.C. Kiedrowski, C.M. Perfetti, Nucl. Sci. Eng. **185**(3), 384 (2017)
12. N. Tsoulfanidis, *Measurement and Detection of Radiation* (CRC press, Boca Raton, 2013)
13. T.K. Ho, Multi. Classif. Syst. **2096**, 53 (2001)
14. I.F. Akyildiz, W. Su, Y. Sankarasubramaniam, E. Cayirci, Comput. Netw. **38**(4), 393 (2002)
15. A. Koschan, M. Pollefeys, M. Abidi, *3D Imaging for Safety and Security*, vol. 35 (Springer Science & Business Media, Berlin, 2007)
16. R.C. Runkle, D.L. Chichester, S.J. Thompson, Nucl. Instrum. Methods Phys. Res. Sect. A Accel. Spectrom. Detect. Assoc. Equip. **663**(1), 75 (2012)
17. G. Zentai, Int. J. Signal Imaging Syst. Eng. **3**(1), 13 (2010)
18. N.G. Cutmore, Y. Liu, J.R. Tickner, in *2010 IEEE International Conference on Technologies for Homeland Security (HST)* (IEEE, Piscataway, 2010), pp. 330–336
19. J. Allison, K. Amako, J. Apostolakis, H. Araujo, P.A. Dubois, M. Asai, G. Barrand, R. Capra, S. Chauvie, R. Chytracek, et al., IEEE Trans. Nucl. Sci. **53**(1), 270 (2006)

20. G. Chen, Nucl. Instrum. Methods Phys. Res. Sect. B Beam Interact. Mater. Atoms **241**(1), 810 (2005)
21. Y. Xing, L. Zhang, X. Duan, J. Cheng, Z. Chen, IEEE Trans. Nucl. Sci. **58**(2), 537 (2011)
22. S.R. Deans, *The Radon Transform and Some of Its Applications* (Courier Corporation, North Chelmsford, 2007)
23. A. Katsevich, Adv. Appl. Math. **32**(4), 681 (2004)
24. K. Takei, Hitachi Rev. **53**(2), 97 (2004)
25. A. Manickavasagan, N. Yasasvy (eds.), in *Imaging with Electromagnetic Spectrum* (Springer, Berlin, 2014), pp. 17–31
26. K. Vetter, M. Burks, L. Mihailescu, Nucl. Instrum. Methods Phys. Res. Sect. A Accel. Spectrom. Detect. Assoc. Equip. **525**(1), 322 (2004)
27. Z. Zhu, L. Zhao, J. Lei, in *IEEE Computer Society Conference on Computer Vision and Pattern Recognition-Workshops, 2005. CVPR Workshops* (IEEE, Piscataway, 2005), pp. 126–126
28. Y. Eisen, A. Shor, I. Mardor, Nucl. Instrum. Methods Phys. Res. Sect. A Accel. Spectrom. Detect. Assoc. Equip. **428**(1), 158 (1999)
29. J. Eberhardt, S. Rainey, R. Stevens, B. Sowerby, J. Tickner, Appl. Radiat. Isot. **63**(2), 179 (2005)
30. Y. Liu, B. Sowerby, J. Tickner, Appl. Radiat. Isot. **66**(4), 463 (2008)
31. B. Sowerby, J. Tickner, Nucl. Instrum. Methods Phys. Res. Sect. A Accel. Spectrom. Detect. Assoc. Equip. **580**(1), 799 (2007)
32. L.W. Alvarez, J.A. Anderson, F. El Bedwei, J. Burkhard, A. Fakhry, A. Girgis, A. Goneid, F. Hassan, D. Iverson, G. Lynch, et al., Science **167**(3919), 832 (1970)
33. C. Morris, C. Alexander, J. Bacon, K. Borozdin, D. Clark, R. Chartrand, C. Espinoza, A. Fraser, M. Galassi, J. Green, et al., Sci. Glob. Secur. **16**(1–2), 37 (2008)
34. J. Armitage, D. Bryman, T. Cousins, G. Gallant, A. Jason, G. Jonkmans, S. Noël, G. Oakham, T.J. Stocki, D. Waller, in *AIP Conference Proceedings*, vol. 1194 (AIP, College Park, 2009), pp. 24–35
35. L.J. Schultz, K.N. Borozdin, J.J. Gomez, G.E. Hogan, J. McGill, C. Morris, W. Priedhorsky, A. Saunders, M. Teasdale, Nucl. Instrum. Methods Phys. Res. Sect. A Accel. Spectrom. Detect. Assoc. Equip. **519**(3), 687 (2004)
36. A. Aleksandrov, A. Bagulya, S. Baklagin, M. Chernyavsky, V. Galkin, V. Grachev, N. Konovalova, A. Managadze, N. Polukhina, T. Roganova, et al., in *EPJ Web of Conferences*, vol. 125 (EDP Sciences, Les Ulis, 2016), p. 02022
37. K.N. Borozdin, G.E. Hogan, C. Morris, W.C. Priedhorsky, A. Saunders, L.J. Schultz, M.E. Teasdale, Nature **422**(6929), 277 (2003)
38. R. Hoch, D. Mitra, K. Gnanvo, M. Hohlmann, *Opportunities and Challenges for Next-Generation Applied Intelligence* (Springer, Berlin, 2009), pp. 225–231
39. S. Riggi, V. Antonuccio-Delogu, M. Bandieramonte, U. Becciani, A. Costa, P. La Rocca, P. Massimino, C. Petta, C. Pistagna, F. Riggi, et al., Nucl. Instrum. Methods Phys. Res. Sect. A Accel. Spectrom. Detect. Assoc. Equip. **728**, 59 (2013)
40. M. Hohlmann, P. Ford, K. Gnanvo, J. Helsby, D. Pena, R. Hoch, D. Mitra, IEEE Trans. Nucl. Sci. **56**(3), 1356 (2009)
41. R.A. Johnson, D.W. Wichern, et al., *Applied Multivariate Statistical Analysis*, vol. 4 (Prentice-Hall, Upper Saddle River, NJ, 2014)
42. M. Najim, *Modeling, Estimation and Optimal Filtering in Signal Processing* (Wiley, Hoboken, NJ, 2008), pp. 335–340
43. H. Hoffmann, Pattern Recogn. **40**(3), 863 (2007)
44. M. Alamaniotis, H. Hernandez, T. Jevremovic, in *2013 Fourth International Conference on Information, Intelligence, Systems and Applications (IISA)* (IEEE, Piscataway, 2013), pp. 1–4
45. M. Alamaniotis, S. Lee, T. Jevremovic, Nucl. Technol. **191**(1), 41 (2015)
46. C.M. Bishop, *Pattern Recognition and Machine Learning* (Springer, New York, 2006)
47. K. Fukunaga, *Introduction to Statistical Pattern Recognition* (Academic Press, Cambridge, 2013)

48. A.S. Kapadia, W. Chan, L.A. Moyé, *Mathematical Statistics with Applications* (CRC Press, Boca Raton, 2017)
49. C.E. Rasmussen, C.K. Williams, *Gaussian Processes for Machine Learning*, vol. 1 (MIT Press, Cambridge, 2006)
50. L.H. Tsoukalas, R.E. Uhrig, *Fuzzy and Neural Approaches in Engineering* (John Wiley & Sons, Hoboken, 1997)
51. V. Kurkova, Neural Netw. **5**(3), 501 (1992)
52. T. Burr, M. Hamada, Algorithms **2**(1), 339 (2009)
53. J. Mattingly, D.J. Mitchell, IEEE Trans. Nucl. Sci. **57**(6), 3734 (2010)
54. M. Alamaniotis, J. Mattingly, L.H. Tsoukalas, IEEE Trans. Nucl. Sci. **60**(3), 2222 (2013)
55. M. Alamaniotis, A. Ikonomopoulos, T. Jevremovic, L.H. Tsoukalas, Nucl. Technol. **175**(2), 480 (2011)
56. C. Sullivan, M. Martinez, S. Garner, IEEE Trans. Nucl. Sci. **53**(5), 2916 (2006)
57. M. Alamaniotis, A. Heifetz, A.C. Raptis, L.H. Tsoukalas, IEEE Trans. Nucl. Sci. **60**(4), 3014 (2013)
58. M. Kamuda, J. Stinnett, C. Sullivan, IEEE Trans. Nucl. Sci. **64**(7), 1858–1864 (2017)
59. D. Culler, D. Estrin, M. Srivastava, Computer **37**(8), 41 (2004)
60. P. Boltryk, C. Harris, N. White, J. Phys. Conf. Ser. **15**(1), 155–160 (2005)
61. R.J. Nemzek, J.S. Dreicer, D.C. Torney, T.T. Warnock, IEEE Trans. Nucl. Sci. **51**(4), 1693 (2004)
62. A. Sundaresan, P.K. Varshney, N.S. Rao, in *IEEE International Conference on Acoustics, Speech and Signal Processing, 2009. ICASSP 2009* (IEEE, Piscataway, 2009), pp. 2901–2904
63. N.S. Rao, Measurement-based statistical fusion methods for distributed sensor networks, in *Distributed Sensor Networks*, ed. by S.S. Iyengar, R.R. Brooks (Chapman and Hall/CRC Publishers, Boca Raton, 2005)
64. M. Wieneke, W. Koch, H. Friedrich, S. Chmel, in *Future Security* (Springer, Berlin, 2012), pp. 376–387
65. A. Sundaresan, P.K. Varshney, N.S. Rao, in *2007 10th International Conference on Information Fusion* (IEEE, Piscataway, 2007), pp. 1–7
66. R. Coulon, V. Kondrasovs, K. Boudergui, S. Normand, IEEE Trans. Nucl. Sci. **61**(4), 2189 (2014)
67. D.S. Hochbaum, B. Fishbain, Ann. Oper. Res. **187**(1), 45 (2011)

Chapter 9
Examples of Active Interrogation Systems

Dennis Slaughter, Anna S. Erickson, and Igor Jovanovic

Abstract In this chapter we briefly describe some of the prototype active interrogation systems that have reached a reasonable level of integration and testing readiness, and we summarize their performance. The focus is on the Nuclear Carwash system developed by Lawrence Livermore National Laboratory, but we also include the Pulsed Neutron Fast Analysis system developed by Rapiscan Systems, the advanced scanner developed by Passport Systems, and the prototype laser-based scanner based on inverse Compton scattering.

9.1 Introduction and Range of Application

A primary national security goal is to detect and identify special nuclear material (SNM) being smuggled into the US via cargo containers arriving at US seaports. As discussed in Chap. 2, while the radiation signal from plutonium is strong and readily detected in many cases, the radiation signal from highly enriched uranium (HEU) is very weak and does not penetrate cargo easily. It is necessary to amplify that signal substantially to make HEU detectable in normal commerce. In this discussion, the goal is to detect and identify a SNM (Pu or ^{235}U) with mass greater than 5 kg embedded in a fully loaded cargo container.

D. Slaughter
Lawrence Livermore National Laboratory, Livermore, CA, USA
e-mail: slaughter1@llnl.gov

A. S. Erickson
Georgia Institute of Technology, Atlanta, GA, USA
e-mail: erickson@gatech.edu

I. Jovanovic (✉)
Department of Nuclear Engineering and Radiological Sciences, University of Michigan, Ann Arbor, MI, USA
e-mail: ijov@umich.edu

I. Jovanovic, A. S. Erickson (eds.), *Active Interrogation in Nuclear Security*, Advanced Sciences and Technologies for Security Applications,
https://doi.org/10.1007/978-3-319-74467-4_9

Detection of SNM signatures is strongly affected by shielding due to cargo. Standard cargo containers have inner length of 5.87 and 12.03 m (frequently referred to as 20-ft and 40-ft container, respectively) and rectangular cross sections of 2.35 m by 2.39 m. If the cargo is homogeneous and loaded to the container net weight limit of $m = 28.2$ MT (characteristic of 20-ft containers), the mean cargo density is $\langle \rho \rangle = 0.86$ g/cm^3. For an SNM object located at the most inaccessible place in the container (on centerline), the corresponding areal density along the $L = 118$ cm attenuation path to the wall, floor or ceiling is at least $\rho L \geq 100$ g/cm^2. This represents the worst case scenario and is the baseline assumption for the discussion and analyses that follow in this chapter. But, for context, we should note that the majority of containers (80%) are 12.03-m long, and their maximum load implies density $\langle \rho \rangle = 0.39$ g/cm^3 with $\rho L = 48$ g/cm^2. The actual cargo loads in commercial traffic [1] have mean density $\langle \rho \rangle = 0.1 - 0.3$ g/cm^3.

9.2 Nuclear Carwash

Recently, a new technology suggested by Prussin and Norman has been developed to not only detect but identify HEU and distinguish it from other radioactive materials [2] by utilization of the high-energy fission product γ-ray radiation. Figure 9.1 shows strong yield of high-energy (3–5 MeV range) γ-ray radiation from fission products and much lower yield of radiative capture products from (n,γ)

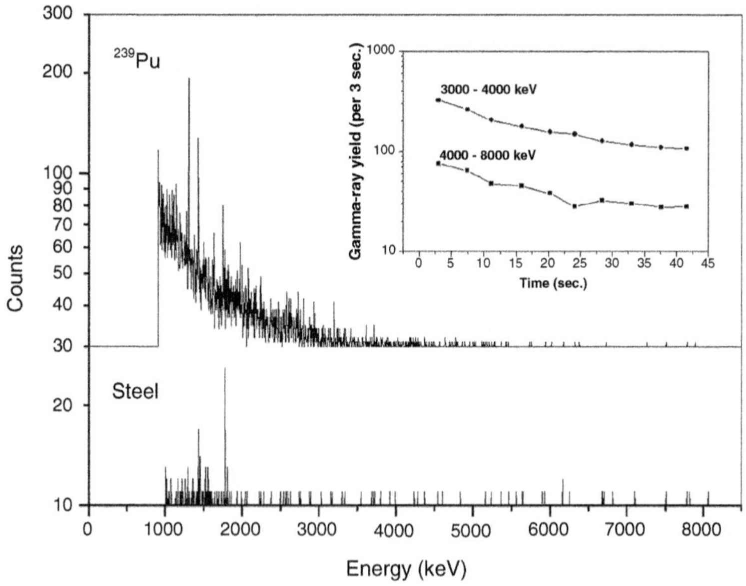

Fig. 9.1 High-energy fission product γ rays compared to capture γ rays when ^{239}Pu is irradiated with neutrons. The fission product yield of γ-ray radiation above $E_\gamma > 3$ MeV is large compared to other reactions. Reproduced from Ref. [2]

and (n,n′γ) reactions. Compared to neutron activation of materials, for example steel, the production of fission products is orders of magnitude more intense. In addition, low-resolution detectors may be employed with energy discriminators set to $E_\gamma > 3$ MeV, making them effectively *blind* to the terrestrial background radioactivity that does not extend higher than 2.6-MeV γ rays. Finally, high-energy γ rays are highly penetrating in both wood and steel cargo, so that signal attenuation is not prohibitive. Practical signal to background ratios are potentially very large in this technique. Utilization of high-energy fission product γ rays in combination with their characteristic decay times with half-lives of 5–100 s unique for fission products is the basis of the new SNM detection concept [3], sometimes referred to as the *Nuclear Car Wash* (NCW). The name was suggested by the architecture shown in Fig. 9.2 that resembles an ordinary car wash where vehicles are towed through an array of nozzles and brushes.

Cargo containers to be inspected are towed along a track that passes the interrogation source shown in Fig. 9.1. The interrogation source emits a fan-shaped beam, which is collimated in the vertical plane and is thin in the direction of cargo motion. The design described here employs a neutron source to generate neutron-induced fission, but a strong photon source with significant intensity above the (γ,f)

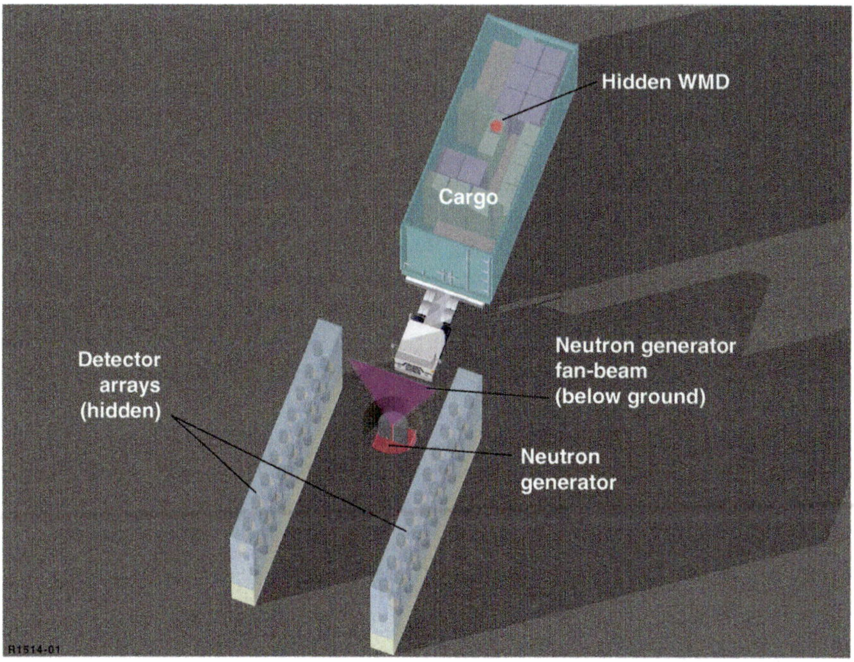

Fig. 9.2 Schematic layout of the Nuclear Car Wash at a US seaport. The interrogation beam is located below the cargo container floor and directed vertically. Two linear detector arrays are arranged to transport the cargo over the interrogation beam and along the detector arrays. Reproduced from Ref. [4]

threshold at $E_\gamma \sim 5.5\,\text{MeV}$ could also be used to induce photofission. In both cases, the interrogation beam generates fission in SNM that might be present and produces fission product radiation that is much more intense than the intrinsic radiation due to HEU radioactive decay. There are roughly 12 fission product γ rays produced per fission and about half of those are β-delayed with lifetimes ranging from $t = 0.1\,\text{s}$ to many minutes. More importantly, the energy spectrum of fission product γ-ray radiation extends to about $E_\gamma \sim 10\,\text{MeV}$ with roughly $0.13\,\gamma$ rays per fission in ^{235}U (0.065 in Pu) emitted at energies above $E_\gamma > 3\,\text{MeV}$ [2, 5–8]. This yield is far more than the 0.015 delayed neutrons per fission in ^{235}U (0.0061 in Pu) previously utilized in some SNM detection technologies. Thus the high-energy part of the fission product γ-ray spectrum provides a very strong fission signal and this is enhanced by low intensity of the terrestrial background at high energy.

The cargo container track down stream from the interrogation source is lined with an array of high efficiency γ-ray spectrometers on one or both sides of the track that detect only the high energy part of the fission spectrum, i.e. the part of the spectrum whose energy exceeds the $E_{th} \sim 3\,\text{MeV}$ threshold set on the spectrometers. In addition, if the interrogation source is pulsed and the detectors time-gated to the source, it is possible to gate-out the γ rays due to (n,γ) elastic and (nn$'$,γ) inelastic scatter reactions in the local environment. A major reduction in detector background count rate may be obtained without invoking high-energy resolution in the detectors. An array of low-resolution liquid scintillators or plastic scintillators provides adequate suppression of response to background radiation and they have very high sensitivity for detection of the high-energy γ-ray radiation characteristic of fission.

As the cargo container is towed past the interrogation source, the fission product radiation peaks when the target to be detected is aligned with a particular detector in the array and thus reveals the axial location of the target to be detected. As the container moves downstream from the source, the product radiation half-lives are determined as the peak count rate decreases from detector to detector. Fission products decay with half-lives in the range from 0.1 s to a few minutes. Typically, several half-lives are observed and their decay characteristics help to identify the target as HEU, DU, or Pu as these species have significantly different fission product yield distributions. Those distributions also differ when a neutron source or γ-ray source is utilized, but that is not the topic of this discussion.

There are a variety of other detection concepts employing active interrogation (AI). Of these, 23 are summarized in a PNNL report [9], and that report employs Monte Carlo simulations to compare the predicted sensitivity of each concept.

9.2.1 Neutron Source Design

The NCW can accommodate a photon source or a neutron source. The latter may produce neutron beams at $E_n = 14\,\text{MeV}$, beams at $E_n = 2.5 - 7\,\text{MeV}$, and low-

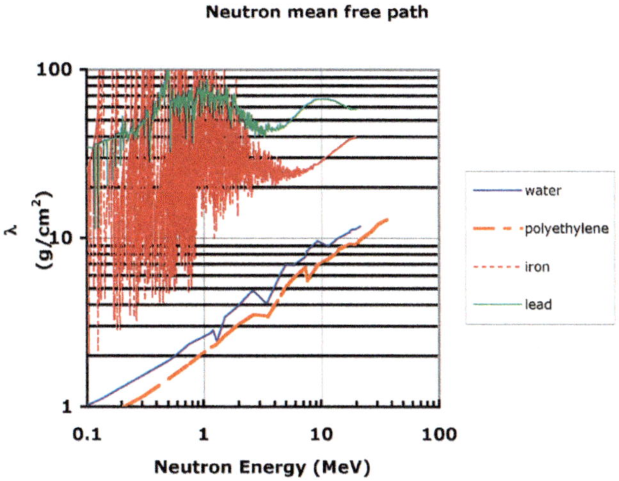

Fig. 9.3 Neutron mean free path in water, polyethylene, iron, and lead. Note that the neutron mean path is $\sim 3\times -9\times$ shorter in hydrocarbons compared to metals. Reproduced from Ref. [11]

energy $E_n = 60\,\mathrm{keV}$ beams by the ^{7}Li(p,n) reaction with proton beams at $E_p = 1.9\,\mathrm{MeV}$ [10]. Only the first two sources will be described here.

9.2.1.1 T(d,n) Neutron Source at $E_n = 14\,\mathrm{MeV}$

The neutron mean free path in several cargos is shown in Fig. 9.3. Examination of Fig. 9.3 shows that a 14-MeV beam produced by an economical electrostatic accelerator via the T(d,n) reaction penetrates well (two mean free paths) to about $\lambda \sim 80\,\mathrm{g/cm^2}$ in metals such as iron or lead but only to about $\lambda \sim 15\,\mathrm{g/cm^2}$ in water or polyethylene. In addition, the source is isotropic and thus the flux at the target is dramatically reduced by divergence of the beam. Nevertheless, an $E_n = 14\,\mathrm{MeV}$ source with high intensity output may be obtained readily from commercial suppliers and it is reasonably compact. Testing described in later sections was carried out with a Kaman Sciences neutron generator that produced a neutron yield $Y_n \leq 4\times 10^{10}\,\mathrm{n/s}$ or a neutron flux $\phi \geq 5\times 10^4\,\mathrm{n/cm^2/s}$ at $R = 2.5\,\mathrm{m}$ and the in-beam dose in air is roughly $D \sim 2.1\,\mathrm{mR/s}$ at that distance. Some applications may be satisfied by the beam available from a T(d,n) source; this will be discussed in a later section.

9.2.1.2 D(d,n) Neutron Source at $E_n = 7\,\mathrm{MeV}$

A radiofrequency quadrupole (RFQ) accelerator can be used to accelerate a deuteron beam to energies in the range $E_d \sim 4\,\mathrm{MeV}$. A gas target of deuterium provides a

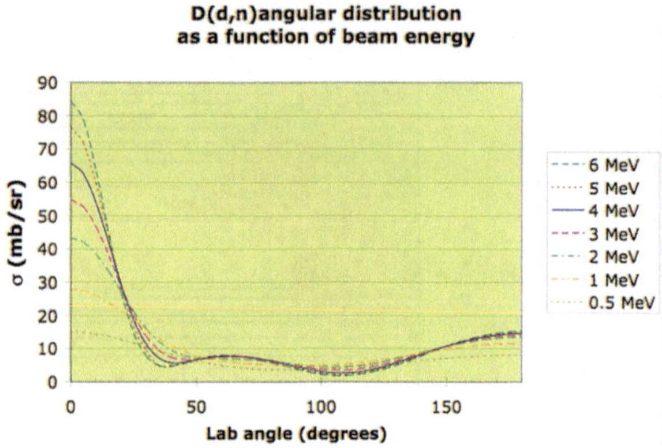

Fig. 9.4 Neutron output from RFQ generator at several deuteron beam energies. Note the narrow beam angle providing sharp *kinematic collimation*. Reproduced from Ref. [11]

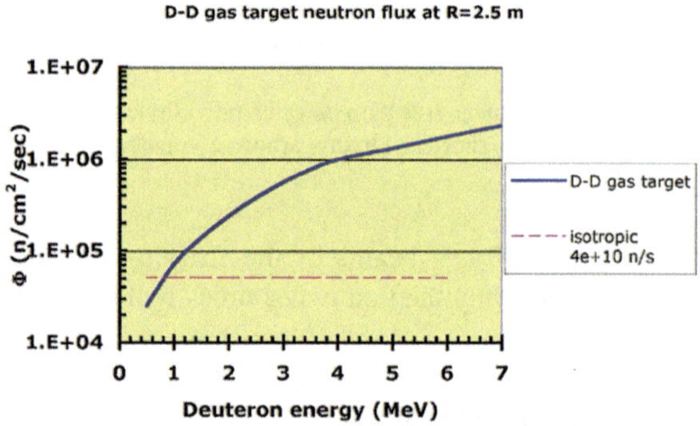

Fig. 9.5 Neutron flux for 14 MeV T(d,n) and 7 MeV RFQ sources. Note the large neutron flux at forward angles for the D-D reaction at MeV energies. Data shown for D(d,n) is for beam current $I_d = 100\,\mu\text{A}$. Reproduced from Ref. [11]

highly efficient beam production mechanism without the beam losses characteristic of a solid target, where most of the beam implants at low energy below neutron production threshold. In the best case most of the deuterons in the beam can produce a neutron. More importantly, the neutron beam is kinematically collimated, i.e. it is produced mostly in the forward direction as shown in Fig. 9.4. Note that a large fraction of the beam is directed within $\pm 15°$ of the axis, so that the flux at forward angles is much larger than that obtained for an isotropic source.

The effect of the neutron flux enhancement can be seen more clearly in Fig. 9.5. The figure shows that a $E_d = 4\,\text{MeV}$ deuteron beam with nominal beam current $I_d = 100\,\mu\text{A}$ into stopping gas target can produce a neutron flux on axis nearly 20× larger than a 4×10^{10} n/s T(d,n) source making the 7 MeV source equivalent to a 10^{12} n/s isotropic source. That is greatly in excess of commercially available T(d,n) sources, which do not quite reach the 10^9 n/s level. The RFQ accelerated deuteron beam with gas deuterium target is the neutron source of choice here, but a photon source could be used as well. Figure 9.6 shows a photo of a typical RFQ system.

The unit shown in the figure is the DL4 deuteron RFQ manufactured by Accsys Technology in Pleasanton, CA. It has a 425 MHz RFQ, producing a pulsed beam

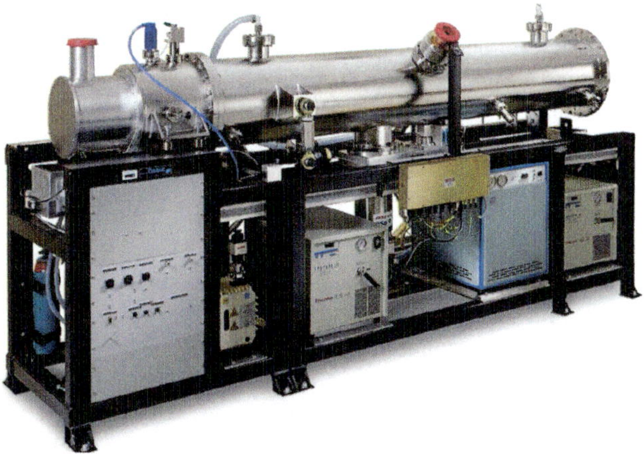

Fig. 9.6 RFQ neutron generator with $E_n = 7\,\text{MeV}$ output. The main accelerator tube on top is roughly 1 m above the floor and the overall system is roughly 5 m long. Reproduced from Ref. [11]

with 1% duty factor with mean current $I_d = 100\,\mu\text{A}$ beam of D^+ ions at $E_d = 4.0\,\text{MeV}$. A gas target produces neutrons by the D(d,n) reaction with energy $E_n = 7.0\,\text{MeV}$. (There is also another model that produces mean beam currents in the range $I_d = 0.4 - 1.0\,\text{mA}$.) Operated at $I_d = 100\,\mu\text{A}$ beam into a deuterium gas target, the in-beam neutron flux is $\phi \leq 7 \times 10^5\,\text{n/cm}^2\text{/s}$ at $R = 2.5\,\text{m}$ and the maximum neutron dose is $D \sim 25\,\text{mR/s}$ at that distance [12]. The neutron flux and dose are nearly two decades greater than the maximum available from commercial off-the-shelf T(d,n) generators and the beam is kinematically collimated in the forward direction, so that less shielding is needed on the back side. The neutron energy is low enough to avoid interference due to ^{16}N without reducing the penetration of the beam into cargo. (See Sect. 9.2.5 for a detailed discussion of ^{16}N interference.)

9.2.1.3 Interrogation Source Pulse Structure

There are at least two modes of operation.

Detection Mode

In the first mode, called the *detection mode*, a suspicious region can be irradiated by a single pulse and then the induced radioactivity detected in a single detector to determine the presence of high-energy fission product γ-ray radiation. This can be used with very short beam pulses followed by short counting time to determine whether fissionable material is present, but does not identify the material isotope as half-lives are not determined. The irradiation pulse length is limited by maximum dose to the cargo and by the maximum time that the threat object is positioned close to the collimated interrogation beam.

Identification Mode

A second mode is available, called the *identification mode*. In this configuration a chain of irradiation pulses illuminates the inspection zone and fission product γ-ray radiation is generated by fissionable material that may be present. The high-energy γ rays are detected in a linear array of detectors and the highest count rate will occur in the detector nearest to the threat location. As the cargo is towed along the track between the detector arrays, the peak count rate moves from one detector to the next as the irradiated threat object passes by. However, the peak count rate decays with time, so that analysis of the peak count vs. detector position can be analyzed to determine the β-decay lifetime of the fission product γ rays. There are several high-yield fission products and their relative yields are distinct from one fissionable species to another, so that the detected fission product radiation will exhibit a mix of half-lives that changes with time in a manner that depends on the fission product yield distribution. Consequently, analysis of the peak count rate decay from one detector to another as the cargo is towed past the detector array can be used to estimate the relative amount of each fissionable isotope present. The radiation times and counting times are increased to facilitate analysis of the various fission product half-lives. The efficacy of this approach is amplified in the results shown in Sect. 9.2.7.

Pulse and Counting Intervals

Given a time limit for detection of the maximum number of signal counts, the sensitivity of the method is greatest when the beam pulse length is equal to the counting interval. There is little benefit when either the pulse length or counting interval is longer than the longest of the β-decay half lives of the fission products generating the high-energy γ rays. For detection mode, both time intervals can be arbitrarily short to reduce interrogation time and to reduce dose to the cargo. In the second, *identification mode*, the available time is again divided into half for irradiation and half for radiation counting, but the pulse length and counting time are long enough to determine the several decay times for fission product decay. This can be as long as 2 min for irradiation and 2 min for observation of the decay as the object is towed along the detector array

9.2.2 Detector Design

The detection of high-energy γ-ray radiation requires a detector that is highly efficient at high energy and capable of counting at high rates. Both liquid and solid scintillators meet this requirement. Large volume scintillators are also relatively cheap, facilitating their deployment in a large array of units along a detection array. In addition, large plastic and liquid scintillators detect the high-energy γ-ray radiation through its Compton interaction. An energy threshold can be set, where the energy deposition by Compton scattering in the detector must exceed $E_{dep} > 2.5\,\text{MeV}$ for the count to be recorded as part of the signal. While liquid and plastic scintillators have generally poor resolution their spectral performance is adequate to establish a useful threshold for detection and to suppress terrestrial background.

Two versions of the detector array have been constructed and performance evaluated.

9.2.2.1 Liquid Scintillators

Figure 9.7 shows elements of the liquid scintillator array that was constructed. The detector cells were filled with liquid scintillator spanning dimensions $D = 20\,\text{cm}$ diameter by $L = 200\,\text{cm}$ length. There was a single $D = 20\,\text{cm}$ photomultiplier at

Fig. 9.7 Photo of liquid scintillator array component. Two PMTs are coupled to the ends and their outputs summed. Reproduced from Ref. [11]

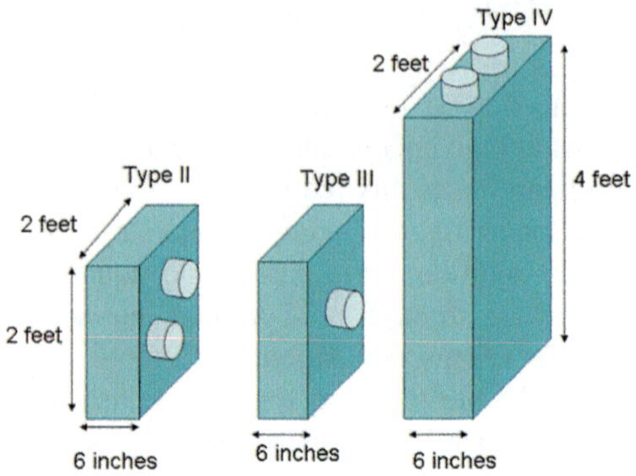

Fig. 9.8 Solid plastic scintillators. In all cases the thickness of scintillator material was $L = 25$ cm. Side dimensions were $L = 61$ cm and $L = 122$ cm. PMTs were installed as shown and PMT outputs were summed. Reproduced from Ref. [11]

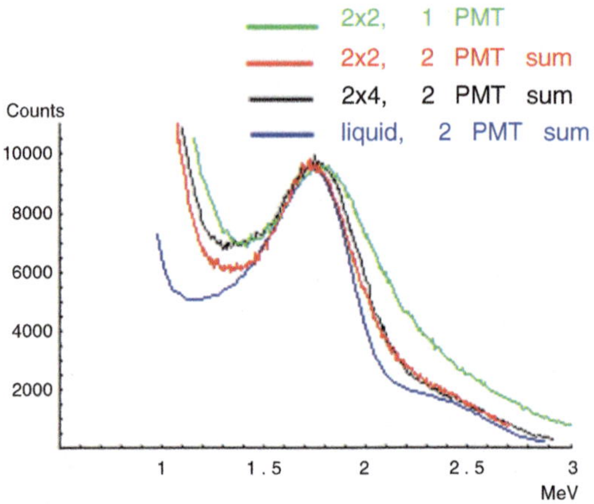

Fig. 9.9 Pulse height spectra for ^{88}Y peak at 1.86 MeV in three plastic scintillators and one liquid scintillator. The solid scintillators were used in pairs and/or in arrays of 2×4 detectors, each component 61 cm long by 61 cm wide by 25 cm thick. PMT outputs were summed except in the first (green) case. Reproduced from Ref. [11]

each end of the detector. The liquid scintillator provided 25–35% energy resolution, as shown in Fig. 9.9. However, there were chronic leaks of the scintillator requiring large catch-basins to trap the leaked material.

9.2.2.2 Solid Plastic Scintillators

As an alternative to liquid scintillators that tended to leak, an array of solid plastic scintillators was developed. Several geometries were chosen and they are shown in Fig. 9.8.

Energy resolution with a ^{88}Y source was determined from the pulse height distributions shown in Fig. 9.9. In most cases the energy resolution for Compton energy deposition was ~33–50% (FWHM) and the detectors were thick enough to provide multiple scattering and thus a suggestion of a full energy peak due to multiple Compton scattering. It is clear that the liquid scintillators with PMTs at each end whose outputs were summed provided the best energy resolution. This is important as it sets the variability of the energy threshold for γ-ray detection and, thus, it determines the background count rate and detection sensitivity.

9.2.3 *Threat Configurations of Interest*

9.2.3.1 Fissionable Target Characteristics

The performance envelope of the NCW was determined by interrogating disks of uranium or even stacks of assorted disks made of HEU. There were also two cylindrical containers of natural uranium oxide beads. All of the targets were utilized both alone and in combination to assess the performance of the NCW. A later section summarizes performance results with two of the targets listed. A summary of target pieces available is given in Table 9.1.

It is interesting to note that the fission rate in the target depends on surface area but not on enrichment or even its mass. In ^{235}U the fission mean free path (mean path length to cause fission) is only $\lambda_f \sim 0.037$ cm for thermal neutrons. Even if the enrichment is reduced to 3%, the thermal neutron mean free path increases to only $\lambda_f = 1.2$ cm. So that a uranium medium with dimensions of cm is essentially *black* to thermal neutrons, and competing capture reactions in other materials have cross sections small compared to ^{235}U fission. Consequently, for uranium enriched to more than 3% ^{235}U, nearly all thermal neutrons entering the surface undergo ^{235}U(n,f) and the enrichment or total mass does not matter. The effect of higher enrichment serves only to focus the thermal fissions closer to the entry surface, where there is less attenuation of the escaping fission signal and increasing mass increases the fission rate only to the extent that the exposed surface area increases.

Table 9.1 Cylindrical uranium targets for performance testing [13]

Target	Type	Total[a]	^{235}U (%)	^{235}U	Diameter	Length	Density	Form
1	Natural uranium	12,000	0.7	84	12	18	6	UO_2 beads
2	Natural uranium	24,000	0.7	168	36	12	6	UO_2 beads
3	HEU	469.7	94.7	376.5	8.64	3.34	2.4	UO_2 powder
4	HEU	280.4	93.2	221.1	8.64	1.96	2.4	UO_2 powder

[a]Total mass includes oxygen

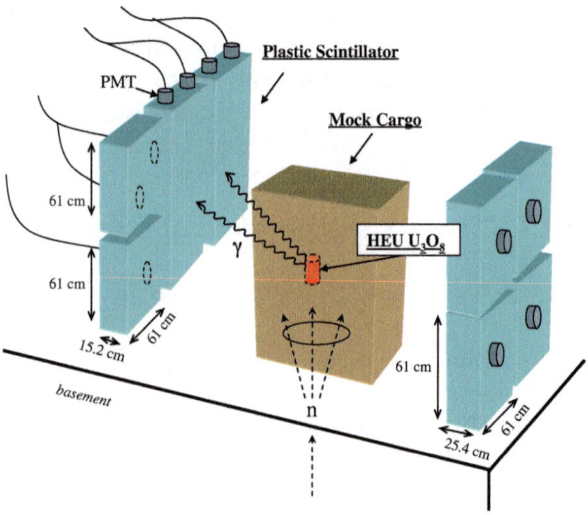

Fig. 9.10 Schematic layout of wood cargo and plastic detectors. The neutron interrogation beam comes up through the floor and into a $L = 1.8$ m high stack of plywood ($\rho = 0.55$ g/cm^3). Detectors were stacked in an array to one side of the wood. Some detectors had frontal area 2×2 ft and others 2×4 ft. Reproduced from Ref. [16]

9.2.3.2 Cargo Characteristics

A recent report [1] indicates that the majority of cargo coming into US ports consists of low density hydrocarbons and metal (mostly steel). Cargo containers have weight limits that prohibit loading with mean density greater than $\rho \leq 0.9$ g/cm^3. For testing purposes, two cargo configurations were employed, i.e. (1) the container was filled the full width and height with dry plywood sheets with spacers so that the density $\langle\rho\rangle \sim 0.55$ g/cm^3, and (2) steel pipes where the mean density averaged over the pipe walls and empty centers was $\langle\rho\rangle = 0.6$ g/cm^3. The U and HEU targets were lowered into holes in the cargo material so that the target was immersed in cargo on all sides. The container with the installed cargo simulants was moved past the collimated neutron beam to simulate scanning of cargo at an inspection station. The steel cargo is shown in Fig. 9.10.

The targets described above were embedded in the two homogeneous cargos described above and were scanned past the interrogation source or, in some cases, count rates were determined with the configuration fixed in space (Fig. 9.11).

9.2.4 Predicted Fission Rates

Monte Carlo modeling has been utilized to predict fission rates for the $E_n = 14$ MeV irradiation and they predict rates in wood cargo of 6.5×10^{-6} fissions/source neutron [15] for a $m = 750$ g HEU oxide target buried so that the intervening cargo thickness is $\rho L \sim 70$ g/cm^2. When the cargo is steel at $\rho \sim 0.6$ g/cm^3 the fission rate is $\sim 1.9 \times 10^{-6}$ fissions/source neutron.

Fig. 9.11 Steel pipes generating mean density $\langle\rho\rangle = 0.6\,\text{g/cm}^3$. Pipe wall thickness was chosen to obtained the design mean density. Reproduced from Ref. [14]

The situation for the D(d,n) source described earlier is quite different. The kinematics of the reaction provides a strong collimation in the forward direction so that this source produces flux on target comparable to an isotropic source $Y_n \sim 10^{12}$ n/s. Monte Carlo simulations have been utilized to estimate the fission production with the D(d,n) source and these have been supported by measurements, indicating that the signal to background ratio is $\geq 100\times$ even at early times after the end of irradiation [3] and is greater than $> 2\times$ even after penetrating $L = 1.2$ m of solid wood cargo.

The simulations indicate that the penetration in cargo is lower at $E_n = 7$ MeV compared to $E_n = 14$ MeV. For a 470 g cylinder of HEU (94% enriched) embedded $L = 122$ cm deep in plywood ($\rho = 0.6\,\text{g/cm}^3$) and at $R = 2.5$ m from the source, the fission rate per source neutron varies from 3.5×10^{-3} at $E_n = 14$ MeV to 1.1×10^{-3} at $E_n = 7$ MeV [13].

9.2.5 *Interferences*

When the T(d,n) reaction is utilized to generate a $E_n = 14$ MeV neutron interrogation beam, there are subsidiary reactions in oxygen due to the ^{16}O(n,p)^{16}N reaction whose reaction energy threshold is $E_n = 10$ MeV. The reaction product,

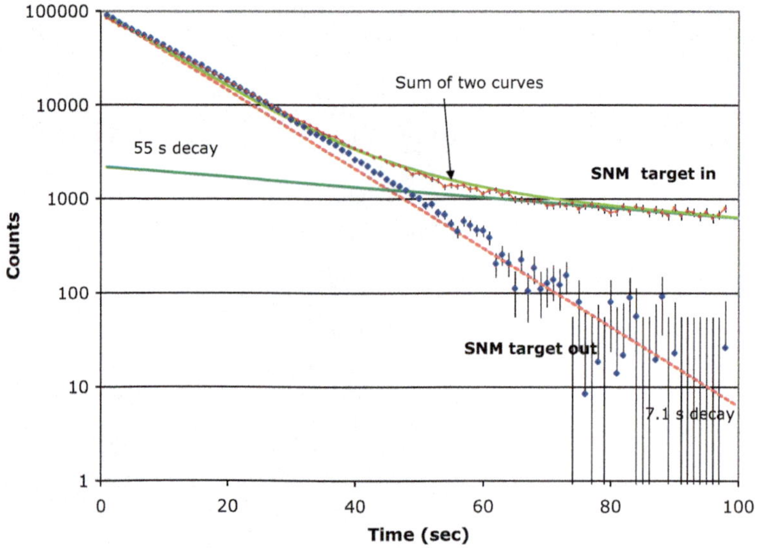

Fig. 9.12 Decay of high-energy γ-ray flux above $E_\gamma \geq 3$ MeV when cargo is irradiated at $E_n = 14$ MeV. Interference is due to the production of ^{16}N is observed. Reproduced from Ref. [4]

^{16}N, decays primarily by emitting a $E_\gamma = 6.1$ MeV γ ray whose decay half-life is $T_{1/2} = 7.1$ s. This generates a flux of γ rays above the $E_\gamma \geq 3$ MeV detection threshold, as shown in Fig. 9.12 [4].

The data was obtained with a $E_n = 14$ MeV neutron source producing 4×10^{10} n/s for 30 s (neutron dose at the entry wall was $D \sim 165$ mR at $R \sim 100$ cm) and that source was located 200 cm from the buried target, which consisted of a $m = 380$ g cylinder of 93% enriched HEU oxide buried 61 cm deep in a stack of plywood sheets 1.8 m high. The injected beam passed from the floor through 40 g/cm^2 of plywood to reach the HEU test object and the detected γ rays passed through 40 g/cm^2 of plywood to reach the detectors alongside the wood stack for a total path length of $\rho L = 80$ g/cm^2. The count rate vs. time is shown and is the sum of counts for four cells of liquid scintillation detectors, as described earlier. Data in Fig. 9.12 includes a green curve, showing detector counts with an SNM target buried in the cargo (*target in*) and a red curve showing detector counts without the SNM present (*target out*). The latter curve has been fitted to a 7.1 s half-life, while the former has been fitted to a 55 s half-life characteristic of one of the prominent high-energy fission product decays. The count rate is dominated by the ^{16}N decay due to environmental oxygen (in the wood and cooling water) until nearly $t > 55$ s, reducing the sensitivity for detection of SNM. The ratio of signal to background ^{16}N radiation is $\sim 10\times$ at $t = 70$ s and increases to $100\times$ at $t = 100$ s.

While interrogation with a 14 MeV beam is feasible, the sensitivity is reduced by environmental oxygen. That interference is generated in the cargo and in the water cooling system for the neutron source. This problem is overcome by employing a lower energy interrogation source. The D(d,n) source described in an earlier section generates a well-collimated beam of $E_n = 7\,\text{MeV}$ neutrons which is well below the reaction threshold for ^{16}N production.

There is a second interference due to fluorine, i.e. major component of Teflon. ^{19}F(n,α)^{16}N has a reaction threshold at $E_n > 1.6\,\text{MeV}$, but the cross section is small for $E_n \leq 5\,\text{MeV}$. Nevertheless, cargo that contains Teflon or other fluoridated products can produce a ^{16}N signal that may trigger false alarms. Detection of the decay curve can quickly identify this interference.

9.2.6 *Test Results*

9.2.6.1 Detection Probability for Small SNM Targets

The $E_n = 7\,\text{MeV}$ interrogation beam was installed in the floor below a stack of plywood sheets or steel pipes 1.8 m tall and its beam collimated to a narrow angle both by reaction kinematics and by a polyethylene/steel collimator. The arrangement is shown in Fig. 9.13. In Fig. 9.13a the plywood has been moved to the side to reveal the neutron source below the floor. The collimator aperture is the white polyethylene square within the larger black square. Four plastic scintillators are configured behind the neutron source. On the right are the liquid scintillators covered by a white curtain to the right of the neutron source. Figure 9.13b shows the plywood moved over the neutron source and it shows the four plastic scintillators to the right of the plywood stack.

Fig. 9.13 (**a**) Experimental layout of the neutron source, collimation, cargo simulation, and detectors. The source is in the floor under small white square within larger black square. The plywood stack on left is cargo simulation, which is moved over the source for testing. The four black squares are plastic scintillation detectors and the liquid scintillators are enclosed by the white curtain on the right; (**b**) plywood stack and rear/external view of plastic scintillators and their PMT connections. Reproduced from Ref. [15]

During the tests described below the neutron interrogation beam was operated at reduced beam current $I_d = 25\,\mu A$, the gas target produced in-beam neutron flux $\phi \leq 1.8 \times 10^5$ n/cm^2/s at $R = 2.5$ m, and was accompanied by an in-beam neutron dose rate $D \leq 6.3$ mR/s with the *beam on*. Beam irradiations were $t = 30$ s for a total dose to the cargo $D = 188$ mR and the detector counting interval was $T_c =$ 30 s.

In the results below a SNM target was assembled from available discs of HEU and embedded in one of several holes at various depths in the simulated cargo and each data set consisted of an irradiation with the *beam on* followed by a counting period, during which four plastic scintillators were summed to provide the total γ-ray count rate vs. time following interrogation.

Typical results are shown in Fig. 9.14. During data measurements there were variations from run to run in the beam current. The data shown in Fig. 9.14 and in Table 9.1 have been scaled up or down to correspond to $I_d = 25\,\mu A$ for purposes of comparison. In all of the data detector background was determined by irradiating the assembly described above without the insertion of an SNM target. Two count rates were recorded, i.e. the background during the 100 s immediately prior to irradiation (*background #1*) and the background immediately after irradiation (*background #2*). As observed in Fig. 9.14, the two backgrounds are the same, indicating that there are no neutron activation products generating γ-ray radiation above the detector energy threshold. Table 9.2 summarizes the analysis of data displayed in Fig. 9.14.

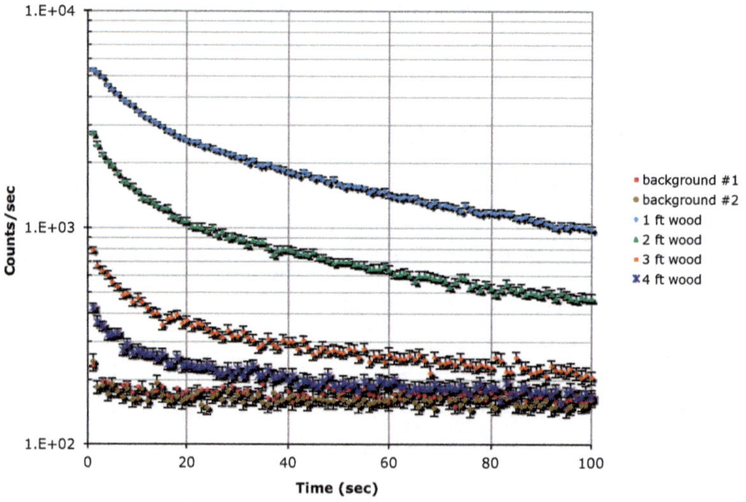

Fig. 9.14 Detector count rate following a $t = 30$ s irradiation of $m = 376$ g ^{235}U (HEU) at mean beam current $I_d = 25\,\mu A$. The SNM object was a cylinder buried in wood ($\rho = 0.6$ g/cm^3) at depths where the neutron path through the wood was $L = 4$ ft (purple, $\rho L = 73$ g/cm^2), $L = 3$ ft (orange, 55 g/cm^2), $L = 2$ ft (green, 37 g/cm^2), and $L = 1$ ft (blue, 18 g/cm^2). The pathway for detected γ rays at the HEU object to the detector arrays was $L = 61$ cm and 91 cm (37 g/cm^2 and 55 g/cm^2) and those detector count rates were summed in the figure. Reproduced from Ref. [12]

Table 9.2 Signals (N_{sig}) and backgrounds (N_{bkd}) accumulated for $t = 30$ s irradiation at $I_d = 25\,\mu$A (dose $D = 188$ mR) followed by detector counting over the interval $T_c = 1 - 30$ s after the end of irradiation for HEU embedded in cargo [16]

Signal window	Cargo	^{235}U mass (g) ρL (g/cm^2)	Beam attenuation $N_{sig} - N_{bkg}$	Net signal counts	Detection confidence $(N_{sig} - N_{bkg})/\sigma$
$\gamma = 3 - 4$ MeV	Wood	221	18[a]	27679	350
		221	35[a]	11156	141
		376	53[a]	2356	32
		376	71[a]	908	12
$\gamma = 3 - 5$ MeV	Steel	376	27[a]	3219	23
		376	46[a]	2157	16.5
		376	73[a]	3053	25
		376	91[a]	348	2.9
Delayed neutron	Steel	376	27[a]	604	11
		376	46[a]	376	7.2
		376	73[a]	141	2.1
		376	91[a]	800	15[b]

Sigma is twice the Poisson standard deviation of the background, $\sigma = 2N_{bkd}^{1/2}$

[a]The areal densities quoted are for the neutron path length from neutron generator to SNM target and do not include the additional $\rho L = 40$ g/cm^2 for the γ-ray path from SNM target to the detectors

[b]Background was not available for delayed neutron counts at $\rho L = 91$ g/cm^2 steel and the value assumed was taken from the $\rho L = 73$ g/cm^2 run and could be in error

The cargo thickness along the signal path from SNM target to container wall is $\sim \rho L = 40$ g/cm^2 and the thickness from neutron source to the SNM target ranges from zero to $L = 122$ cm ($\rho L = 91$ g/cm^2). It is noteworthy that the data in wood cargo with $I_d = 25\,\mu$A and $t = 30$ s irradiation produced a detected signal at least twice the background level (at early times) even for the deepest burial, where the neutrons penetrate up to $\rho L \leq 71$ g/cm^2 of wood cargo and $\rho L \leq 91$ g/cm^2 in steel. The target mass was only $m = 221$ g or $m = 376$ g of ^{235}U. Generally, the detections over this range of attenuation were made with a confidence level > 2.0× standard deviation of the background in both wood and steel cargo. In these data the detection probability is very high and the detections are high confidence, i.e. not false positives. Detection at twice the standard deviation of the background is usually considered $P_d \sim 98\%$ confidence, i.e. probability that these detections are not false positives.

9.2.6.2 Extrapolation to Performance with Goal Quantities of SNM

Attenuation was not tested at $\rho L = 100$ g/cm^2 as no cargo was available at the needed high density. Instead, the available results may be extrapolated to the detection of the goal quantity of SNM, i.e $m = 5\,\text{kg}$ is the goal. As stated earlier, the fission rate in the target is expected to increase linearly with target surface area. If the surface area increases with target mass, as $A \sim m^{2/3}$, a factor of 5.6× from $m = 376$ g to $m = 5\,\text{kg}$, the fission signal in the $m = 5\,\text{kg}$ target is also larger by that amount. The data in Table 9.2 shows increasing signal levels by factors roughly ~ 2.5–$5\times$ in wood when ρL is decreased by $\sim 20\,\text{g/cm}^2$. Similarly, fission signals increase by factors of a few when ρL decreases by $\sim 20\,\text{g/cm}^2$ in steel. It is reasonable to expect that a target object with $m = 5\,\text{kg}$ and exposing a surface area $\sim 5.6\times$ larger than the surface area of targets irradiated in these tests would generate a confidently detectable signal in steel at the goal $\rho L \sim 100\,\text{g/cm}^2$. The overall conclusion is that a full-size test target ($m = 5\,\text{kg}$) would be confidently detectable in steel cargo at thickness $\rho L \sim 100\,\text{g/cm}^2$ with radiation dose $D \sim 200\,\text{mR}$ and a $T_c = 30\,\text{s}$ counting interval following irradiation. A signal in wood cargo may be detectable at $\rho L \leq 90\,\text{g/cm}^2$, but that has not been demonstrated here.

As a backup detection mode, delayed neutron emission was monitored by an array of 14 moderated ^{3}He tubes [16]. The tubes ($D = 5.08\,\text{cm}$ diameter and $L = 91\,\text{cm}$ long) were embedded in $L = 10\,\text{cm}$ thick polyethylene moderator and filled with $p = 4$ atm ^{3}He. The background levels and background-subtracted signal levels for the neutron monitors are given in Table 9.2. Just as with the γ-ray signal, the delayed neutron signal registers confident detections at $\rho L = 90\,\text{g/cm}^2$ for the small targets, though there is a possible error in the background level in that measurement. As with γ-ray detection, the $m = 5\,\text{kg}$ target provides confident delayed neutron detections for the baseline target.

9.2.6.3 Concept of Operations for Field Implementation

The data shown was based on $T_c = 30\,\text{s}$ counting time. A long counting time will generally be required to identify fissionable isotopes through analysis of their fission product decay times. However, in the case where it is only necessary to detect the presence of fissionable material, the detection process can be shortened using a higher intensity pulsed interrogation beam and a truncated detection period. At higher beam current ($I_d = 1\,\text{mA}$), the dose $D \sim 250\,\text{mR}$ can be delivered in a pulse as short as $t \geq 1.0\,\text{s}$. This facilitates the same localization accuracy determined by the width of a detector module ($L = 120\,\text{cm}$) with faster scanning, i.e. at $V \sim 1.2$ m/s and scanning the entire 20 ft container in roughly $t \sim 5\,\text{s}$ and the 40 ft container in $t \sim 10\,\text{s}$. However, the signal level is dependent on the counting time, which depends on the length of the detector array and the scanning speed. For example, irradiating the inspection zone for only $t \sim 4\,\text{s}$ provides a dose $D \sim 1000\,\mathrm{mR}$ and a 5.3× increase in count rate compared to the measured data. A $T_c = 4\,\mathrm{s}$ counting interval provides nearly the same counting statistics as were obtained in the measured data. A pulsed format is then implemented, where the beam is alternately *on* and *off* with 50% duty cycle and $t = 8\,\mathrm{s}$ period. The container is drawn through the scanning system at speed up to $V \sim 3.6\,\mathrm{m/s}$ (8 mph).

9.2.6.4 Isotopic Identification

The previous section described a simple system to detect SNM by observing neutron-induced fission, where a very brief irradiation is followed by counting in a linear array of detectors, where the peak count rate is compared among detectors to determine the axial location of the SNM to be detected.

There are only a few fission products that produce high-energy γ-ray radiation at high yield. Their fission product yields are distinct for each fissionable isotope, so that the γ-ray spectrum evolves with time depending on which fissionable isotopes are present. Simulations can be developed that provide a basis set of pulse height spectra vs. time for each fissionable isotope of interest. This basis set can be used to apportion the contributions of each isotope to a measured 2-D pulse height spectrum vs. time. The fitting function then determines the relative contribution of each fissionable isotope to the measured spectrum over the first 100 s or so following the end of irradiation. Very rudimentary fitting functions have been developed and their fit on ^{235}U spectra validated [13]. An example is shown in Fig. 9.15.

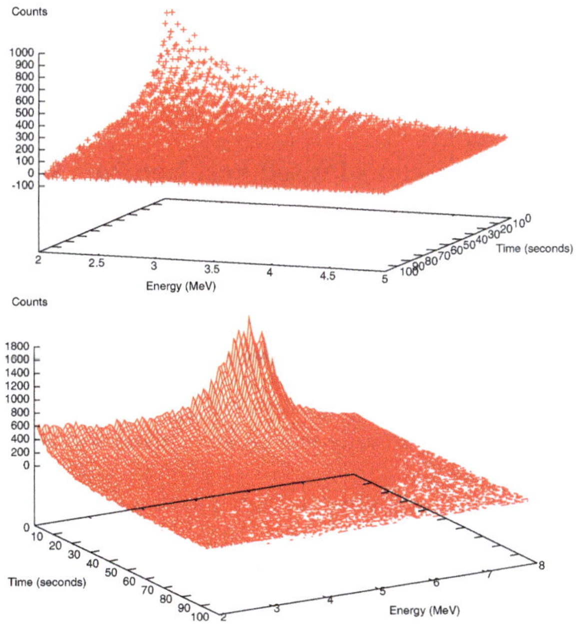

Fig. 9.15 Time-dependent pulse height data obtained with HEU buried in wood cargo. Two views are shown with axes interchanged for clarity. Reproduced from Ref. [17]

However, this work is in its earliest stages and its efficacy has yet to be demonstrated. An experimental study is required to determine the minimum data quality that leads to correct identification of the relative contributions of each isotope in a given experimental measurement as well as the error rate when significant activity is attributed to an isotope not present.

9.2.7 Efficacy and Limitations

Results shown in the previous sections establish that detection of $m = 0.4$ kg quantities of HEU embedded behind $\rho L = 91$ g/cm^2 of shielding material can be made with high confidence when an interrogation beam of $E_n = 7$ MeV is permitted to subject the cargo to collateral dose $D \sim 200$ mR. This supports the projected ability to detect small amounts of SNM deeply buried in cargo loaded to the weight limit of a standard cargo container, and to do this in short times (seconds). Extrapolation to $m = 5$ kg suggests that detection meets the goal for steel cargo at $\rho L \leq 100$ g/cm^2, but results with wood cargo are harder to predict. The architecture works best when there is space to employ a linear array of plastic scintillation detectors extending $L \sim 15$ m along a track leading from the irradiation portal. There is a potential to infer the isotopic identification of SNM that may be present, but this requires a scan of duration $T_c = 30$–60 s in order to analyze the decay times of the fission products that are present and to assign their unique mixture to a particular fissionable isotope. Primitive software has been developed for this purpose. But, this application has not yet been demonstrated. It is important to also note that this technology may produce false alarms in the presence of fluorine, though the isotopic ID capability may be used to identify these false alarms.

9.3 Rapiscan's Pulsed Fast Neutron Analysis

A cargo inspection system to detect explosives based on Pulsed Fast Neutron Analysis (PFNA) technique was deployed by Rapiscan at the George Bush Intercontinental Airport in Houston, Texas. A tandem Van de Graaff accelerator operating at a frequency of 3.5 MHz outputs deuterons with a full width at half maximum of 1.5 ns in duration. The pulsed 8.5-MeV neutrons are produced as a result of these deuterons interacting in a deuterium gas target. Oscillating collimator near the gas target creates vertically-moving neutron beam spot 9-cm wide by variable (typical) 12-cm tall.

Neutrons interacting in cargo produce γ rays collected in NaI detector arrays. The system utilizes a tandem Van de Graaff accelerator operating at a frequency of 3.5 MHz that produces pulses of deuterons with a Full Width at Half Maximum of 1.5 ns in duration. Neutrons of several nanosecond duration are created through the (d,D) reaction at an energy of around 8 MeV at a deuteron beam current of up to 140 μA. A neutron collimator near the deuteron gas target produces a neutron beam spot 9-cm wide by variable (typical) 12-cm tall at the center of the container. This neutron beam oscillates vertically by moving the collimator. Translational motion of the air cargo is provided by a constant-velocity conveyor system. The inspection volume is surrounded by a large array of 14-cm cube NaI detectors to collect the γ rays from the neutron inelastic scattering reactions occurring within the volume. Using a time-of-flight technique to determine the position in the container in which the neutron inelastic scattering reactions occur, the data acquisition system and the image reconstruction engine produce a three-dimensional image of the cargo contents. The images have a typical volume element granularity of 6.3 cm wide, 6.3 cm tall, and 8 cm deep. This technology employs personnel from a variety of engineering disciplines. Electrical engineers designed the particle accelerator, mechanical engineers designed the cargo conveyance system, and chemical engineers fabricated the explosive threat simulants.

9.4 EURopean Illicit TRAfficking Countermeasures Kit (EURITRACK)

The EURopean Illicit TRAfficking Countermeasures Kit (EURITRACK) was a special detection kit developed by an international collaboration to protect European's ports from entering illicit materials. The focus of the project was nonintrusive identification of chemical composition of objects in cargo containers. The system design was based on existing X-ray equipment producing radiographic images of the contents, but also included innovative Tagged Neutron Inspection System (TNIS) to provide chemical information about any objects flagged on the image.

TNIS is based on 14-MeV neutron interrogation, which produced characteristic γ rays utilized in identification of material in question. These characteristic γ rays are measured by several $5'' \times 5'' \times 10''$ NaI(Tl) detectors, which are time-tagged to the DT neutron generator via alpha detection in YAP:Ce array. As shown in Fig. 9.16, the reflection block houses the neutron source and two cylindrical $5'' \times 5''$ NaI(Tl) units which act as associated particle detectors. The top set of detectors contains 16 shielded units allowing for transverse scan of the container, while the transmission set is placed on the opposite side from the source and consists of four $5'' \times 5''$ NaI(Tl) detectors and one $5'' \times 5''$ liquid scintillator capable of PSD, which is used for neutron detection.

The principle of operation of EURITRAC is based on differential response of various materials to 14-MeV interrogating neutron beam. The γ-ray spectra obtained with the 22 NaI(Tl) detectors is unfolded to analyze C/O and C/N (carbon-to-oxygen and carbon-to-nitrogen, respectively) ratios. These ratios are further compared to values characteristic of explosives or benign materials.

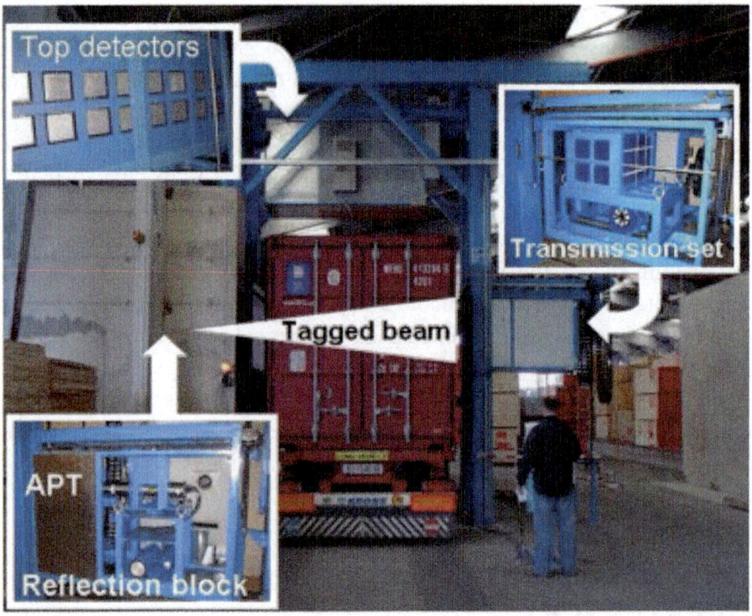

Fig. 9.16 EURITRACK portal during a truck inspection in the seaport of Rijeka, Croatia. Source and detectors are mounted on sliding rails and can be moved up and down to align the tagged neutron beam with the target position. Reproduced from Ref. [18]

9.5 Photon Interrogation Prototype by Passport Systems

The Passport Systems' philosophy in design of SMARTSCAN 3D AI system for cargo scanning has been speed and complete 3-D mapping of contents. The system includes EZ-3D technology based on nuclear resonance fluorescence (NRF) imaging, which aims to provide the map of the entire interior of the container in 15 s. If any questionable object is detected as a result of the first scan, additional investigations can be conducted through secondary inspections. Figure 9.17 shows the layout of the overall system, with the radiography based on a 9-MeV electron source. As the truck moves through the system, a X-ray scanning is conducted top-down using high-resolution imaging detectors (shown on the bottom of the system). EZ-3D detector arrays provide further information about atomic number of mass of cargo materials based on photon interactions via NRF. Fissionable material detectors look for high-energy neutrons resulting from photofission when the X ray interacts with materials, providing ultimate confirmation of SNM presence. Finally, if a secondary scan is required based on the primary scan information, material/isotope ID detector array inspects specific voxels and provides signatures without opening the container.

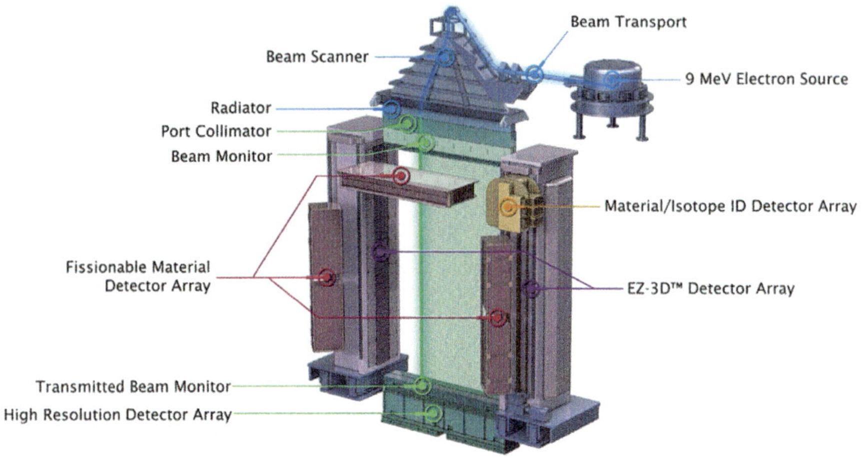

Fig. 9.17 Design of the core of the Smartscan 3D system by Passport Systems (https://www.passportsystems.com/mod/file/view.php?file_guid=33608, 2017)

9.6 Inverse Compton Scattering Prototype Systems

There has been significant interest in developing a new class of AI systems in which the probe beams are derived from interactions of intense, short laser pulses with relativistic electrons [19]. In the interaction process, referred to *inverse Compton scattering* (ICS), the low-energy photon from a laser (typically of order eV) is Doppler upshifted to high energy. The resulting photons with energies on the order of 1–10 MeV can then be used to perform AI measurements. There are several special properties such sources which make them of interest for AI applications. The first property is that the ICS-produced beams can emerge from the production process as highly collimated (*pencil beams*). The second property, which has generated significant attention in the community, is the projected ability of the ICS process to produce small energy spreads ($\sim$1% and even smaller) with continuous tunability over a wide energy range, which can be realized by tuning the electron energy or by adjusting the wavelength of the laser pulse. Finally, it has been predicted that the brightness of such sources should increase rapidly with the energy of X rays they produce [20].

The relatively small expected energy spread and continuous tunability of ICS sources may be of particular interest to detection of specific nuclides, such as ^{235}U or ^{239}Pu by nuclear resonance fluorescence (NRF) [21]. While the broad bremsstrahlung sources could also be used for NRF measurements, the narrow spectrum of ICS sources could offer higher contrast through a less prominent *notch refilling* effect [21] and lower dose, since a greater fraction of the incident photons would be overlapped with the relatively narrow NRF resonance. It has also been

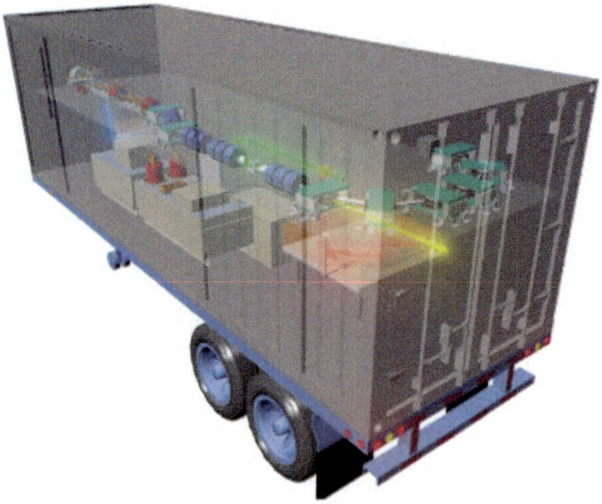

Fig. 9.18 Conceptual rendering of a MEGA-Ray source [25]. US DOE public domain

recognized that the shape of the energy spectrum from an ICS source may be advantageous for AI via photofission, where the goal is for the majority of the incident photon spectrum to overlap with the giant dipole resonance in the range of ~10–15 MeV [22].

The researchers at Lawrence Livermore National Laboratory has invested a major effort to develop a concept for an AI system that could be used to detect ^{235}U by tuning an ICS source to its NRF resonance at 1.733 MeV [23]. The predecessor of this work was the PLEIADES experiment at LLNL, which has reached the energy of 78 keV, energy spread of 0.2%, and the total X-ray production rate of 1.3×10^6/shot [24]. While suitable for atomic measurements such as X-ray backlighting and time-resolved material studies, the energy of X rays produced by PLEIADES was too low for AI, in which MeV-class beams are sought.

An effort to scale the energy of ICS sources to MeV range was initially termed the Thomson-radiated extreme X ray (T-REX) project for production of monoenergetic gamma-rays (MEGa-RAY, Fig. 9.18) [25, 26]. The technical approach in PLEAIDES and in T-REX/MEGA-RAY projects was common: a conventional linear accelerator is used to accelerate electrons, while a separate short-pulse laser system, such as Ti:sapphire (in PLEAIDES) or Nd:YAG (in T-REX/MEGA-RAY) produces the low-energy photons. The laser photons may be first frequency upshifted by second- or third-harmonic generation, after which they are collided with the electron bunches. Synchronization between the electron bunches and laser pulses was achieved by generating the electrons on a photocathode using laser pulses derived from the laser that also drives the ICS interaction. The conceptual layout of the T-REX experiment is shown in Fig. 9.19.

The T-REX source has reached the energy of 478 keV by colliding the second-harmonic Nd:YAG laser pulses (532 nm) with relativistic electrons from an S-band 120-MeV linac at a repetition rate of 10 Hz, which has allowed the detection of a prominent NRF line in ^{7}Li (Fig. 9.20) [28]. It has been recognized that a major

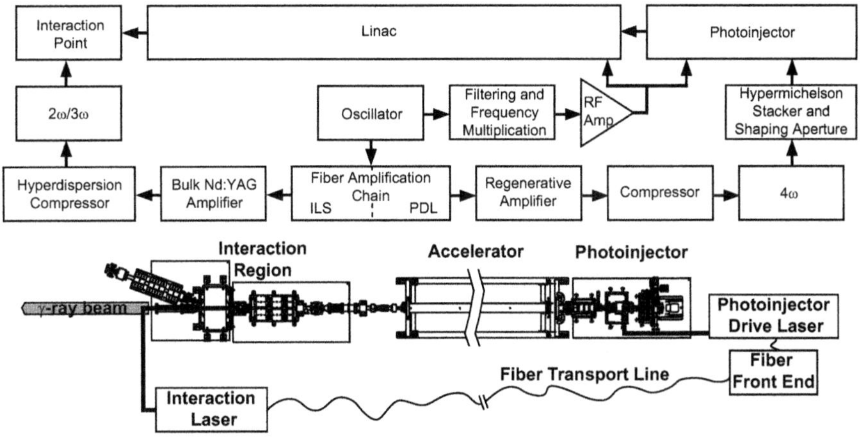

Fig. 9.19 Design of the T-REX source at LLNL [27]. US DOE public domain

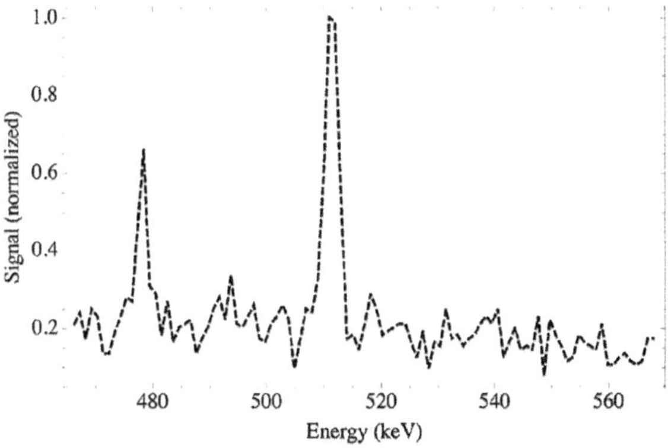

Fig. 9.20 Detection of an NRF line at 478 keV in ^{7}Li. Reproduced from Ref. [28]

technological bottleneck of those sources is the low utilization of laser photons; because of the very small effective cross section for the ICS process, only one in $>10^{10}$ incident laser photons is scattered. The efficiency of these sources can therefore be greatly improved by reusing (recirculating) laser photons using cavities, either as resonant cavities [29] or through a scheme termed recirculation injection by nonlinear gating (RING) [30].

One of the most challenging aspects of the approach taken by sources such as T-REX/MEGA-RAY and Extreme Light Infrastructure-Nuclear Physics (ELI-NP) under construction in Romania [31] is the complexity in interfacing the two technologies (linear accelerator and a short-pulse laser). As a result of this complexity, it has been difficult to develop a technical path for transitioning these sources to practical applications, where they would ultimately need to be deployed in the field and operated by non-experts. With the development of laser wakefield accelerators, it has become possible to simplify the system design and reduce the system footprint by taking advantage of a short-pulse laser as both the ICS driver and a driver for electron acceleration. An ICS source of this type has been constructed at the University of Nebraska-Lincoln [32] and generated X-ray energies of up to 9 MeV [33]. This source has been used to perform X-ray transmission imaging [34].

References

1. M.A. Descalle, D. Manatt, D. Slaughter, Analysis of recent manifests for goods imported through us ports. Technical Report UCRL-TR 225708, Lawrence Livermore National Laboratory, Livermore, CA (2006)
2. E.B. Norman, S.G. Prussin, R.M. Larimer, H. Shugart, E. Browne, A.R. Smith, R.J. McDonald, H. Nitsche, P. Gupta, M.I. Frank, T.B. Gosnell, Nucl. Instrum. Methods Phys. Res. Sect. A Accel. Spectrom. Detect. Assoc. Equip. **521**(2), 608 (2004). https://doi.org/10.1016/j.nima.2003.10.097. http://www.sciencedirect.com/science/article/pii/S0168900203028948
3. D. Slaughter, M. Accatino, A. Bernstein, P. Biltoft, J. Church, M. Descalle, J. Hall, D. Manatt, G. Mauger, T. Moore, E. Norman, D. Petersen, J. Pruet, S. Prussin, Nucl. Instrum. Methods Phys. Res. Sect. A Accel. Spectrom. Detect. Assoc. Equip. **579**(1), 349 (2007). https://doi.org/10.1016/j.nima.2007.04.058. http://www.sciencedirect.com/science/article/pii/S0168900207006456. Proceedings of the 11th Symposium on Radiation Measurements and Applications
4. D. Slaughter, M. Accatino, A. Bernstein, J. Church, M. Descalle, T. Gosnell, J. Hall, A. Loshak, D. Manatt, G. Mauger, T. Moore, E. Norman, B. Pohl, J. Pruet, D. Petersen, R. Walling, D. Weirup, S. Prussin, M. McDowell, Nucl. Instrum. Methods Phys. Res. Sect. B. Beam Interact. Mater. Atoms **241**(1), 777 (2005). https://doi.org/10.1016/j.nimb.2005.07.236. http://www.sciencedirect.com/science/article/pii/S0168583X05013212. The Application of Accelerators in Research and Industry
5. M. Gmar, J. Capdevila, Nucl. Instrum. Methods Phys. Res. Sect. A Accel. Spectrom. Detect. Assoc. Equip. **422**(1), 841 (1999). https://doi.org/10.1016/S0168-9002(98)01121-8. http://www.sciencedirect.com/science/article/pii/S0168900298011218
6. D. Slaughter, M. Accatino, A. Bernstein, J. Candy, A. Dougan, J. Hall, A. Loshak, D. Manatt, A. Meyer, B. Pohl, S. Prussin, R. Walling, D. Weirup, Detection of special nuclear material in cargo containers using neutron interrogation. Technical Report UCRL-ID 155315, Lawrence Livermore National Laboratory (2003)
7. R.B. Firestone, L.P. Ekstrom, S.Y.F. Chu, in *APS Division of Nuclear Physics Meeting Abstracts* (1999)
8. T.R. England, B.F. Rider, End-349: evaluation and compilations of fission product yields. Technical Report LA-UR-94-3106, Los Alamos National Laboratory (1993)
9. L. Brodzinski, D.D. Dietrich, N.R. Hilton, J.L. Jones, G.W. McKinney, B.J. Micklich, J.T. Mihalczo, C.E. Moss, C.E. Okada, H.M. Sanger, D.R. Slaughter, M. Todosow, Evaluation of active interrogation techniques for detecting special nuclear material in maritime cargo containers. Technical Report PNNL-15722, Pacific Northwest National Laboratory (2006)
10. P. Kerr, M. Rowland, D. Dietrich, W. Stoeffl, B. Wheeler, L. Nakae, D. Howard, C. Hagmann, J. Newby, R. Porter, Nucl. Instrum. Methods Phys. Res. Sect. B. Beam Interact. Mater. Atoms **261**(1), 347 (2007). http://dx.doi.org/10.1016/j.nimb.2007.04.190. http://www.sciencedirect.com/science/article/pii/S0168583X07009202. The Application of Accelerators in Research and Industry
11. D.R. Slaughter, S.J. Asztalos, P.J. Biltoft, J.A. Church, M.A. Descalle, J.M. Hall, T.C. Luu, D.R. Manatt, G.J. Mauger, E.B. Norman, D.C. Petersen, J.A. Pruet, S.G. Prussin, Optimized performance for neutron interrogation to detect SNM. Technical Report UCRL-TR 228924, Lawrence Livermore National Laboratory (2007)

12. J. Hall, S. Asztalos, P. Biltoft, J. Church, M.A. Descalle, T. Luu, D. Manatt, G. Mauger, E. Norman, D. Petersen, J. Pruet, S. Prussin, D. Slaughter, Nucl. Instrum. Methods Phys. Res. Sect. B. Beam Interact. Mater. Atoms **261**(1), 337 (2007). http://dx.doi.org/10.1016/j.nimb.2007.04.263. http://www.sciencedirect.com/science/article/pii/S0168583X07007355. The Application of Accelerators in Research and Industry
13. D.R. Slaughter, S.J. Asztalos, P.J. Biltoft, J.A. Church, M.A. Descalle, J.M. Hall, T.C. Luu, D.R. Manatt, G.J. Mauger, E.B. Norman, D.C. Petersen, J.A. Pruet, S.G. Prussin, A 7 MeV neutron interrogation system to detect well shielded SNM: the final report. Technical Report UCRL-TR 236521, Lawrence Livermore National Laboratory (2007)
14. D. Slaughter, M. Accatino, A. Bernstein, J. Church, M.A. Descalle, T. Gosnell, J. Hall, A. Loshak, D. Manatt, J. Mauger, T. Moore, E. Norman, B. Pohl, D. Petersen, J. Pruet, S. Prussin, R. Walling, S. Asztalos, in *IEEE Nuclear Science Symposium* (San Juan, Puerto Rico, 2005)
15. S. Prussin, M.A. Descalle, J. Hall, J. Pruet, D. Slaughter, M. Accatino, O. Alford, S. Asztalos, A. Bernstein, J. Church, T. Gosnell, A. Loshak, N. Madden, D. Manatt, G. Mauger, A. Meyer, T. Moore, E. Norman, B. Pohl, D. Petersen, B. Rusnak, T. Sundsmo, W. Tenbrook, R. Walling, Nucl. Instrum. Methods Phys. Res. Sect. A Accel. Spectrom. Detect. Assoc. Equip. **569**(3), 853 (2006). http://dx.doi.org/10.1016/j.nima.2006.08.002. http://www.sciencedirect.com/science/article/pii/S016890020601357X
16. J. Church, D. Slaughter, S. Asztalos, P. Biltoft, M.A. Descalle, J. Hall, T. Luu, D. Manatt, J. Mauger, E. Norman, D. Petersen, S. Prussin, Nucl. Instrum. Methods Phys. Res. Sect. B. Beam Interact. Mater. Atoms **261**(1), 351 (2007). http://dx.doi.org/10.1016/j.nimb.2007.04.264. http://www.sciencedirect.com/science/article/pii/S0168583X07007434. The Application of Accelerators in Research and Industry
17. T. Luu, P. Biltoft, J. Church, M. Descalle, J. Hall, M.J. Manatt, D.E. Norman, D. Petersen, J. Pruet, S. Prussin, D. Slaughter, Description of ALARMA: the alarm algorithm developed for the nuclear car wash. Technical Report UCRL-TR–227515, Lawrence Livermore National Laboratory (2006)
18. A. Donzella, G. Bonomi, E. Giroletti, A. Zenoni, Radiat. Phys. Chem. **81**(4), 414 (2012). https://doi.org/10.1016/j.radphyschem.2011.12.038. http://www.sciencedirect.com/science/article/pii/S0969806X11005202
19. P. Sprangle, A. Ting, E. Esarey, A. Fisher, J. Appl. Phys. **72**(11), 5032 (1992). http://dx.doi.org/10.1063/1.352031
20. F.V. Hartemann, W.J. Brown, D.J. Gibson, S.G. Anderson, A.M. Tremaine, P.T. Springer, A.J. Wootton, E.P. Hartouni, C.P.J. Barty, Phys. Rev. ST Accel. Beams **8**, 100702 (2005). https://link.aps.org/doi/10.1103/PhysRevSTAB.8.100702
21. J. Pruet, D.P. McNabb, C.A. Hagmann, F.V. Hartemann, C.P.J. Barty, J. Appl. Phys. **99**(12), 123102 (2006). http://dx.doi.org/10.1063/1.2202005
22. S. Boucher, P. Frigola, A. Murokh, M. Ruelas, I. Jovanovic, J. Rosenzweig, G. Travish, Nucl. Instrum. Methods Phys. Res. Sect. A Accel. Spectrom. Detect. Assoc. Equip. **608**(1), S54 (2009). http://dx.doi.org/10.1016/j.nima.2009.05.035. http://www.sciencedirect.com/science/article/pii/S0168900209009681. Compton sources for X/γ rays: Physics and applications
23. F.V. Hartemann, S.G. Anderson, D.J. Gibson, C.A. Hagmann, M.S. Johnson, I. Jovanovic, M.J. Messerly, J.A. Pruet, M.Y. Shverdin, A.M. Tremaine, D.P. McNabb, C.W. Siders, C.P.J. Barty, in *2007 16th IEEE International Pulsed Power Conference*, vol. 2 (2007), pp. 1382–1386. https://doi.org/10.1109/PPPS.2007.4652445
24. D.J. Gibson, S.G. Anderson, C.P.J. Barty, S.M. Betts, R. Booth, W.J. Brown, J.K. Crane, R.R. Cross, D.N. Fittinghoff, F.V. Hartemann, J. Kuba, G.P.L. Sage, D.R. Slaughter, A.M. Tremaine, A.J. Wootton, E.P. Hartouni, P.T. Springer, J.B. Rosenzweig, Phys. Plasmas **11**(5), 2857 (2004). http://dx.doi.org/10.1063/1.1646160
25. K. Walter, Going deep with MEGa-rays (2011). https://str.llnl.gov/AprMay11/pdfs/4.11.3.pdf

26. D.J. Gibson, S.G. Anderson, S.M. Betts, F.V. Hartemann, I. Jovanovic, D.P. McNabb, M.J. Messerly, J.A. Pruet, M.Y. Shverdin, C.W. Siders, A.M. Tremaine, C.P.J. Barty, in *2007 IEEE Particle Accelerator Conference (PAC)* (2007), pp. 1248–1250. http://dx.doi.org/10.1109/PAC.2007.4441045
27. D.J. Gibson, F. Albert, S.G. Anderson, S.M. Betts, M.J. Messerly, H.H. Phan, V.A. Semenov, M.Y. Shverdin, A.M. Tremaine, F.V. Hartemann, C.W. Siders, D.P. McNabb, C.P. Barty, Design and operation of a tunable MeV-level Compton-scattering-based (gamma-ray) source design and operation of a tunable MeV-level compton-scattering-based (gamma-ray) source. Technical Report LLNL-JRNL-414643, Lawrence Livermore National Laboratory (2009)
28. F. Albert, S.G. Anderson, G.A. Anderson, S.M. Betts, D.J. Gibson, C.A. Hagmann, J. Hall, M.S. Johnson, M.J. Messerly, V.A. Semenov, M.Y. Shverdin, A.M. Tremaine, F.V. Hartemann, C.W. Siders, D.P. McNabb, C.P.J. Barty, Opt. Lett. **35**(3), 354 (2010). https://doi.org/10.1364/OL.35.000354. http://ol.osa.org/abstract.cfm?URI=ol-35-3-354
29. G. Klemz, K. Mönig, I. Will, Nucl. Instrum. Methods Phys. Res. Sect. A Accel. Spectrom. Detect. Assoc. Equip. **564**(1), 212 (2006). http://dx.doi.org/10.1016/j.nima.2006.03.017. http://www.sciencedirect.com/science/article/pii/S0168900206005389
30. I. Jovanovic, M. Shverdin, D. Gibson, C. Brown, Nucl. Instrum. Methods Phys. Res. Sect. A Accel. Spectrom. Detect. Assoc. Equip. **578**(1), 160 (2007). http://dx.doi.org/10.1016/j.nima.2007.04.168. http://www.sciencedirect.com/science/article/pii/S0168900207007887
31. N.V. Zamfir, in *2017 10th International Symposium on Advanced Topics in Electrical Engineering (ATEE)* (2017), pp. 358–362. https://doi.org/10.1109/ATEE.2017.7905166
32. S. Banerjee, S. Chen, N. Powers, D. Haden, C. Liu, G. Golovin, J. Zhang, B. Zhao, S. Clarke, S. Pozzi, J. Silano, H. Karwowski, D. Umstadter, Nucl. Instrum. Methods Phys. Res. Sect. B. Beam Interact. Mater. Atoms **350**, 106 (2015). https://doi.org/10.1016/j.nimb.2015.01.015. http://www.sciencedirect.com/science/article/pii/S0168583X15000282
33. C. Liu, G. Golovin, S. Chen, J. Zhang, B. Zhao, D. Haden, S. Banerjee, J. Silano, H. Karwowski, D. Umstadter, Opt. Lett. **39**(14), 4132 (2014). https://doi.org/10.1364/OL.39.004132. http://ol.osa.org/abstract.cfm?URI=ol-39-14-4132
34. S. Chen, G. Golovin, C. Miller, D. Haden, S. Banerjee, P. Zhang, C. Liu, J. Zhang, B. Zhao, S. Clarke, S. Pozzi, D. Umstadter, Nucl. Instrum. Methods Phys. Res. Sect. B. Beam Interact. Mater. Atoms **366**, 217 (2016). https://doi.org/10.1016/j.nimb.2015.11.007. http://www.sciencedirect.com/science/article/pii/S0168583X15011258

Chapter 10
Radiation Dose in Active Interrogation

Shaheen Dewji and Nolan Hertel

Abstract The fundamental concepts and quantities used to describe radiation dose continue to be refined to represent the best scientific understanding to protect humans and the environment. The safe implementation of active interrogation systems requires trade-offs between optimizing the use of radiation-generating technology to the benefit of mankind while minimizing the hazards and risks. Predicting the effects of the source and target radiation as it interacts is critical to optimizing the system's ability to identify the unknown target material with high confidence in an accurate, economical, and time-efficient manner, while ensuring the safe operation of such systems in either open or closed environments. The radiation protection risks posed by the deployment of such systems are varied and potentially dangerous.

10.1 Introduction

The concept of active interrogation (AI) dates back to Wilhelm Röntgen, whose work using penetrating x-ray radiation on his wife's hand pioneered the concept of employing radiation for inspection [1, 2]. As the beneficial applications of radiation for diagnostic medical radiography were evolved the detrimental consequences of radiation exposure were also observed. These physical consequences led to the

This submission was written by the author(s) acting in (his/her/their) own independent capacity and not on behalf of UT-Battelle, LLC, or its affiliates or successors.

S. Dewji (✉)
Oak Ridge National Laboratory, Oak Ridge, TN, USA
e-mail: dewjisa@ornl.gov

N. Hertel
Georgia Institute of Technology, Atlanta, GA, USA
e-mail: nolan.hertel@me.gatech.edu

I. Jovanovic, A. S. Erickson (eds.), *Active Interrogation in Nuclear Security*, Advanced Sciences and Technologies for Security Applications,
https://doi.org/10.1007/978-3-319-74467-4_10

concept of *radiation dose* which, in simple terms, is defined as the amount of energy imparted per unit mass. This field of radiological physics evolved as the science analyzing the interactions of ionizing radiation, with focus on the energy absorbed in matter. Radiation dosimetry is the quantitative determination of this interaction [3].

Radiation dosimetry addresses how ionizing radiation interacts with matter and the effects of energy deposited, notably within tissues and organs of the human body. Ionizing radiation is characterized by the ability to excite and ionize atoms in matter. The ionization energy of an electron required to cause a valence electron to escape an atom ranges from 4 to 25 eV, thus requiring radiation that deposits energy in excess of this range to be classified as ionizing radiation. Ionizing radiation can cause damage or break up a molecule, such as deoxyribonucleic acid (DNA), which would ordinarily not occur. If the region affected by ionizing radiation is small, the risk of delayed effects such as cancer, remains low; however, if the damage successively accumulates (*chronic exposure*), or if high-energy radiation interacts with tissue in a short period of time (*acute exposure*), the consequences can be more serious.

10.2 Radiation of Concern in Active Interrogation

A primary concern in the deployment of AI systems is accounting for the radiation dose received by both the interrogated target (such as package, vehicle, cargo) and the receptors in various roles in the vicinity of the system such as system operators, members of the public, or stowaways. In the assessment of dose in AI systems, both the primary interrogating radiation and the secondary induced radiation must be considered. AI mechanisms can be summarized as follows and captured in Table 10.1 [4–7].

Fast Neutron Interrogation Mechanisms Fast neutron mechanisms include systems that employ either steady-state (spontaneous/continuous) or pulsed sources, monoenergetic or broad spectral energies, and directional or isotropic distributions. The induced (secondary) radiation from the target object inspected could be signature gamma (γ)-rays from excited nuclear states caused by inelastic scatter; prompt γ-rays from neutron thermalization and capture; γ-rays from decay of neutron activation products; and prompt or delayed fission neutrons from fissionable materials, which could include multiplication of neutrons and γ-rays due to multiple fission interactions by the secondary fission neutrons.

Thermal Neutron Interrogation Mechanisms Thermal neutrons are produced by thermalization of fast neutrons in the AI system, usually via a series of inelastic and elastic scattering reactions. Secondary radiation observed includes prompt γ-rays from neutron thermalization and capture, γ-rays from decay of thermal neutron activation, and prompt or induced fission neutrons that yield potential fission chains with significant multiplication of neutrons and γ-rays.

Table 10.1 Summary of sources used in active interrogation

Source	Interrogation reaction	Source energy	Direction	Source frequency	Radiation protection
^{252}Cf	Fission	Broad spectrum neutrons Mean: 2.1 MeV Probable: 0.7 MeV	Isotropic	Continuous	Broad neutron spectrum isotropic
AmBe PuBe AmLi	(α,n)	Spectrum dependent on α-particle energy	Isotropic	Continuous	Broad neutron spectrum isotropic
^{9}Be, ^{2}H	Photonuclear (γ from ^{24}Na, ^{72}Ga, ^{123}Sb, ^{140}La, ^{226}Ra)	Neutron spectrum dependent on γ-ray energy, ^{9}Be: 1.67 MeV ^{2}H: 2.23 MeV	Isotropic	Continuous	High-energy incident γ-rays resulting in monoenergetic neutrons
D-D D-T	^{2}H(d,n)^{3}He ^{3}H(d,n)^{4}He	Monoenergetic 14.1 MeV 2.5 MeV	Anisotropic	Controlled/ pulse	High-energy neutron flux
Brems-strahlung (electron accelerator)	NRF	Dependent on electron energy	Anisotropic (bremsstrahlung cone)	Controlled/ pulse	Electron accelerator beam; high-energy incident γ-rays

High-Energy Photon Interrogation Mechanisms High-energy photon sources (x-ray or radionuclide), as well as accelerator-produced monoenergetic γ-rays, provide target signature information by nuclear resonance fluorescence (NRF), nuclear resonance absorption, or photonuclear fission. Prompt or delayed neutrons or γ-rays from induced fission are signatures of the interrogated target. Similar to fast and thermal interrogation, induced fission neutrons yield potential fission chains with significant multiplication of neutrons and γ-rays.

10.3 Regulations of Dose Exposure

The objectives of radiation protection have revolved around the principle of *As Low as Reasonably Achievable* (ALARA). To this extent, the benefits of activities that involve radiation must balance the risk, notably in limiting the incidence of *deterministic* effects (such as radiation sickness and cataracts), as well as *stochastic* effects (such as cancer and hereditary effects).

The growth of the widespread use of ionizing radiation has necessitated the creation of organizations and standards for developing the scientific and technical basis of the safe application of ionizing radiation. The International Commission on Radiation Units and Measurements (ICRU) was established in 1925 [8] to

> ... develop and promulgate internationally accepted recommendations on radiation-related quantities and units, terminology, measurement procedures, and reference data for the safe and efficient application of ionizing radiation to medical diagnosis and therapy, radiation science and technology, and radiation protection of individuals and populations.

Shortly thereafter in 1928, the International Commission on Radiological Protection (ICRP) was established, whose mission focused on providing recommendations and guidance on all aspects of protection against ionizing radiation using the science of radiological protection to the benefit of the public [9]. The National Council on Radiation Protection and Measurements (NCRP) was chartered by the US Congress in 1964 to collect, analyze, develop, and disseminate information and recommendations about protection against radiation and about radiation measurements, quantities and units, particularly those concerned with radiation protection in the public interest [10]. These three organizations form the core trifecta for radiation protection science and policy in the United States and internationally. Based on research and recommendations of the ICRU, ICRP, and NCRP, the following standards, federal codes of conduct/codes of federal regulations (CFR), and recommendations (including their updates and revisions) are most relevant to the regulation of dose in AI systems in the United States.

10.3.1 ICRU/ICRP/NCRP Recommendations

Recommendations relevant to radiation dose to human exposure from ionizing radiation, inclusive of AI systems, are provided in a plurality of ICRU, ICRP, and NCRP references:

- ICRU Report 57/ICRP Publication 74: Conversion Coefficients for Use in Radiological Protection Against External Radiation [11]
- ICRU Report 85a: Fundamental Quantities and Units for Ionizing Radiation [12]
- ICRP Publication 103: Recommendations of the International Commission on Radiological Protection [13]
- ICRP Publication 116: Conversion Coefficients for Radiological Protection Quantities for External Radiation Exposures [14]
- NCRP Report 116: Limitation of Exposure to Ionizing Radiation [15]
- NCRP Commentary 16: Screening of Humans for Security Purposes using Ionizing Radiation Scanning Systems [16]
- NCRP Commentary 17: Pulsed Fast Neutron Analysis System Used in Security Surveillance [17]
- NCRP Statement 10: Recent Applications of the NCRP Public Dose Limit Recommendation for Ionizing Radiation [18]

10.3.1.1 NCRP Recommendations

The NCRP Commentary 17, "Pulsed Fast Neutron Analysis System Used in Security Surveillance," provides a detailed radiation protection assessment for pulsed fast neutron analysis (PFNA) systems, which is further applicable to AI systems. NCRP Commentary 17 addresses the following concerns [17]:

- description of relevant concepts of radiation protection that should be applied to PFNA systems;
- application of ALARA to the PFNA system;
- recommendation of the appropriate dose limit for a person inadvertently irradiated by the PFNA system, including recommendations of the proper methods to determine the dose received;
- recommendation of the specific methods and instruments for the measurement and determination of the radiation dose (effective dose) that an individual would receive by inadvertent exposure to radiation from the PFNA system; and
- determination of whether the use of the PFNA system could result in levels of activation products in pharmaceuticals and medical devices that could be of concern to public health.

The most recent recommendations made by the NCRP at the time of authorship of Commentary 17 were derived on the basis of from guidance in the NCRP Report 116, "Limitation of exposure to ionizing radiation" [15], which recommends that an inadvertently exposed person should receive no more than 100 mrem (1 mSv) but that the dose can be raised to 500 mrem (5 mSv) to achieve national security objectives. This guidance was reinforced in an update of NCRP's recommendations in Statement 10, "Recent Applications of the NCRP Public Dose Limit Recommendation for Ionizing Radiation" [18].

For occupational exposures related to operators of PFNA systems (and applicable to all occupational radiation workers), NCRP Commentary 17 cites recommendations of NCRP Report 116, which recommends that the cumulative lifetime effective dose does not exceed the age of the individual in years times 1,000 mrem (10 mSv) and the limit on annual effective dose as 5,000 mrem (50 mSv) [15]. This limit excludes exposure from natural background radiation and exposure associated with medical procedures. The NCRP, under Commentary 17 and Report 116, further recommends an administrative control limit of 25 mrem (0.25 mSv) from a single exposure event.

10.3.1.2 ICRP Recommendations

The ICRP recommendations are supported by a series of documents that reflect the latest scientific information on the intake, distribution, retention, and elimination of radioactive material, as well as doses to organs and tissues. At the time of publication of NCRP Commentary 17, the latest guidance available from ICRP was given in Publication 60, "1990 Recommendations of the International Commission

Table 10.2 Recommended dose limits in planned exposure situations given in ICRP Publication 103[a] (Reproduced from [13])

Type of limit	Occupational	Public
Effective dose	20 mSv per year, averaged over defined periods of 5 years[b]	1 mSv in a year[c]
Annual equivalent in:		
Lens of the eye[d]	150 mSv	15 mSv
Skin[e,f]	500 mSv	50 mSv
Hands and feet	500 mSv	–

[a]Limits on effective dose are for the sum of the relevant effective doses from external exposure in the specified time period and the committed effective dose from intakes of radionuclides in the same period
[b]With the further provision that the effective dose should not exceed 50 mSv in any single year. Additional restrictions apply to the occupational exposure of pregnant women
[c]In special circumstances, a higher value of effective dose could be allowed in a single year, provided that the average over 5 years does not exceed 1 mSv per year
[d]This limit is currently being reviewed by an ICRP Task Group
[e]The limitation on effective dose provides sufficient protection for the skin against stochastic effects
[f]Averaged over 1 cm^2 area of skin regardless of the area exposed

on Radiological Protection" [19]. ICRP Publication 60 has been superseded by Publication 103, "Recommendations of the International Commission on Radiological Protection" [13]. In Publication 103, ICRP derived the limit of an average of 2,000 mrem (20 mSv) per year averaged over 5 years (100 mSv in 5 years) for the occupational dose limit, with the further provision that the dose should not exceed 5,000 mrem (50 mSv) in any single year. The recommendation of the annual public dose limit was 100 mrem (1 mSv) per year (Table 10.2).

Dose concepts and the methodologies underpinning recommendations by ICRU and ICRP are explored further in Sect. 10.4.

10.3.2 *Code of Federal Regulations*

Drawing upon the recommendations of ICRU, ICRP, and NCRP listed previously, the following sections of the US Code of Federal Regulations were developed largely in alignment regarding regulation of dose limits to occupational workers and members of the public, which applies to the operation of AI systems:

- Code of Federal Regulations, Title 10, Part 20 (10 CFR 20): Standards for Protection Against Radiation [20]
- Code of Federal Regulations, Title 10, Part 50 (10 CFR 50): Domestic Licensing of Production and Utilization Facilities [21]
- Code of Federal Regulations, Title 21, Part 179 (21 CFR 179): Irradiation in the Production, Processing and Handling of Food [22]

- Code of Federal Regulations, Title 29, Part 1910 (29 CFR 1910): Occupational Safety and Health [23]

  Regulation 10 CFR 20 outlines the occupational dose limits, which could be applicable to system operators with dosimeters defined as radiological workers operating an AI system, in Subpart C—20.1201, "Occupational Dose Limits for Adults" [20]:

1. *The licensee shall control the occupational dose to individual adults, except for planned special exposures under 20.1206, to the following dose limits.*
   a. *An annual limit, which is the more limiting of—*
      i. *The total effective dose equivalent being equal to 5 rem (0.05 Sv); or*
      ii. *The sum of the deep-dose equivalent and the committed dose equivalent to any individual organ or tissue other than the lens of the eye being equal to 50 rem (0.5 Sv).*
   b. *The annual limits to the lens of the eye, to the skin of the whole body, and to the skin of the extremities, which are:*
      i. *A lens dose equivalent of 15 rem (0.15 Sv), and*
      ii. *A shallow-dose equivalent of 50 rem (0.5 Sv) to the skin of the whole body or to the skin of any extremity.*
2. *Doses received in excess of the annual limits, including doses received during accidents, emergencies, and planned special exposures, must be subtracted from the limits for planned special exposures that the individual may receive during the current year (see 20.1206(e)(1)) and during the individual's lifetime (see §20.1206(e)(2)).*

10 CFR 20 outlines dose limits to members of the public in Subpart D—20.1301,"Occupational Dose Limits for Adults" [20]:

1. *Each licensee shall conduct operations so that—*
   a. *The total effective dose equivalent, which is the sum of the effective dose equivalent (for external exposures) and the committed effective dose equivalent (for internal exposures), to individual members of the public from the licensed operation does not exceed 0.1 rem (1 mSv) in a year, exclusive of the dose contributions from background radiation, from any administration the individual has received, from exposure to individuals administered radioactive material and released under §35.75, from voluntary participation in medical research programs, and from the licensee's disposal of radioactive material into sanitary sewerage in accordance with §20.2003, and*

   b. *The dose in any unrestricted area from external sources, exclusive of the dose contributions from patients administered radioactive material and released in accordance with §35.75, does not exceed 0.002 rem (0.02 mSv) in any 1 h.*

2. *If the licensee permits members of the public to have access to controlled areas, the limits for members of the public continue to apply to those individuals.*
3. *Notwithstanding paragraph (a)(1) of this section, a licensee may permit visitors to an individual who cannot be released, under §35.75, to receive a radiation dose greater than 0.1 rem (1 mSv) if—*

   a. *The radiation dose received does not exceed 0.5 rem (5 mSv); and*
   b. *The authorized user, as defined in 10 CFR Part 35, has determined before the visit that it is appropriate.*

4. *A licensee or license applicant may apply for prior NRC authorization to operate up to an annual dose limit for an individual member of the public of 0.5 rem (5 mSv). The licensee or license applicant shall include the following information in this application:*

   a. *Demonstration of the need for and the expected duration of operations in excess of the limit in paragraph (a) of this section;*
   b. *The licensee's program to assess and control dose within the 0.5 rem (5 mSv) annual limit; and*
   c. *The procedures to be followed to maintain the dose as low as is reasonably achievable.*

Current recommendations under 10 CFR 20 are primarily based on recommendations of ICRP Publication 26 [26], which was superseded by ICRP Publication 60 [19] and the latest guidance given in ICRP Publication 103 [13, 24]. 10 CFR 20 does not reflect the latest recommendations in ICRP Publication 103 (or its predecessor, ICRP Publication 60) for members of the public or occupational workers but is the governing standard in the United States for annual dose limits to occupational workers and members of the public.

10.3.3 Standards

Standards directly relevant to the operational and performance requirements for AI systems align with established NCRP recommendations and the Code of Federal Regulations titles listed previously, with further details on operational and performance requirements of the AI systems:

- American National Standards Institute (ANSI) N42.23-1996 American National Standard Measurement and Associated Instrument Quality Assurance for Radioassay Laboratories [25]

- ANSI NA42.41-2007 American National Standard Minimum Performance Criteria for Active Interrogation Systems Used for Homeland Security [4]
- ANSI NA42.41-2007, "American National Standard Minimum Performance Criteria for Active Interrogation Systems Used for Homeland Security" outlines the operational and performance requirements for AI systems for homeland security applications [4]. This standard focuses on systems employing penetrating ionizing radiation to simulate secondary radiation or resonances that detect and identify elemental or radionuclide compositions of hidden chemical, nuclear, or conventional explosives. The standard further accounts for inspection zones of various sizes and locations (indoor/outdoor), as well as sizes of containers (packages, briefcases, suitcases, air/sea cargo containers, passenger vehicles, trucks/semi-trailers, and rail cars). ANSI NA42.41-2007 provides the requirements for minimally acceptable performance criteria of candidate systems. The standard cites that at the time of its production few commercially available systems employing AI systems were available, none with broad scale deployment either domestically or internationally, which, arguably, continues to be the case to date.
- ANSI NA42.41-2007 cites that in contrast to x-ray and γ-ray passive/ transmission systems, which are limited largely to densitometry measurements, AI systems are limited by attenuation of interrogating radiation and of the induced, or secondary, emissions. A further limiting factor with AI systems is that if the constituents of the composition of the interrogated object are similar to substances of interest (such as explosives), then this could lead to distorted stoichiometry and hence false positive or negative results.

  This standard identifies general characteristics of AI systems, including inspection zone characteristics (dimensions, fixed or scanned), interrogating radiation type and physical principles (fast neutron interrogation, thermal neutron interrogation, high-energy photon interrogation), and physical configurations (portable, mobile or fixed systems); performance requirements for explosives, chemical warfare agents, and special nuclear materials and nuclear weapons components; inspection times for cargo by container category; and specification of masses by container category.

In §6.11 of ANSI NA42.41-2007, radiation dose guidance is provided for workers without dosimetry, workers with dosimetry, stowaways, prescreened inspection zones, and activation [4]:

- **Workers without monitored dosimetry and passersby.** Per 10 CFR 20 [20], workers without dosimeters and passersby shall not exceed the total effective dose equivalent limit to the general public of 100 mrem (1 mSv) per year.
- **Workers with monitored dosimetry.** 10 CFR 20 [20] for occupational workers and 29 CFR Part 1910 [23].

- **Stowaways.** Per NCRP Statement 10 [18] and 10 CFR 20 [20], inadvertent exposure of stowaways shall not exceed the effective dose limit of 100 mrem (1 mSv), or 500 mrem (5 mSv) in exceptional circumstances of national security.
- **PreScreened Inspection Zones.** Per 21 CFR Part 179 [22], if the container or vehicle has undergone prescreening that could expose a stowaway to a maximum of 100 mrem (1 mSv), and if the screening system was capable of detecting human occupancy with at least 95% confidence, then the vehicle or container may be exposed to up to 20 rem (0.2 Sv) by the AI system.
- **Activation Limits.** Per 21 CFR Part 179 [22] and 10 CFR Part 20 [20].

10.4 External Dose Assessment

AI systems result in the production of complex radiation fields comprising a distribution of photons (γ-rays) and neutrons (thermal, fast). The incident type, energy, and direction of the interrogating source and the secondary radiations produced by the interrogated target must all be considered in the dose assessment of radiation fields of AI systems.

10.4.1 Fundamental Dose Quantities

The absorbed dose, D, is the fundamental unit describing the deposition of ionizing radiation in matter as the quotient of the mean energy imparted by the ionizing radiation into a target volume divided by the mass of the matter in the target volume. The unit of absorbed dose is joules per kilogram (J/kg), which is given the name gray (Gy) [13]. In complex radiation fields interacting with specified organs or tissues in the body (T), the absorbed dose (DT) for said tissue must be modified to reflect greater biological effects for the different types of radiation, notably that of neutrons, as well as differentiating radiation risk among organs or tissues in the body. The application of a radiation-weighting factor (w_R), representing the relative biological effectiveness of different radiation types (photons and neutrons in AI systems), yields a quantity of equivalent dose (H_T) for the organ or tissue in units of J/kg with the special name, sievert (Sv). The application of a tissue-weighting factor (w_T) is the same for all radiations but different for each organ based on its associated radiation risk. The sum of the products of H_T and w_T yield the quantity of effective dose (E), which is used to quantify the radiation dose received by an exposed person in units of J/kg with the special name, sievert (Sv) [13]. The computation of these quantities using the methods prescribed by ICRP is given in Sect. 10.4.3.

10.4.2 Relating Physical, Operation, and Protection Quantities Using Dose Coefficients

It is the convention in radiation protection to employ dose coefficients to relate the protection quantities to physical quantities. The ICRP defines protection (limiting) quantities as dosimetric quantities specified in the human body [13]. Protection quantities are not measurable but can be determined by calculation or a combination of calculations and measurements [13, 14]. Operational quantities, developed by the ICRU, are meant to demonstrate compliance with dose limits by providing calibration quantities for instrumentation measuring or monitoring dose to estimate the protection quantities [11].

The computation of fluence-to-dose coefficients enables relating a radiation field (operational) quantity, such as the conversion of fluence (Φ) to dose. The ICRU defines physical quantities such as fluence, kerma, and absorbed, which are used in radiation dosimetry. The fluence (Φ) is the quotient of the number of particles (dN) incident on a sphere of cross-sectional area (da) [12]. As fluence is a radiometric quantity characterizing a radiation field, it does not describe the interaction of radiation with matter. The kerma (K) for ionizing uncharged particles is the quotient of (dE_{ptr}) by $\mathrm{d}m$, where $\mathrm{d}E_{ptr}$ is the mean sum of the initial kinetic energies of all the charged particles liberated in a mass $\mathrm{d}m$ of a material by the uncharged particles incident on dm [12]. The relationship between physical, operational, and protection quantities is captured in Fig. 10.1.

In AI systems, the application of fluence-to-dose coefficients might not necessarily be applicable in a specialized setup, such as inside a cargo container, which warrants further study and computation since they were computed for fixed radiation

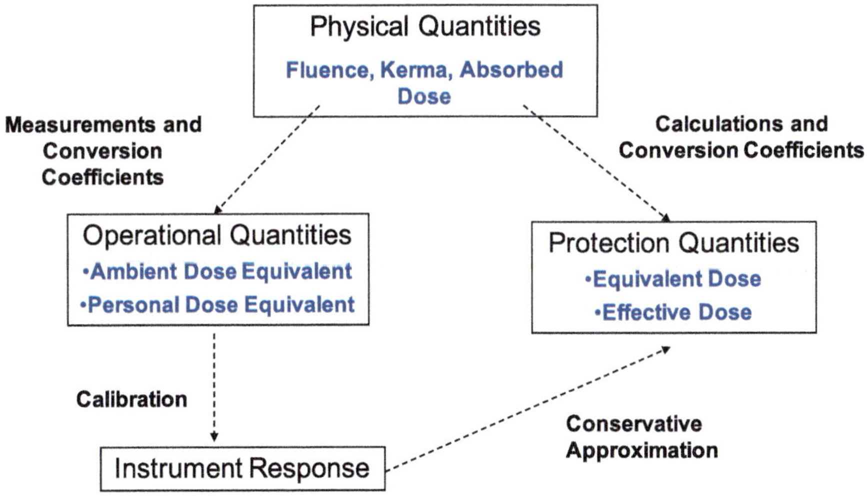

Fig. 10.1 Relationship between operational and protection quantities (Reproduced from [32])

field geometries, as recommended by NRCP Commentary 17 [17]. However, fluence-to-dose coefficients might be applicable in broader environmental exposures and remain the core data relating radiation protection operational quantities with protection quantities. The recommendations of the ICRP and ICRU, in addition to joint efforts between the two organizations, have led to the computation of dose coefficients aligned with the latest scientific recommendations. The quantities and methodologies for producing fluence-to-dose coefficients, per the latest international scientific recommendations [11–14], are described subsequently.

10.4.3 Protection Quantities

Protection quantities provide dose limits for both occupational workers and members of the public and are derived using computational reference anthropomorphic phantoms. Protection quantities are defined by ICRP in Publications 26, 60, and 103 [13, 19, 26].

10.4.3.1 Equivalent and Effective Dose

Dose limits and administrative controls for delayed stochastic effects are expressed in effective dose (E). The latest methodology recommended in ICRP Publication 103 for the computation of effective dose and its constituent quantities [13], introduced in Sect. 10.4.1, is summarized in Fig. 10.2.

ICRP Publication 60 introduced the term effective dose, with its recommended set of organs to be used in its computation. In addition, the radiation quality factor was further redefined as the radiation-weighting factor, based on radiation type and energy incident on the body.

As reinforced in ICRP Publication 103, the radiation-weighted absorbed dose in the organ or tissue is referred to as the equivalent dose, H_T, and is found from

$$H_T = \sum_R w_R D_{T,r}, \tag{10.1}$$

where w_R is defined as a function of the energy of the incident radiation, $D_{T,R}$ is the absorbed dose averaged over the organ or tissue (T) from radiation (R).

The computation of effective dose is the sum of the product of the tissue-weighting factor (w_T) with the sex-averaged value of the equivalent dose (H_T)

$$E = \sum_T w_T \left(\frac{H_T^M + H_T^F}{2} \right), \tag{10.2}$$

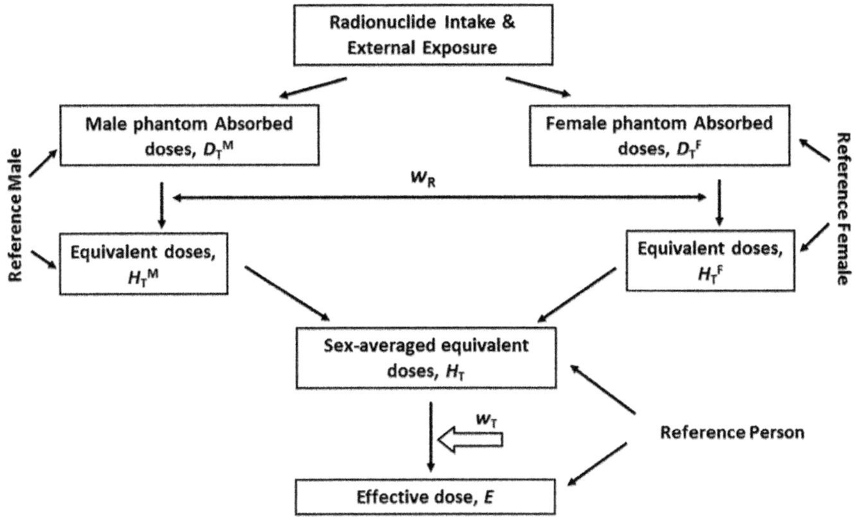

Fig. 10.2 Methodology recommended by ICRP Publication 103 for the computation of equivalent and effective dose (Reproduced from [13])

where H_T^M and H_T^F are equivalent doses to organ or tissue T in the male and female phantoms, respectively.

10.4.3.2 Radiation-Weighting Factors

The recommendations made by ICRP for the radiation weighting factor associated with the type of incident radiation changed with the latest available science. Recommendations from ICRP Publication 60 and ICRP Publication 103 are juxtaposed in Table 10.3 and graphed in Fig. 10.3.

The continuous function representing the neutron w_R was developed to serve as an approximation as a function of neutron energy (E) in megaelectronvolts. ICRP Publication 60 defines this function as [19]

$$w_R = 5 + 17\exp\left(-\ln^2(2E)/6\right). \tag{10.3}$$

The continuous function representing the neutron w_R was developed to serve as an approximation as a function of neutron energy (E) in megaelectronvolts. ICRP Publication 103 defines this function as [13]

Table 10.3 Comparison of radiation weighting factors from ICRP Publication 60 [19] and ICRP Publication 103 [13]

Organ/tissue	ICRP Publication 60	ICRP Publication 103
Photons	1	1
Electrons and muons	1	1
Protons and charged pions	5	2
Alpha particles, fission fragments, and heavy ions	20	20
Neutrons[a] < 10 keV	5	
10 keV to 100 keV	10	
>100 keV to 2 MeV	20	Continuous function
>2 MeV to 20 MeV	10	
>20 MeV	5	

[a]A continuous function serves as an approximation

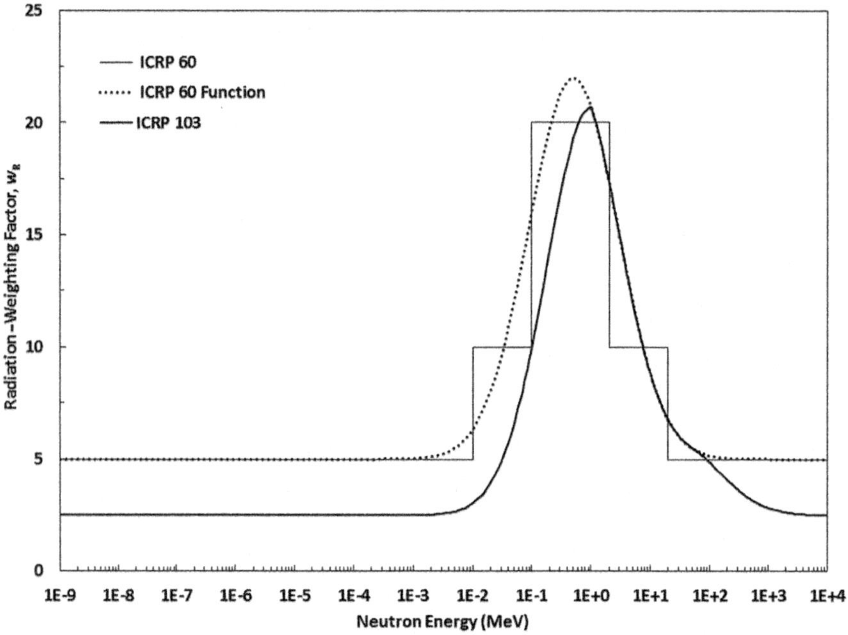

Fig. 10.3 Radiation-weighting factors given in ICRP Publication 60 and ICRP Publication 103 (Reproduced from [13])

$$w_R = \begin{cases} 2.5 + 18.2 \exp\left(-\ln^2(E_n)/6\right) & E_n < 1 \text{ MeV} \\ 5 + 17 \exp\left(-\ln^2(2E)/6\right) & 1 \text{ MeV} \leq E_n \leq 50 \text{ MeV} \\ 2.5 + 3.24 \exp\left(-\ln^2(0.04E)/6\right) & E_n > 50 \text{ MeV} \end{cases} \tag{10.4}$$

Table 10.4 Comparison of tissue-weighting factors per ICRP recommendations

Tissue/organ[a]	ICRP Publication 26[b]	ICRP Publication 60[c]	ICRP Publication 103[d]
Gonads	0.25	0.20	0.08
Bone marrow (red)	0.12	0.12	0.12
Colon		0.12	0.12
Lungs	0.12	0.12	0.12
Stomach		0.12	0.12
Bladder		0.05	0.04
Breast	0.15	0.05	0.12
Liver		0.05	0.04
Esophagus		0.05	0.04
Thyroid	0.03	0.05	0.04
Skin		0.01	0.01
Bone surface	0.03	0.01	0.01
Remainder	0.3[e]	0.05[f]	0.12[g]
Brain			0.01
Salivary glands			0.01

[a]Organs without tissue-weighting factors defined comprise the remainder
[b][26]
[c][19]
[d][13]
[e]The five most highly irradiated other organs and tissues
[f]Adrenals, brain, upper large intestine, small intestine, kidneys, muscle, pancreas, spleen, thymus, and uterus (female)
[g]Adrenals, extrathoracic tissue, gall bladder, heart, kidneys, lymphatic nodes, muscle, oral mucosa, pancreas, prostate (male), small intestine, spleen, thymus, uterus/cervix (female)

10.4.3.3 Tissue-Weighting Factor

The recommendations made by ICRP for radiation risk associated with the tissues and organs of the human body have changed with the latest available science. The recommendations for tissue-weighting factors (w_T), which are independent of indecent radiation type, are provided per the recommendations of ICRP Publication 26, ICRP Publication 60, and ICRP Publication 103 in Table 10.4.

10.4.4 Operational Quantities

While protection quantities define dose limits to occupational workers and members of the public through computation of radiation and tissue-risk weighted absorbed doses, these quantities can neither be measured nor applied to calibration or radiation detection instrumentation. As a result, operational quantities were developed by the ICRU to provide quantities reflective of the corresponding protection quantities and dose limits. Operational quantities have been developed by ICRU in Reports 39,

43, 47, 51, 60, and 85a [12, 27–31]. Just as protection quantities employ the radiation-weighting factor, operational quantities employ a quality factor (Q) to differentiate the biological effectiveness of radiation.

The operational quantities applied to area monitoring are the ambient dose equivalent and directional dose equivalent; the operational quality for individual monitoring is the personal dose equivalent. These quantities are used to correlate the protection and operational quantities for recommendations of whole body dose, dose to lens of the eye, and dose to skin. It is worthwhile to note that ICRP Publication 26 employs the term "dose equivalent" to represent operational and protection quantities, whereas ICRP Publication 60 and ICRP Publication 103 protection quantities employ "equivalent dose" for organs and tissues and dose equivalent for operational quantities.

10.4.4.1 Area Monitoring

The ambient dose equivalent $H^*(d)$ at a point in a radiation field is the dose equivalent that would be produced by the corresponding expanded and aligned field in the ICRU sphere at a depth d on the radius opposing the direction of the aligned field. As explained in ICRP Publication 74 [32], an expanded and aligned radiation field is a hypothetical field where the fluence and its energy distribution are the same as an expanded field but the fluence is unidirectional. The unit of ambient dose equivalent is joule per kilogram (J/kg) and called Sievert, Sv. The recommended value of d in the ICRU sphere is 10 mm for penetrating radiation, with notation $H^*(10)$ [32].

The directional dose equivalent $H'(d, \Omega)$ is used for assessing dose to the skin and the extremities (hands, wrists, and feet), as well as the dose to the lens of the eye. The directional dose equivalent at a point in a radiation field is the dose equivalent that would be produced by the corresponding expanded field in the ICRU sphere at a depth d on a radius in a specified direction Ω. As reinforced in ICRP Publication 116, for all types of external radiation, the operational quantities for area monitoring are defined on the basis of a dose-equivalent quantity that would exist in the ICRU sphere (30 cm in diameter) as a theoretical construct of tissue-equivalent material of ICRU 4-element tissue with a density of 1 g/cm^3 and a mass composition of 76.2% oxygen, 11.1% carbon, 10.1% hydrogen, and 2.6% nitrogen [14].

For assessing dose to skin and the extremities, a depth of $d = 0.07$ mm is recommended with notation $H'(0.07, \Omega)$. For monitoring dose to the lens of the eye, depth $d = 3$ mm is recommended, with notation $H'(3, \Omega)$ is recommended.

10.4.4.2 Personnel Monitoring

The operational quantity employed when calibrating the dosimeters of personnel is the personal dose equivalent $H_p(d)$ with units of J/kg (Sv). Just as applied with area monitoring in ambient and directional dose equivalent, the same depths are

Table 10.5 Scheme of operational quantities used for dose monitoring in external exposure stations (reproduced from [14])

Task	Area monitoring	Personnel monitoring
Control of effective dose	Ambient dose equivalent $H^{*}(10)$	Personal dose equivalent $H_p(10)$
Control of dose to the lens of the eye	Directional dose equivalent $H'(3, \Omega)$	Personal dose equivalent $H_p(3)$
Control of dose to skin, hands, and feet	Directional dose equivalent $H'(0.07, \Omega)$	Personal dose equivalent $H_p(0.07)$

recommended for personal dose equivalent with 10 mm for penetrating radiation, 3 mm for lens of the eye, and 0.07 mm for skin [14].

In computing personal dose equivalent for dosimeter calibration, the ICRU commonly uses a 30 cm × 30 cm × 15 cm thick slab phantom constructed of the same aforementioned ICRU 4-element tissue substitute, representing the trunk of the body, and denotes the corresponding quantity as $H_{p,slab}(d)$. This material cannot be fabricated, so actual calibrations frequently use polymethyl methacrylate with thermoluminescent detectors.

For the different monitoring tasks of external exposures, area monitoring, and individual monitoring, the scheme shown in Table 10.5 can be used to describe the application of the different operational dose quantities.

10.4.5 Fluence-to-Dose Coefficients

The methodology used to calculate the ambient dose equivalent coefficients has evolved over the past three decades [33]. Ambient dose coefficients are calculated by modeling the energy deposition in a specified volume at various depths of the ICRU sphere (Fig. 10.4). Critical to the calculation of ambient dose equivalent coefficients is how charged particle buildup is handled and the depth at which charged particle equilibrium occurs, which is dependent on the incident radiation type and energy. To simplify the calculations, the kerma approximation is employed, which assumes that at the point of interest in the ICRU sphere phantom, all secondary charged particles are in equilibrium with the primary radiation; consequently, the absorbed dose is equal to the kerma less any energy removed by uncharged radiation, such as neutrons or bremsstrahlung. ICRP Publication 116 stipulates that ambient dose equivalent $H^{*}(10)$ continues to provide a reasonable assessment of the effective dose under charged-particle equilibrium for photons [14]. For electrons, $H^{*}(10)$ gives a reasonable estimate of the effective dose up to 10 MeV. For neutrons, $H^{*}(10)$ overestimates the effective dose or gives a reasonable approximation of this quantity up to ~40 MeV [14].

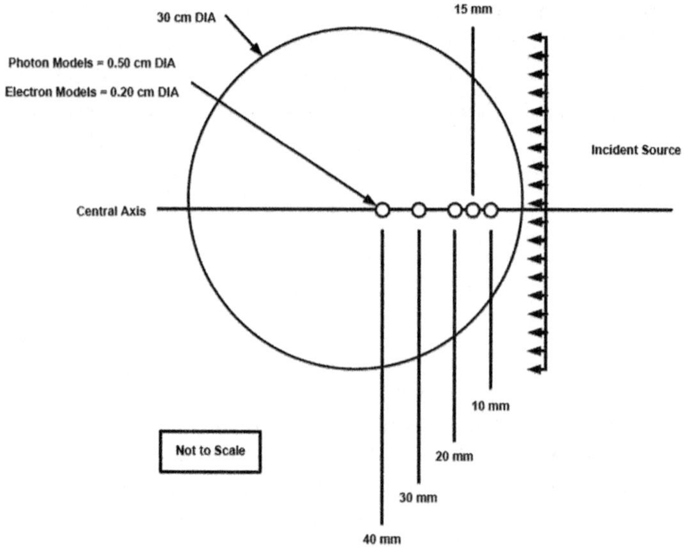

Fig. 10.4 Schematic of ambient dose equivalent conversion coefficient model (reproduced from [5])

Based on joint collaboration with ICRU, ICRP Publication 116 [14] provides a set of photon fluence-to-ambient dose equivalent coefficients succeeding ICRP Publication 74 [32].

The photon fluence-to-ambient dose equivalent coefficients adapted by the ICRP were based on computations using the kerma approximation, which assumes that all secondary electrons generated by photons are in equilibrium and that the kinetic energy of these electrons is transferred to the medium and locally deposited [14]. The kerma approximation is satisfactory for photons up to 3 MeV at 10 cm, where at increasing photon energies, the equilibrium increasingly deviates since charged particle equilibrium is not obtained at this depth due to the longer range of the secondary radiation [5, 14].

Particle fluence-to-ambient dose conversion coefficients relate the particle fluence to the effective dose, calculated for whole-body exposure of the ICRP/ICRU adult reference phantoms for the broad parallel beams in irradiation geometries conducted in ICRP Publication 74 [32] updated in ICRP Publication 116 [14], and being updated (in progress) in ICRU Report 47 [29], of which the details of the computations are explained in extensive detail. Conversion coefficients for ambient dose equivalent and air kerma in free air are summarized for photons in Tables 10.6 and 10.7, respectively; conversion coefficients for ambient dose equivalent for neutrons are summarized in Table 10.8. Fluence-to-organ absorbed dose and fluence-to-effective doses (as pSv cm^2) additionally from electrons, positrons, protons, positive/negative muons, positive/negative pions, and helium ions are given in Appendix A of ICRP Publication 116 [14].

Table 10.6 Conversion coefficients from photon fluence (Φ) to ambient dose equivalent ($H^*(10)$) (Reproduced from [32])

Photon energy (MeV)	$H^*(10)/\Phi$ (pSv cm^2)
0.010	0.061
0.015	0.83
0.020	1.05
0.030	0.81
0.040	0.64
0.050	0.55
0.060	0.51
0.080	0.53
0.100	0.61
0.150	0.89
0.200	1.20
0.300	1.80
0.400	2.38
0.500	2.93
0.600	3.44
0.800	4.38
1.000	5.20
1.500	6.90
2.000	8.60
3.000	11.10
4.000	13.40
5.000	15.50
6.000	17.60
8.000	21.60
10.000	25.60

As an update to ICRU Report 47, the conversion coefficients using the ambient dose value $h^*_{E_{max}}$ from particle fluence-to-ambient are being considered. These data relate the particle fluence to the maximum value of the effective dose, calculated for whole-body exposure of the ICRP/ICRU adult reference phantoms [14] for broad parallel beams incident in irradiation geometries.

10.5 Methods of Dose Reduction in Active Interrogation

As a core tenet of radiation protection, application of the principles of ALARA is foundational to dose management and reduction in AI facilities. The ALARA principle comprises three factors: time, distance, and shielding. By decreasing the time of exposure, increasing the distance between the source and receptor, and by applying appropriate shielding to minimize the radiation reaching the receptor, whether an occupational worker or member of the public, all are equally applicable

Table 10.7 Conversion coefficients from photon air kerma to ambient dose (Reproduced from [32])

Photon energy (MeV)	$H^*(10)/K_a$ (Sv/Gy)
0.010	0.008
0.015	0.26
0.020	0.61
0.030	1.10
0.040	1.47
0.050	1.67
0.060	1.74
0.080	1.72
0.100	1.65
0.150	1.49
0.200	1.40
0.300	1.31
0.400	1.26
0.500	1.23
0.600	1.21
0.800	1.19
1.000	1.17
1.500	1.15
2.000	1.14
3.000	1.13
4.000	1.12
5.000	1.11
6.000	1.11
8.000	1.11
10.000	1.10

in I systems. Consideration of primary and secondary radiations, (i.e., neutrons and photons), requires special attention to shielding, as photons are best shielded by high-Z materials, which can be ineffective against neutron penetration depending on the material; neutrons are effectively moderated to low energies by low-Z materials, which are ineffective for photon attenuation. For example, tungsten can be used for high-energy neutrons since it has a high (n,n′) and (n,2n) cross section. However, reduce to thermal energies, a moderator is required. Neutron shields are often laminated materials to gain both of these effects.

ANSI N42.41-2007 addresses exposure duration by providing guidance on allowed average inspection time for full inspection zones containing only normal cargo packages [4], with average inspection times ranging from 90 to 900 s, depending on the container category as defined by the standard ranging from small articles to rail cars in size. Average allowed inspection times, for example, as for clearing a single, externally specified subzone containing normal cargo packages, ranging from 60 to 300 s.

Table 10.8 Conversion coefficients from neutron fluence (Φ) to ambient dose equivalent ($H^*(10)$) (reproduced from [32])

Neutron energy (MeV)	$H^*(10)/\Phi$ (pSv cm^2)
1.00×10^{-9}	6.60
1.00×10^{-8}	9.00
2.53×10^{-8}	10.60
1.00×10^{-7}	12.90
2.00×10^{-7}	13.50
5.00×10^{-7}	13.60
1.00×10^{-6}	13.30
2.00×10^{-6}	12.90
5.00×10^{-6}	12.00
1.00×10^{-5}	11.30
2.00×10^{-5}	10.60
5.00×10^{-5}	9.90
1.00×10^{-4}	9.40
2.00×10^{-4}	8.90
5.00×10^{-4}	8.30
1.00×10^{-3}	7.90
2.00×10^{-3}	7.70
5.00×10^{-3}	8.00
1.00×10^{-2}	10.5
2.00×10^{-2}	16.6
3.00×10^{-2}	23.7
5.00×10^{-2}	41.1
7.00×10^{-2}	60.0
1.00×10^{-1}	88.0
1.50×10^{-1}	132
2.00×10^{-1}	170
3.00×10^{-1}	233
5.00×10^{-1}	322
7.00×10^{-1}	375
9.00×10^{-1}	400
1.00×10^{0}	416
1.20×10^{0}	425
2.00×10^{0}	420
3.00×10^{0}	412
4.00×10^{0}	408
5.00×10^{0}	405
6.00×10^{0}	400
7.00×10^{0}	405
8.00×10^{0}	409

(continued)

Table 10.8 (continued)

Neutron energy (MeV)	$H^*(10)/\Phi$ (pSv cm^2)
9.00×10^0	420
1.00×10^1	440
1.20×10^1	480
1.40×10^1	520
1.50×10^1	540
1.60×10^1	555
1.80×10^1	570
2.00×10^1	600
3.00×10^1	515
5.00×10^1	400
7.50×10^1	330
1.00×10^2	285
1.25×10^2	260
1.50×10^2	245
1.75×10^2	250
2.01×10^2	260

In addition to these four key areas of ALARA, in Table 3.1 of the NCRP Commentary 17 classifies four types of areas (ranging from 1 to 4) with associated dose limits or administrative controls [17]. NCRP Commentary 17 recommends that the shielding walls of the radiation-generating device area be designed to prevent exposure to personnel from exceeding 0.25 mSv per year when the voltage is on and dosage in the restricted access area from exceeding 0.25 mSv per year when neutrons are being produced.

NCRP Commentary 17 identifies four aspects specific to PFNA systems where ALARA could be applied. These include tritium production and release, inadvertently exposed persons, neutron activation of foodstuffs, and radiation levels outside the facility [17]:

- **Management of tritium production.** In managing tritium, NCRP recommends that the facility housing the PFNA system is sufficiently equipped to manage tritium produced in a beam line, minimize the risk of rupture of the target accumulating tritium, and manage tritium produced in the target via venting or capture.
- **Doses due to inadvertent exposures.** Doses to inadvertently exposed persons are directly related to the neutron fluence of the PFNA system. Reduction of neutron fluence can be accomplished by increasing the sensitivity of γ-ray detection or the acceptance of reduced resolution of measurements, which could compromise the performance and confidence of the PFNA system to identify the constituents of its interrogated target.
- **Activation of food and medical devices.** The production of trace amounts of radioisotopes via neutron activation of foodstuffs should emulate doses due to radiopharmaceuticals and medical devices/implants, contributing a maximum

effective dose contribution of 1×10^{-6} mGy [15]. Tables 2.1 and 2.2 of Commentary 17 provide an excellent reference in the extensive tabulation of significant thermal and fast neutron activation products of elements with atomic numbers (Z) between 1 and 60, in addition to high-Z materials that occur in pharmaceuticals and medical devices including gold, platinum, iridium, and bismuth. These data assume fast neutron activation with a narrow energy spectrum peaking at 8.5 MeV and thermal neutron activation following a Maxwellian energy distribution, both cases assuming total neutron fluence of 6.4×10^5 neutrons-cm^{-2} in the tabulations [17]. Reactions identified as yielding the highest absorbed doses from activation products with long half-lives or high interaction cross-section probabilities include ^{24}Mg(n,p)^{24}Na (milk of magnesia) in fast neutron activation, and ^{23}Na(n,γ)^{24}Na (saline) in thermal neutron activation. Absorbed doses of concern from activation products of ^{64}Zn and ^{59}Co are discussed in further detail in Commentary 17.

References

1. W.C. Röntgen, Science **3**(59), 227 (1896). https://doi.org/10.1126/science.3.59.227. http://science.sciencemag.org/content/3/59/227
2. W.C. Röntgen, Ann. Phys. **300**(1), 1 (1898)
3. F. Attix, *Introduction to Radiological Physics and Radiation Dosimetry* (Wiley, London, 2008). https://books.google.com/books?id=WN8LTw3S5CEC
4. American National Standards Institute (ANSI), Minimum performance criteria for active interrogation systems used for homeland security. Technical Report N42.41 (2007)
5. M.P. Shannon, The dosimetry of a highly-collimated bremsstrahlung source in air. Ph.D. thesis, Georgia Institute of Technology (2009)
6. Z.D. Whetstone, K.J. Kearfott, J. Radioanal. Nucl. Chem. **301**(3), 629 (2014). https://doi.org/10.1007/s10967-014-3260-5
7. E. Morse, *Analytical Methods for Nonproliferation*. Advanced Sciences and Technologies for Security Applications (Springer International Publishing, Berlin, 2016). https://books.google.com/books?id=vjHuCwAAQBAJ
8. ICRU at a Glance (2017). https://icru.org/icru-at-a-glance-pdf/uncategorised/icru-at-a-glance
9. B. Lindell, H.J. Dunster, Swedish radiation protection institute SE 171:16 (1998)
10. NCRP Mission (2017). https://icru.org/icru-at-a-glance-pdf/uncategorised/icru-at-a-glance
11. ICRU Report 57: Conversion coefficients for use in radiological protection against external radiation. Technical Report (1998)
12. ICRU Report 85a: Fundamental quantities and units for ionizing radiation. Technical Report (2011)
13. ICRP Publication 103: Recommendations of the international commission on radiological protection. Ann. ICRP **37**(2–4), 1–332 (2007). Technical Report
14. ICRP Publication 116: Conversion coefficients for radiological protection quantities for external radiation exposures. Ann. ICRP **40**(2–5), 1–257 (2010). Technical Report
15. NCRP Report 116: Limitation of exposure to ionizing radiation. Technical Report (1993)
16. NCRP Commentary 16: Screening of humans for security purposes using ionizing radiation scanning systems. Technical Report (2003)
17. NCRP Commentary 17: Pulsed fast neutron analysis system used in security surveillance. Technical Report (2003)

18. NCRP Statement 10: Recent applications of the NCRP public dose limit for ionizing radiation. Technical Report (2004)
19. ICRP Publication 60: 1990 Recommendations of the International Commission on Radiological Protection. Ann. ICRP **21** (1–3), 1–167 (1991). Technical Report
20. Code of federal regulations title 10, Part 20: Standards for protection against radiation. Technical Report 10 CFR 20, US Nuclear Regulatory Commission (1993)
21. Code of federal regulations title 10, Part 50: Domestic licensing of production and utilization facilities. Technical Report 10 CFR 50, US Nuclear Regulatory Commission (2004)
22. Irradiation in the production, processing and handling of food. Technical Report 21 CFR 179, Food and Drug Administration - Health and Human Services (2017)
23. Occupational safety and health standards. Technical Report 29 CFR 1910, Food and Drug Administration - Health and Human Services (2017)
24. Issue Paper 1. Technical Report, US Nuclear Regulatory Commission (2014)
25. American national standard measurement and associated instrument quality assurance for radioassay laboratories. Technical Report ANSI N42.23-1996, American National Standards Institute (1999)
26. ICRP Publication 26: Recommendations of the ICRP. Ann. ICRP **1**(3) (1977). Technical Report
27. ICRU Report 39: Determination of dose equivalents resulting from external radiation sources. Technical Report (1985)
28. ICRU Report 43: Determination of dose equivalents resulting from external radiation sources - Part 2. Technical Report (1988)
29. ICRU Report 47: Measurement of dose equivalents from external photon and electron radiations. Technical Report (1992)
30. ICRU Report 51: Quantities and units in radiation protection dosimetry. Technical Report (1993)
31. ICRU Report 60: Fundamental quantities and units for ionizing radiation. Technical Report (1998)
32. ICRP Publication 74: Conversion coefficients for use in radiological protection against external radiation. Ann. ICRP **21** (1–3), 1–167 (1997). Technical Report
33. M. Pelliccioni, Radiat. Prot. Dosim. **88**(4), 279 (2000). https://doi.org/10.1093/oxfordjournals.rpd.a033046

Chapter 11
Active Interrogation Testing Standards

Richard Kouzes

Abstract Active interrogation systems for cargo inspection are designed to automatically determine the presence of special nuclear material (SNM) in transport by observing the radiation emitted by an object when exposed to an external radiation source. Active interrogation systems are contrasted with passive detection systems, such as radiation portal monitor systems, that detect the neutron and gamma radiation spontaneously emitted by SNM. Operational limitations to not interfere with commerce can restrict the use any interdiction system, including passive detection, active-interrogation and imaging systems. Because of their cost and complexity, active interrogation systems are intended for applications where SNM may be in shielded configurations that may not normally be detectable by passive systems. Active interrogation systems range from those that only indicate the presence of high-Z materials, to fissionable material detection, to those that detect specific SNM materials. The decision to deploy an AI system will depend on its ability to meet standards and specifications, its effectiveness, and its ability to fit into the operational environment. To ensure AI systems are designed and tested to a consistent level, minimum performance standards have been developed for evaluating these systems. Development of an active interrogation system that has the sensitivity to SNM that is needed while also being deployable is a challenge. It is the aim of standards to define a set of tests that can be performed on a system in an economic manner while challenging the capability of the system.

Richard Kouzes; with contributions from Edward Siciliano, Glen Warren (Pacific Northwest National Laboratory) and Peter Chiaro (Oak Ridge National Laboratory)

R. Kouzes (✉)
Pacific Northwest National Laboratory, Richland, WA, USA
e-mail: RKouzes@pnnl.gov

I. Jovanovic, A. S. Erickson (eds.), *Active Interrogation in Nuclear Security*, Advanced Sciences and Technologies for Security Applications,
https://doi.org/10.1007/978-3-319-74467-4_11

11.1 Introduction

Active interrogation refers to techniques that use an external source of radiation to induce an interaction with or excitation of the material of interest and detect any resulting outgoing radiation signature, noting that the type of ingoing and outgoing radiation may be different. AI systems for cargo inspection are designed to automatically determine the presence of SNM in transport (e.g., an intermodal cargo container) by observing the radiation emitted by an object when exposed to an external radiation source.[1] The term SNM includes plutonium, especially ^{239}Pu, or uranium enriched in the isotopes ^{233}U or ^{235}U. The term highly enriched uranium (HEU) refers to uranium enriched to contain more than 20% ^{235}U. Active interrogation systems are contrasted with passive detection systems, such as radiation portal monitor (RPM) systems, that detect the neutron and gamma radiation spontaneously emitted by SNM. There are over 3000 such passive RPM systems deployed worldwide for interdiction applications since they are inexpensive, robust and effective for many interdiction scenarios. Passive detection systems are often paired with imaging systems that can detect contraband and shielding material that could be used to attempt to hide SNM. Operational limitations to not interfere with commerce can restrict the use of any interdiction system, including passive detection and imaging systems.

Because of their cost and complexity, AI systems are intended for applications where SNM may be in shielded configurations that may not normally be detectable by passive systems. Active interrogation systems range from those that only indicate the presence of high-Z materials, to fissionable material detection, to those that detect specific SNM materials [1]. These inspection systems use photon or neutron interrogating beams, where outgoing photons and/or neutrons are detected. Radiography systems that produce images of cargo may complement the capabilities of AI and passive systems. Cosmic-ray muon based systems, where muon scattering is detected, can be used for SNM detection, and may be considered active though no external man-made interrogating beam is used.

A convenient method to categorize AI systems is to consider the interrogation particle (photon, neutron or muon) and the detected particle (photon, neutron, or muon). For example, a technology that exploits photofission may use a photon source as the interrogation beam with the intent to induce fission in the object and to examine either outgoing photon or neutron signatures to identify material as SNM. Further categorization is possible by subdividing the observed particles as either prompt or delayed (Chap. 2). These distinctions are important when considering materials for use in testing and shielding material since SNM is often not available and very difficult to manage for testing.

[1]While AI systems are also used for other contraband detection, the focus here is only on standards for SNM detection.

An AI system should also localize any detected suspect material. The localization requirement is meant to aid follow-on interdiction efforts. This localization is thus at a coarser resolution than is tested by image-quality metrics in other standards (e.g., standards for imaging systems [2]). It is possible that an AI system is part of a cascade or linear grouping of subsystems in which an early step is the rapid determination of a suspect region of a container, with a follow-on step to determine if SNM is present in the suspect region. All of these considerations are applied to the development of standardized testing requirements for AI systems. The decision to deploy an AI system will depend on its ability to meet standards and specifications, its effectiveness, and its ability to fit into the operational environment. To ensure AI systems are designed and tested to a consistent level, performance standards have been developed for evaluating these systems. Standards typically provide a minimal set of requirements for acceptance.

11.2 Specific AI Systems

Specific realizations of AI systems based on probing and detected radiation have been demonstrated, though none are currently deployed. These AI systems are generally differentiated from non-intrusive inspection systems (single or dual-energy) that produce only radiography images. There are many categories of AI systems based on the interrogating particle (photon, neutron, or muon), the detected particle (photon, neutron, or muon), whether detection is of prompt or delayed particles, and the nature of the physics process being observed [1]. However, only a limited number of AI systems have been implemented. The AI systems, categorized by physics process, considered for standard based testing are:

- *High-Z* detection involves systems that utilize photons of one or more endpoint energies to interrogate cargo, generally associated with radiography. The material can be differentiated based on its atomic number through absorption and/or backscatter detection of photons. Such systems can localize the presence of high atomic number (Z) materials.
- *Photofission (PF)* is the process of photon-induced fission of fissionable material, followed by the detection of the resulting prompt or delayed fast-neutron or delayed photon signature.
- *Nuclear resonance fluorescence (NRF)* is based on the resonant nuclear absorption and reemission of photons, where the resonant energy for the photons is indicative of each specific nuclide. The NRF method can be applied to detection of various types of contraband, including SNM.
- *Differential die-away (DDA)* uses a pulsed neutron interrogation source directed into inspected cargo. The neutrons are thermalized and absorbed, decaying with a time constant on the order of hundreds of microseconds. If fissile material is present, the thermalized neutrons from the source cause fissions that produce a new, delayed source of fast neutrons.

- *Muon scattering* systems rely upon the high-energy cosmic ray muons that are naturally present impinging on a transport. The muons undergo scattering within any material, with much larger angle scattering occurring for high-Z materials, such as SNM.

Active interrogation systems also vary depending on the type of object to be scanned and the scanning geometry. Different test configurations and procedures may be required for different geometries. Scanning geometries include:

- *Portals*, in which the object that is scanned while driven or pulled through a stationary measurement device;
- *Gantries*, where the measuring device either moves past the object under interrogation or is large enough to fully enclose the object and the object remains stationary;
- Steerable *Point-and-shoot beam* directed at a region of the object under interrogation. This category might be usable for objects too large for a portal or gantry.

Developing standards for testing of AI systems is challenging because of this variety of modalities and input and output particles is so varied. However, one common theme among testing standards is that any AI system must meet the same minimum detection criteria in order to be included for acceptance as a system for consideration.

11.2.1 Targets

The targeted mass of SNM for detection is the most crucial specification of an AI standard. The targeted mass needs to be small enough that any and all threat quantities of serious impact can be detected. Physics can limit the ability to detect very small, shielded masses, so the mass must be large enough to detect in meaningful scenarios. Any mass of SNM can be shielded from passive or active detection with a large enough shield. However, such a shield itself can be detected with imaging technology, so there is some range of shielding size that may not be obvious, but would be sufficient to shield some useful mass of SNM. The AI systems that are considered in standards development need to be sensitive to the size, material composition, and, to some extent, the shape of the test objects. Test objects that have been developed for passive-inspection systems focus on radioactivity and spectral features. It is this difference that drives the need for unique test objects for AI systems. For AI systems used to exploit signatures of SNM, the cross section of the test item that is presented to the beam, the elemental composition, and sometimes the isotopic composition of the test item impact the magnitude of the signature.

Table 11.1 lists some possible target quantities of SNM. Target masses of SNM could be the DOE quantities [3] "sufficient for a nuclear explosive device", the

Table 11.1 Potential SNM targeted masses

Agency	HEU mass	^{239}Pu mass
DOE: sufficient for a nuclear explosive device [3]	25 kg HEU	4 kg ^{239}Pu
IAEA: Significant quantity [4]	25 kg HEU with >20% ^{235}U	8 kg ^{239}Pu
IAEA: Category 1 [4]	5 kg HEU with >20% ^{235}U	2 kg ^{239}Pu

International Atomic Energy Agency (IAEA) "Significant Quantity," or the IAEA "Category 1" quantities [4]. The absolute minimum level of performance for all AI systems should be better than the IAEA "Significant Quantities" of SNM [4]. Even better would be the IAEA Category 1 values as the target for detection by an AI system.

Plutonium and HEU are both fissile materials and, thus, most AI systems are approximately equally sensitive to equal masses of these materials. Therefore, it may be possible to pick the lesser of the masses of SNM as the targeted threat quantity to be the goal for detection by AI systems.

Since handling of large masses of SNM can be problematic for testing, surrogates are usually defined that replace the SNM for testing purposes in standards. Surrogates may be fissile materials like low enriched uranium (LEU), defined as uranium enriched to less than 20% ^{235}U, fissionable materials like depleted uranium (DU) with less than the 0.7% enrichment of natural uranium, or simply high atomic number materials like Pb or W for systems that only detect the presence of high-Z materials.

11.3 Testing of AI Systems

Testing involves target materials to verify the detection capability, false-alarm items to assure that the system can discriminate target materials from other materials that might appear in commerce, and shielding and/or cargo scenarios to evaluate performance in likely scenarios to be encountered during deployment. Standards specify the conditions under which measurements are to be made, and the specific tests to be performed. Standards usually try to minimize the number and complexity of tests in order to allow the tests to be performed in an economical manner without going to extraordinary steps. Thus, surrogates are used instead of actual materials; shielding scenarios are kept simple and limited to a few typical cases; and the number of repeated measurements may be kept small, which limits the statistical significance of measurements. The overall goal is to provide a minimal, but meaningful, set of requirements that must be met by systems that aspire to be used in homeland security applications. Shielding of the input and/or output radiation, either by design or due to the presence of cargo, will affect AI system performance. In some AI implementations, shielding and matrix material (cargo) are used to create a secondary source of interrogating radiation (e.g., thermalization of fast neutrons by hydrogenous material to increase the cross section of an interrogating neutron beam). Thus, some form of shielding and surrogate-cargo is needed for testing AI systems. Testing has sometimes been done with a wide range of "cargo" configurations intended to simulate real world situations, but this is not usually done for standards since repeatability by independent testers is required of a standard and economy of testing is a consideration. Testing to standards is often performed at a facility, such as a national laboratory or commercial test organization, which has experience performing such tests. Since AI systems are so large and few will be built, standards testing may have to be performed on a site with limited access to target materials. This again means that standards need to use surrogate materials that do not require special handling, and a limited number of shielding or cargo scenarios.

11.4 Existing Related Standards

There is a need to develop performance requirements and test standards for AI systems that detect SNM in unshielded and shielded configurations within a container or conveyance. Previous American National Standards Institute (ANSI) standards [2, 5] and International Electrotechnical Commission (IEC) standards [6] have considered testing requirements for radiography and AI systems.

11.4.1 ANSI N42.41

The *American National Standard Minimum Performance Criteria for Active Interrogation Systems Used for Homeland Security* (ANSI N42.41) standard [5] was approved in 2008 with the following scope:

> "This standard specifies the operational and performance requirements for active interrogation systems for use in homeland security applications. These systems employ penetrating ionizing radiation (e.g., neutrons, high-energy x-rays, gamma-rays) to detect and identify hidden chemical, nuclear, and explosive agents by detection of stimulated secondary radiations or by nuclear resonance contrast, giving elemental and/or nuclidic identification of the composition of the substances-of-interest. These inspection systems may be designed for open inspection zones of various sizes or for various sizes of containers such as small packages, briefcases, suitcases, air cargo containers, passenger vehicles, two-axle trucks, intermodal cargo containers, semi-trailers/tractor rigs, or rail cars. The systems may be designed for operation in indoor, outdoor, or mobile facilities."

This ANSI N42.41 standard thus has a broad reach in terms of the size of targets and transports considered (packages to containers), referred to as *container category*, and the range of materials to be identified (chemicals, explosives, and nuclear material). It includes neutron (fast and thermal) and high-energy photon interrogation in fixed, mobile and portable systems. Simulants (surrogates) for the targeted materials are defined because of the difficulty of handling actual target materials. For the SNM threat, LEU with 19.5% ^{235}U (surrogate for HEU) and tungsten carbide spherical shells (surrogate for Pu) are used as simulants. Inspection times of 90–900 s are defined, depending on the container category, with the longest times being for rail inspection. Four loading configurations are used for simulated cargo at specified densities, including bare, newsprint, aluminum and steel. Testing with a minimum of ten trials for each configuration is required.

The mass of the simulant for each container category is specified. For trucks and intermodal cargo container (IMCC) configurations, the targeted masses are 200 kg for explosives, 200 kg for chemical agents, 25 kg for fissionable material, and 16 kg for weapons shells. The target SNM value is currently considered to be too massive for large conveyance screening with AI systems, as discussed in the previous section.

In addition to these threat related specifications, like all ANSI standards, there are many additional requirements for environmental and electromagnetic effects. There are additional specifications for radiation exposure to workers and to stowaways. For unmonitored workers, the general public and stowaways, the radiation dose limit is no more than 100 mrem (1 mSv) per year.

11.4.2 ANSI N42.46

The *American National Standard for Determination of the Imaging Performance of X-Ray and Gamma-Ray Systems for Cargo and Vehicle Security Screening* (ANSI N42.46) standard [2] was approved in 2008 with the following scope:

> "This standard is intended to be used to determine the imaging performance of x-ray and gamma-ray systems utilized to inspect loaded or empty vehicles, including personal and commercial vehicles of any type; marine and air cargo containers of any size; railroad cars; and palletized or unpalletized cargo larger than 1 meter by 1 meter in cross-section."

This standard is not specific to active interrogation, though similar single or multiple energy photon sources are used. Both transmission and backscatter detection are included. The system requirements are primarily for imaging but these systems also may have complementary features such as material discrimination and automatic active or passive threat alerts. Such features include identification of high-Z elements, so the standard does overlap with the AI standards. The standard focuses on image quality, resolution and object localization.

11.4.3 IEC 62523

The IEC Radiation Protection Instrumentation—Cargo/Vehicle Radiographic Inspection System (IEC 62523) standard [6] was approved in 2010 with the following scope:

> "This international standard applies to radiographic inspection systems with photon radiation energy of at least 500 keV for inspection of cargo, vehicles and cargo containers. Such inspection systems generally consist of radiation source(s), detectors, control system, image processing system, radiation safety system and other auxiliary devices/facilities. The object of this standard is to define the tests and the relevant testing methods for determining the performance characteristics of the radiographic inspection systems. This standard is not applicable to those cargo/vehicle inspection systems using neutron source radiography, computed tomography or backscatter technology."

This standard is not specific to active interrogation, though similar single or multiple energy photon sources are used. Only transmission and detection is included. The standard focuses on image quality, resolution and object localization.

11.5 Development of an AI Technical Capability Standard

A Technical Capability Standard (TCS) is a government unique standard that establishes targeted performance requirements for radiation detection and non-intrusive imaging systems. The purpose of a TCS is to establish, where practical, requirements and applicable test methods that are based on threat-informed unclassified source materials and test configurations that are not addressed in consensus standards. Technical Capability Standards are developed by an inter-agency Technical Capability Standard Working Group, which includes representatives from the Department of Homeland Security Domestic Nuclear Detection Office (DNDO), National Institute of Standards and Technology (NIST), Customs and Border Protection (CBP), the Nuclear Regulatory Commission (NRC), the Department of Energy (DOE), the Federal Bureau of Investigation (FBI), Office of Assistant Secretary of Defense for Homeland Defense and America's Security Affairs, Defense Threat Reduction Agency (DTRA), and several national laboratories (Los Alamos National Laboratory, Oak Ridge National Laboratory, Savannah River National Laboratory, Sandia National Laboratories, and Pacific Northwest National Laboratory). The DNDO works within the consensus standards arena to ensure that future ANSI N42 series consensus standards reflect the capabilities described by the TCS benchmarks, where applicable.

The proposed Technical Capability Standard for Special Nuclear Materials Detection and Localization by Active Interrogation [7] (still under development in 2018) has the following scope:

> "This TCS establishes performance requirements for systems that detect special nuclear materials (SNM) in unshielded (bare) and shielded configurations within a container or conveyance using radiation from a source that is external to that container or conveyance,

often referred to as active interrogation (AI). This TCS includes a test against a bare configuration of SNM to verify system functionality. Systems considered in this TCS include those that detect high atomic number (High-Z) materials, fissionable materials, or specific special nuclear materials. These systems automatically evaluate signatures generated by the interaction of the interrogating radiation with the material in the container under interrogation to determine the presence of SNM. This TCS applies to systems that can provide automated detection (i.e., not requiring human interpretation) and localization of SNM. The required localization accuracy is intended to assist in physical inspection, if deemed safe, in accordance with the end-users' respective safety protocols, manuals, guidelines, and/or directives. This TCS applies to systems that inspect large conveyances such as cargo containers or truck-borne cargo using an external source of radiation."

The specific instantiations of AI systems considered in the TCS are photofission, nuclear resonance fluorescence, differential die-away, and High-Z detection. These systems are to be tested independently of any other systems (e.g., passive detection or radiography). A group of system-dependent surrogate test objects (cylinders or spheres of DU or HEU) sufficient for use when testing using the TCS was determined. Radiation transport modeling was used extensively to explore the detection capabilities of each of these AI approaches and to determine the specific surrogates for each detection modality.

These test objects are sufficient surrogates for SNM for the purposes of this testing for AI systems considered in the TCS. However, in order to minimize the number of different DU surrogate masses, the difference in the strength of the signature between SNM and the surrogate material requires the use of lead attenuators to achieve comparable signal strengths depending on the interrogation modality and cargo configuration for some approaches. Each lead attenuator is a uniform shell of lead surrounding the DU surrogate, which vary in thickness depending on the AI modality. Testing is performed with the surrogates and appropriate lead shielding for ten trials, and requires ten out of ten detections. Cargo may also shield the interrogation and resultant signature radiation. While the complexity of heterogeneous cargo may allow streaming paths for radiation, it is not the intent of this TCS to test cargo complexity. Instead, only two uniform "cargo" distributions, one of mild steel and one of high-density polyethylene (HDPE) as a substitute for wood, are used, configured as shown schematically in Fig. 11.1.

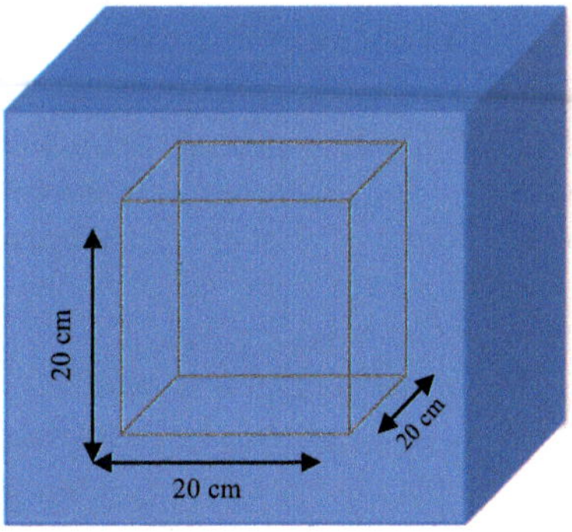

Fig. 11.1 Configuration of "cargo" for testing of surrogates and false-alarm items. The cube has an access cover, allowing for placement of the test object inside. The dimensions shown are nominal

The density of HDPE and mild steel are approximately 0.96 and 7.86 g/cm^3, respectively. This "cargo" is configured as a cube that can be located within an intermodal cargo container for testing. Testing is performed with the surrogates with the appropriate lead shielding in each of the cargo configurations for ten trials.

In addition to the SNM surrogates, systems are also tested with "false-alarm" objects that vary by AI modality. False-alarm items include W and Pb, which are high-Z materials that can be confused with threat objects in AI systems that are not specific to SNM. Materials such as heavy water and Be can cause large numbers of neutrons that can be mistaken for fissile material in neutron detection modalities. Some systems should not respond to these false-alarm objects, and the test is to verify such non-response. Some systems will respond to these false-alarm objects because they do not have the discrimination capability to separate threats from the false-alarm objects. Tests with these false-alarm items evaluate system performance with respect to these materials. The testing with false-alarm items follows the same testing approach as the SNM surrogates.

11.6 Modeling of AI Modalities

In the development of a standard for AI systems, an analytical approach was used to predict AI system response to targeted SNM, but further analysis was required for PF and DDA approaches. In order to determine the response of the various DDA and PF modalities to SNM, surrogates, cargo, and false-alarm items, extensive computer simulation was performed using GEANT4 [8], MCNP6 [9] and MCNPX [10].

11.6.1 Differential Die-Away Models

For DDA, an intense pulsed neutron beam is used as the interrogation source, and the resulting time-dependent neutron signal is observed. The approach to modeling DDA was validated against measurements discussed in references [11] and [12]. In the measurements, a sample of LEU enriched to 19.5% was placed in three locations within a cube of copy paper. The neutron source was a deuterium-tritium neutron generator capable of delivering 10^8 n/s with a 3-ms cycle starting with a 250-μs pulse of neutrons. The epithermal neutron detector was positioned in-line with the center of the beam and referenced as having a measured 23 μs thermal die-away time. Figure 11.2 shows a comparison of measurements from reference [12] and simulations for three different measured target positions, as well as a target-free background measurement.

The experimental and modeled signal shows the characteristic long decay time (relative to no target) of a DDA measurement. There is good agreement between the simulations and data for all measurements, except for the background measurement,

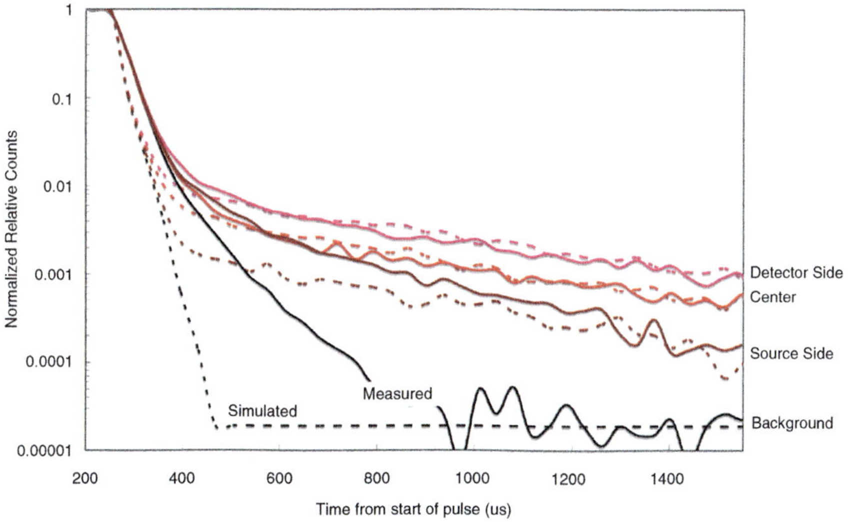

Fig. 11.2 Comparison of modeling results for benchmark measurement, as reported in [12]. The solid lines are the measurements and dashed lines the simulations

where the model did not include all elements of the room scatter. This benchmark model was used to validate the MCNP evaluation methods against the published results, and to define DDA performance metrics to be used for comparison of different surrogate targets.

There are many ways to record results from radiation transport modeling, called *tallies* in MCNP. All tally values in MCNP are normalized per starting particle, and in that sense can be considered as *efficiencies.* Three different types of tallies were used in this work: particle current, volume-averaged particle flux, and total capture efficiency. Of these tallies, the particle current is the least dependent upon on the details of the detection system, easiest to interpret and most computationally efficient. To use the particle currents in DDA analysis, one must demonstrate that the particle currents behave in a similar manner as the observed detector response. This validation was performed, and it was found that the neutron currents at the detectors for neutrons with kinetic energies greater than 0.5 eV provide an accurate analog to the rate of neutrons detected in a DDA measurement. As a result, these currents were used to evaluate the DDA surrogates. It was also shown that using a cylindrical neutron beam, rather than a more realistic isotropic neutron beam, improves computational performance without impacting the reliability of the results. For DDA, the surroundings can significantly impact the measurement, so a higher fidelity geometry (compared to photofission modeling) is appropriate.

The geometry for the DDA surrogate evaluation used a validated model of a 6.1 m (20-ft) IMCC mounted on a standard steel chassis, as seen in Fig. 11.3. The targets and shielding materials were centered 1 m above the floor of the IMCC and were pulsed with a 15-cm diameter circular beam of 14-MeV neutrons. The beam

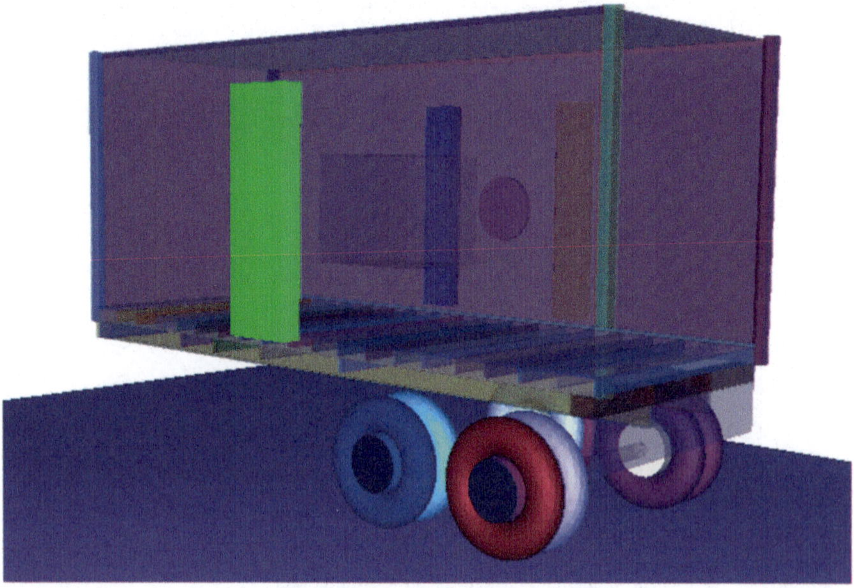

Fig. 11.3 Projection of DDA Cargo Model used for surrogate evaluations. The in-beam detector volume is the green box. The neutron source is shown as the magenta disk behind the IMCC.Two off-beam detectors are on either side of the source disk. The cargo is the gray box floating in the middle of the IMCC

was directed perpendicular to one side of the IMCC, and the currents were tallied entering an In-Beam detector on the opposite side of the IMCC and two Off-Beam detectors that straddle the beam port.

The results of the modeling effort indicate that the ideal mass of an LEU surrogate for DDA varies only slightly for each of the three cargo scenarios (bare, HDPE, and iron), which simplifies choosing an appropriate surrogate mass.

11.6.2 Photofission Models

For AI systems using photon beams, a bremsstrahlung beam such as the simulation shown in Fig. 11.4 was used. Such a photon source, extending to about 9 MeV maximum energy, is currently the only option for a high flux source. A high intensity monoenergetic photon source would be desirable, but not currently technologically feasible.

For PF, an interrogating photon beam of sufficient energy induces fission in SNM, and the observed signal can be prompt (within 1 μs of the photon striking that target material) neutrons (prompt gamma rays from photofission cannot be distinguished from the source photons scattered in the target through non-fission

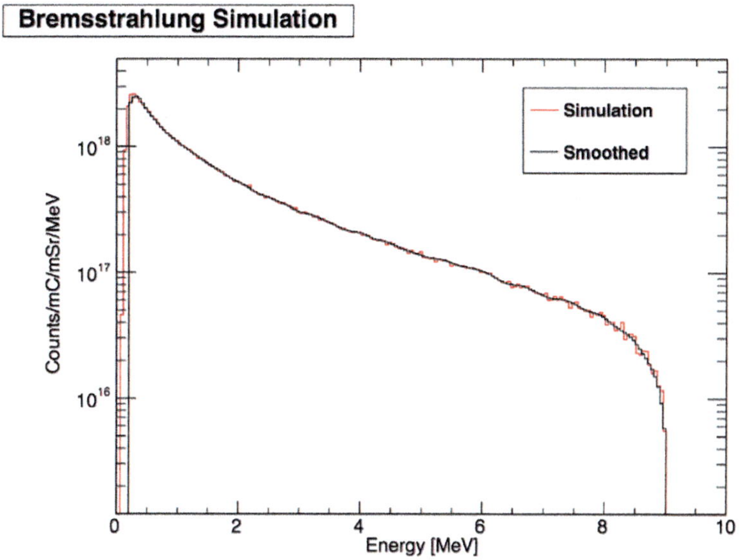

Fig. 11.4 Simulated bremsstrahlung photon energy distribution. The red line shows the spectrum as generated by the simulation and the black line is the smoothed spectrum

related processes), or delayed neutrons or gamma rays. The product particles can be binned in both energy ($E > 0.5\,\text{eV}$, $E > 1\,\text{MeV}$, $> 3\,\text{MeV}$) and the angle of their momentum with respect to the initial photon beam. A description of the photofission model used in MCNPX/MCNP6 is given in reference [13], along with associated photonuclear data and libraries.

Modeling showed that, for a DU-only surrogate to reasonably reproduce the SNM target rates in each of the test conditions (bare, HDPE cargo, Fe cargo), it would take up to nine different DU surrogates. A large number of DU surrogates would be difficult and expensive to manage operationally. Instead, modeling showed that only two different mass surrogates were required if various amounts of lead shielding was used around the DU to reduce the signal from the DU to the appropriate SNM signal being evaluated for the AI system under test.

Modeling was performed for SNM (HEU and Pu) targets in the three test conditions, and this was compared to model results for the DU surrogates in the same three test conditions, with various amounts of lead attenuator around the DU in order to match the SNM cases for each modality of the PF system. The ideal situation would be to have a unique lead attenuator for each of the neutron energy ranges and cargo loading scenarios. However, nine lead attenuators would be burdensome from an operational perspective without providing a significant benefit. Instead, modeling showed three different lead attenuators adequately span the range of attenuations necessary.

11.6.3 Conclusion

Development of an active interrogation system that has the sensitivity to SNM that is needed while also being deployable is a challenge. It is the aim of standards to define a set of tests that can be performed on a system in an economic manner while challenging the capability of the system.

Previous ANSI and IEC standards provide basic requirements, but do not require detection of the small target SNM masses that are meaningful when compared to masses of SNM that can produce potential harm. To justify the expense and operational impact that an AI system would have on a port-of-entry, the systems would have to be effective at detecting the demanding threat for which they are designed.

A Standard is being developed for AI systems in order to set specific requirements for detection of SNM across the differing capabilities of such systems. The purpose of the standard is to detail specific radiation detection requirements for a variety of AI systems using surrogate materials instead of SNM for testing. Both bare and shielded configurations were considered. Extensive modeling of the various AI modalities allowed for the definition of a limited number of surrogates to be used for testing.

The need now is to test specific AI implementations against the standards, which requires that one or more such systems be developed into a complete, robust system designed to meet the requirements of the standards that have been created.

References

1. R.C. Runkle, D.L. Chichester, S.J. Thompson, Nucl. Instrum. Methods Phys. Res. Sect. A Accel. Spectrom. Detect. Assoc. Equip. **663**(1), 75 (2012). http://dx.doi.org/10.1016/j.nima.2011.09.052. http://www.sciencedirect.com/science/article/pii/S016890021101847X
2. A.N.S.I. (ANSI), American national standard for determination of the imaging performance of x-ray and gamma-ray systems for cargo and vehicle security screening. Technical Report N42.46-2008 (2008)
3. U.D. of Energy, Restricted data declassification decisions 1946 to the present. Technical Report, U.S. Department of Energy (2001)
4. I.A.E. Agency, The physical protection of nuclear material and nuclear facilities. Technical Report INFCIRC/225/Rev.5 (Corrected) (1998)
5. A.N.S.I. (ANSI), Minimum performance criteria for active interrogation systems used for homeland security. Technical Report N42.41 (2007)
6. I.E.C. (IEC), Radiation protection instrumentation – cargo/vehicle radiographic inspection system. Technical Report 62523 (2010)
7. U.D. of Homeland Security Domestic Nuclear Detection Office, Technical capability standard for special nuclear materials detection and localization by active interrogation. Technical Report (2017)
8. S. Agostinelli, J. Allison, K. Amako, J. Apostolakis, H. Araujo, P. Arce, M. Asai, D. Axen, S. Banerjee, G. Barrand, F. Behner, L. Bellagamba, J. Boudreau, L. Broglia, A. Brunengo, H. Burkhardt, S. Chauvie, J. Chuma, R. Chytracek, G. Cooperman, G. Cosmo, P. Degtyarenko, A. Dell'Acqua, G. Depaola, D. Dietrich, R. Enami, A. Feliciello, C. Ferguson, H. Fesefeldt,

G. Folger, F. Foppiano, A. Forti, S. Garelli, S. Giani, R. Giannitrapani, D. Gibin, J.G. Cadenas, I. González, G.G. Abril, G. Greeniaus, W. Greiner, V. Grichine, A. Grossheim, S. Guatelli, P. Gumplinger, R. Hamatsu, K. Hashimoto, H. Hasui, A. Heikkinen, A. Howard, V. Ivanchenko, A. Johnson, F. Jones, J. Kallenbach, N. Kanaya, M. Kawabata, Y. Kawabata, M. Kawaguti, S. Kelner, P. Kent, A. Kimura, T. Kodama, R. Kokoulin, M. Kossov, H. Kurashige, E. Lamanna, T. Lampén, V. Lara, V. Lefebure, F. Lei, M. Liendl, W. Lockman, F. Longo, S. Magni, M. Maire, E. Medernach, K. Minamimoto, P.M. de Freitas, Y. Morita, K. Murakami, M. Nagamatu, R. Nartallo, P. Nieminen, T. Nishimura, K. Ohtsubo, M. Okamura, S. O'Neale, Y. Oohata, K. Paech, J. Perl, A. Pfeiffer, M. Pia, F. Ranjard, A. Rybin, S. Sadilov, E.D. Salvo, G. Santin, T. Sasaki, N. Savvas, Y. Sawada, S. Scherer, S. Sei, V. Sirotenko, D. Smith, N. Starkov, H. Stoecker, J. Sulkimo, M. Takahata, S. Tanaka, E. Tcherniaev, E.S. Tehrani, M. Tropeano, P. Truscott, H. Uno, L. Urban, P. Urban, M. Verderi, A. Walkden, W. Wander, H. Weber, J. Wellisch, T. Wenaus, D. Williams, D. Wright, T. Yamada, H. Yoshida, D. Zschiesche, Nucl. Instrum. Methods Phys. Res. Sect. A Accel. Spectrom. Detect. Assoc. Equip. **506**(3), 250 (2003). http://dx.doi.org/10.1016/S0168-9002(03)01368-8. http://www.sciencedirect.com/science/article/pii/S0168900203013688
9. J.T. Goorley, M.R. James, T.E. Booth, F.B. Brown, J.S. Bull, L.J. Cox, J.W.J. Durkee, J.S. Elson, M.L. Fensin, R.A.I. Forster, J.S. Hendricks, H.G.I. Hughes, R.C. Johns, B.C. Kiedrowski, R.L. Martz, S.G. Mashnik, G.W. McKinney, D.B. Pelowitz, R.E. Prael, J.E. Sweezy, L.S. Waters, T. Wilcox, A.J. Zukaitis, Initial mcnp6 release overview. Technical Report LA-UR-11-07082, Los Alamos National Laboratory, Los Alamos, New Mexico (2011)
10. D. Pelowitz, Mcnpx user's manual version 2.7.0. Technical Report LA-CP-11-00438, Los Alamos National Laboratory, Los Alamos, New Mexico (2011)
11. K.A. Jordan, T. Gozani, Nucl. Instrum. Methods Phys. Res. Sect. A Accel. Spectrom. Detect. Assoc. Equip. **579**(1), 388 (2007). http://dx.doi.org/10.1016/j.nima.2007.04.083. http://www.sciencedirect.com/science/article/pii/S0168900207006584. Proceedings of the 11th Symposium on Radiation Measurements and Applications
12. K.A. Jordan, T. Gozani, in *Joint International Topical Meeting on Mathematics and Computation and Supercomputing in Nuclear Applications* (American Nuclear Society, Monterey, 2007)
13. J. Verbeke, C. Hagmann, D. Wright, Simulation of neutron and gamma ray emission from fission and photofission. Technical Report UCRL-AR-228518, Lawrence Livermore National Laboratory, Livermore (2014)

Chapter 12
Conclusion

Igor Jovanovic

Abstract The problem of detection of special nuclear material, especially in shielded configurations, is a remarkable technical challenge that has eluded the past efforts that rely exclusively on passive detection. In this book we have addressed the various aspects of active interrogation from the perspective of science, technology, and system design and operation. Here we summarize the major conclusions and insights that have been presented at length in the previous chapters.

In all detection scenarios it is important to understand the detailed characteristics of the signatures of interest, including their magnitudes. The physical origin of characteristic signatures by which materials of interest such as special nuclear material (SNM) can be detected are the energetic processes occurring in the atomic nucleus. In the processes such as radioactive decay and fission, γ rays and neutrons can be emitted with unique energies and/or in a typical sequence that can allow their discrimination from the background radiation that does not share those characteristics. It is equally important to understand the properties of background, including its variability and how the background may be affected by presence of nearby objects with which radiation interacts, or which represent sources of natural radiation.

The simplest method by which SNM and other radioactive materials may be detected is by measuring its spontaneously emitted radiation and detecting the characteristic signatures within the measured radiation data. In the context of this book, only γ rays and neutrons constitute sufficiently penetrating signatures to be viable candidates for detection in passive systems, such as radiation portal monitors. The γ rays can be particularly useful if detected in spectroscopic systems with good energy resolution, which allows them to be better distinguished from

I. Jovanovic (✉)
Department of Nuclear Engineering and Radiological Sciences, University of Michigan, Ann Arbor, MI, USA
e-mail: ijov@umich.edu

I. Jovanovic, A. S. Erickson (eds.), *Active Interrogation in Nuclear Security*, Advanced Sciences and Technologies for Security Applications,
https://doi.org/10.1007/978-3-319-74467-4_12

various backgrounds. The use of imaging in conjunction with spectroscopy can also significantly improve the capabilities in passive detection by better distinguishing the signal from background based on their spatial distributions. However, all of these passive methods are still fundamentally limited by the relatively low rate of spontaneous emission of radiation and the relative ease with which the emitted radiation can be shielded.

The limitations of passive measurements can be overcome by the use of the AI technique, which delivers several benefits, including increasing the intensity of the characteristic radiation signatures and imposing the unique time structure which further allows the characteristic signal to be distinguished. The two major methods of AI that involve stimulating emission of characteristic signatures from an atomic nucleus rely on nuclear fission or nuclear resonance fluorescence. Nuclear fission is induced either using a neutron source or a photon source (photofission). In the recent period there has been a vigorous effort to understand the characteristics of the nuclear resonance fluorescence signature from SNM and to develop practical methods for screening containers using this method. There is also a significant effort to develop the modeling and simulation tools to meet the needs of design of AI systems and understand their performance characteristics.

AI is distinct from passive measurements in one primary aspect—it makes use of an external, typically much more intense source of radiation, to excite the desired characteristic signatures and overcome backgrounds. Various radiation sources, referred to as probes, were introduced in the context of AI and their principle of operation, typical performance, features, and limitations were discussed. The most general principle is that of a charged particle acceleration method and a converter target, which allows the penetrating neutron or photon radiation to be produced. While the probe technologies capitalize on the decades of technological progress in fundamental research such as nuclear and particle physics, important research and development is still ongoing to make them compact, reliable, less expensive, easier to operate, and deliver a desired time structure to use in conjunction with AI. Simultaneously, there is a concerted research in a non-traditional class of sources based on ultrashort pulse lasers, which has a potential to deliver significant benefits in some AI scenarios.

Radiation detectors constitute the second important component of an AI system. Similar to probe technologies, radiation detectors have been used for many decades in fundamental research and in applications such as medicine, power generation, and industrial processes. As a result, most of the radiation detector technologies are relatively mature and well understood from a physics point of view. However, choosing the best detector technology and adapting it for use in AI still requires a careful consideration of the measurement need, appropriate match to the probing radiation, and prudent interface to data acquisition. Of particular interest is the combination of detectors into larger systems, usually referred to as *detector arrays* to realize large absolute detection efficiency or serve a task complementary and important to AI, such as transmission radiography. There is a vigorous development program in detector materials, some which will undoubtedly benefit the future AI systems.

The success of measurements made using one of the AI techniques can be sensitive to the method by which the data is acquired and processed. The special conditions experienced by DAQ systems used in AI include high data rates, typically a large number of data channels, and the need to record time stamps, which allow the time correlations in the data to be exploited. The development of digital DAQ methods has allowed significant advances in digital data analysis, allowing better particle identification through techniques such as pulse shape discrimination. The main challenges in DAQ for AI still remain the typically high data rates leading to pulse pile-up, challenges in baseline estimation, and data reduction to enable more convenient and faster offline or online data analysis.

While no AI system has been deployed to date, there have been several credible efforts to develop a fully functioning prototype AI system by both the national laboratories and the industry. Perhaps the most well-known among them is the Nuclear Carwash at Lawrence Livermore National Laboratory, which relies on energetic neutrons to induce characteristic signatures of SNM. Companies such as Rapiscan and Passport Systems have developed their implementations of neutron and high-energy X-ray AI systems and have conducted substantial testing, with a goal of commercializing the technology in the near future. More exotic systems such as those that use laser-based sources as a probe have not yet reached the level of technological maturity that would allow their testing at a system level, but the pace of technological progress is high and they may be considered in the near future.

One topic that has stimulated significant discussion between the technical and regulatory community is the magnitude of radiation dose received by people and objects when exposed to AI probes. This is a significant concern especially because of the problem of stowaways, but also because of the necessary shielding and the potential for exposure of system operators. In some cases a concern may also exist about the activation of the cargo. The impetus is therefore to develop AI systems which minimize the deposited dose per unit useful information provided.

Essential to the development of future AI systems that could undergo deployment is establishing the AI testing standards. All AI systems designed for SNM detection need to be able to detect a certain minimum amount of material, while being sensitive to the object size, material composition, and shape. These requirements different from typical requirements presented to passive detection systems, where the emphasis is on the magnitude of spontaneous emission of radiation and the presence of unique, identifiable features in the γ-ray spectra. Testing methods must include both the target object under interest but also items that could trigger false alarms, such that the realistic performance under deployment can be predicted. As of 2017, a Technical Capability standard for AI is under development through a working group that involves many US federal agencies and laboratories and addresses both unshielded and shielded configurations.

Despite the substantial engagement of the broad scientific community in the recent period on the subject of AI, many aspects of the problem, especially that of detection of shielded SNM, remain only partially addressed. It is our intent for the content of this book to motivate a growing community of researchers to undertake the effort to detect nuclear threats with a fresh perspective while developing or adopting a growing repertoire of advanced technologies.

Glossary

Actinides Metallic chemical elements with atomic numbers from 89 to 103, including thorium, uranium, neptunium, and plutonium.

Active interrogation Generation and measurement of characteristic nuclear signatures using an external radiation source.

Acute exposure Radiation exposure over a short period of time.

ADONIS Algorithmic Development Framework for Nuclear Instrumentation and Spectrometry.

Analog-to-digital converter (ADC) Part of DAQ that converts an analog signal, such as a pulse produced by a radiation detector, into a digital signal.

Bayesian method Method of statistical inference in which Bayes' theorem is used to update the probability for a hypothesis as more evidence or information becomes available.

Breit-Wigner distribution Continuous probability distribution used to model nuclear resonances.

Bremsstrahlung Electromagnetic radiation produced by the deceleration of a charged particle, also called *braking radiation.*

Bunching Modulating the particle beam in a way that results in production of a train of short pulses.

Characteristic X ray Energetic photon emitted in a transition between two bound states of an electron in atom, having a well-defined energy.

Charge integration window A time period over which the detector current pulse is integrated.

Charge-to-Digital Converter (QDC) Part of DAQ designed to directly integrate the pulse of current from the detector to give a charge value, which is itself proportional to the energy deposited by the radiation particle.

Charge-sensitive preamplifier (CSP) Part of DAQ which integrates the charge pulse and converts it to a voltage signal that can be accepted by the stage that follows.

I. Jovanovic, A. S. Erickson (eds.), *Active Interrogation in Nuclear Security*, Advanced Sciences and Technologies for Security Applications, https://doi.org/10.1007/978-3-319-74467-4

Cherenkov radiation Radiation emitted in the wake of a charged particle propagating in a medium at a velocity that exceeds the speed of light in that medium.

Coded aperture imaging Method of imaging that employs an aperture in front of detectors that blocks the incident radiation with a known pattern.

Compton edge Sharp feature in the Compton spectrum that corresponds to maximum photon energy lost in Compton scattering.

Compton imaging The use of position and energy measurement of two interactions of a single photon to restrict the reconstructed photon momentum to the surface of a cone.

Compton scattering Scattering of an energetic photon on electron.

Cosmic rays Energetic radiation of solar or galactic origin that is incident onto Earth's atmosphere.

Cosmogenic radionuclides Radionuclides that are continuously produced by cosmic rays and are relatively short-lived.

Coulomb scattering Collision of two charged particles in which the Coulomb force is the dominant interaction.

Chronic exposure Radiation exposure which occurs over a long period time.

Current mode Detector operation during which a continuous signal that is a time average of the individual current bursts is produced.

Cyclotron Accelerator in which charged particles accelerate outwards from the centre along a spiral path.

Data acquisition (DAQ) Electronic system required to interpret the electrical signals produced by radiation detectors when a γ ray, neutron or other radiation type interacts with it.

Data smoothing Process that replaces the value at hand with a new value taken as the arithmetic average of a window of values centered at the value at hand.

Delayed signatures Properties of radiation emitted with some delay after the occurrence of fission.

Dead time Minimum amount of time that must separate two events in order that they are recorded as two separate pulses.

Detector A device or system used to sense ionizing radiation.

Detection efficiency System's ability to measure and record pulses, generally composed of intrinsic and geometric terms.

Die-away time Characteristic detector-dependent time constant associated with neutron interaction in detector, typically on the order of 30–100 μs.

Differential die-away self-interrogation Technique developed to compute multiplication and plutonium content in spent fuel.

Direct laser acceleration Interaction of a charged particle with strong laser field without the use of intervening plasma waves that results in particle acceleration.

Drift tube linac Standing-wave structure with increasing length of the drift tubes along beam propagation direction.

Doppler broadening Spreading of spectral lines due to the Doppler effect.

Electron gun Device that produces a narrow, collimated electron beam that has a precise kinetic energy.

Electron Light negatively charged particle that comprises the atomic shell.

Emittance Mean spread of particle coordinates in position and momentum phase space.

Energy resolution Capacity to consistently record the same pulse height in different events in which the same energy is deposited in the detector.

Filtered backprojection Analytic reconstruction algorithm that applies a convolution filter to remove blurring.

Fission products The fragments of a nucleus after it undergoes fission.

FIGARO Fissile interrogation using γ rays from oxygen.

Fission Splitting of a heavy nucleus into two lighter fragments, accompanies by emission of energetic neutrons and γ rays.

Fluorescence A form of luminescence resulting from excitation.

Flux Rate of passage of radiation quanta through unit surface area.

Gamma ray High-energy photon produced in nuclear de-excitation or annihilation process.

Geant4 A platform for "the simulation of the passage of particles through matter," using Monte Carlo methods.

High-pass filter Sets a duration of the pulse by introducing a decay time constant.

Imaging system Information system that forms information by probing the object, records information with detectors, and presents information in the form of images.

Induction accelerator Type of accelerator which uses ferromagnetic cores to provide stored energy to accelerate beams of electrons or ions.

Inverse Compton scattering Interaction of a low-energy photon with an energetic electron, from which photon emerges with increased energy.

KLYNAC Compact linear accelerator that integrates the klystron RF source and a coupled-cavity linear accelerator into a single resonant system.

Klystron Device in which electron beam interacts with RF fields as it passes through resonant cavities in the structure separated by a cylindrical metal drift tube.

Laser wakefield acceleration Process by which an intense laser pulse produces a plasma wave, which can in turn accelerate injected or self-trapped electrons to high energies.

Linear accelerator (linac) Particle accelerator that accelerates particles by providing an oscillating electric potential along a linear beamline.

Low enriched uranium (LEU) Uranium enriched to less than 20% ^{235}U.

Magnetic quadrupole Arrangement of four magnets used for focusing a particle beam.

Magnetron High-powered vacuum tube that uses crossed electron and magnetic fields to produce the high-power RF radiation.

MCNP Monte Carlo N-Particle Transport Code—a software package for simulating nuclear processes.

Mean square voltage (MSV) mode Adaptation of current mode where the direct-current (DC) component of the current mode signal is blocked and the signal of interest becomes the fluctuation of the current mode signal about its mean value.

Medical radionuclides Radionuclides commonly used in medical imaging or treatment.

Minimum detectable activity (MDA) The smallest activity of radiation that can be detected given some level of background while not exceeding the standard false positive and negative probability.

Multichannel analyzer (MCA) Part of DAQ that can be used at the back end of the data acquisition chain to visualize the energy (or any quantity that can be represented as an analog voltage level) and frequency of radiation events.

Multichannel scaler (MCS) Part of DAQ that can be used at the back end of the data acquisition chain to visualize the time profile of radiation events.

Multiplicity Number of emitted radiation quanta following a decay process.

Muon tomography Use of cosmic ray muons to generate 3-D images of volumes using information contained in the Coulomb scattering of the muons.

Muon Light, highly energetic charged particle produced in large quantities in cosmic ray interactions with the atmosphere.

Naturally occurring radioactive material (NORM) Radioactive material present naturally in the environment.

Near-monoenergetic photon source (MPS) A laser-driven source which employs a process such as inverse Compton scattering to produce photons with narrow energy spread.

Neutron generator Typically a compact device that uses the $D(d,n)^3He$ or $T(d,n)^4He$ reaction to produce 2.45 MeV or 14.1 MeV neutrons, respectively.

Neutron multiplicity Number of neutrons emitted in either spontaneous or induced fission.

Neutron scatter camera Two-plane arrangement of fast neutron detectors that enables the localization of incident neutron momentum to the surface of a cone.

Neutron Heavy neutral constituent of a nucleus.

Non-paralyzable detector System with fixed dead-time to avoid saturation of the detector.

Nonproliferation Efforts to curtail proliferation activities through diplomatic, legal, and administrative methods.

Nuclear counterproliferation Is the effort to combat nuclear proliferation by focusing on intelligence and military methods.

Nuclear forensics Development of methods and practices that can be used to determine the provenance (origin) of the nuclear material, whether after its use (post-detonation) or in the cases when it is intercepted (pre-detonation).

Nuclear proliferation The spread of nuclear weapons, fissile material, and the requisite technology and information that could be used to construct them, to the states that are not among the five nuclear weapons states.

Nuclear resonance fluorescence Process in which a nucleus absorbs and emits high-energy γ rays.

Nuclear safeguards Measures that can be used to verify that countries comply with their international obligations not to use nuclear materials for nuclear explosives.

Nuclear treaty verification Development and deployment of measures to ensure verifiable compliance with treaties and other international agreements, implementation of regimes to reduce nuclear weapons, and detection and dismantlement of undeclared nuclear programs.

Nuclear weapons states The five countries that have acquired nuclear weapons before January 1, 1967.

Nuisance alarms Erroneous alarm that consumes unnecessary resources.

Paralyzable detector System affected by dead-time to the point where it can reach a saturation point and is unable record any events.

Passive measurement Measurement that does not involve the use of externally generated radiation.

Penning source Particle source in which strong magnetic field parallel to the electric field of the sheath guides electrons and ions on cyclotron spirals from cathode to cathode.

Phase space Multi-dimensional space in which any state of a system can be assigned a unique set of coordinates.

Photofission Fission induced by photons.

Photoneutrons Neutrons ejected from a nucleus by photons with energies exceeding neutron binding energy in the nucleus.

Photopeak Feature in the detector pulse spectrum that corresponds to full energy deposition in a detector.

Pinhole imaging The simplest imaging concept that uses a small hole in opaque material to reconstruct the image of an object.

Primordial radionuclides Radionuclides that have been present on Earth since its formation and have relatively long half lives.

Principal component analysis (PCA) Use of orthogonal transformation to convert a set of observations of possibly correlated variables into a set of values of linearly uncorrelated variables, referred to as principal components.

Prompt signatures Properties of radiation emitted simultaneously with the fission process.

Proton Heavy positively charged constituent of a nucleus.

Pulse height distribution Variation of pulse magnitude at a constant applied voltage due to charge collection, detector geometry and various interactions of radiation in the counter.

Pulse mode Detector operation during which information on the amplitude and timing of individual radiation events is preserved.

Pulse-shape discrimination (PSD) Technique based on variation in the amount of light produced by delayed fluorescence can be utilized to distinguish different types of particles.

Q-value Energetics of nuclear reaction characterized as difference between of the masses of the initial reactants and the sum of the masses of the final products.

Radiative capture Neutron absorption in a nucleus followed by nuclear de-excitation, in which an energetic γ ray is emitted.

Radioisotope Atom with an excess of energy, making it unstable.

Radiofrequency quadrupole accelerator (RFQ) Compact accelerating structure for proton and ion acceleration that captures, bunches, and simultaneously focuses and accelerates low-energy beams.

Radiography Imaging technique that uses attenuating properties of penetrating radiation to reveal the internal structure of an object.

RF cavity Metallic chamber that contains an electromagnetic field.

Rossi-Alpha distribution Probability distribution of the time intervals between pulses generated in the neutron detection process.

Sampling Theorem States that the uniformly spaced discrete samples are a complete representation of the signal if its bandwidth is less than half the sampling rate.

Sensor network Arrangement of several interconnected sensors known as the network nodes.

Shielding Intervening material that absorbs or scatters the radiation.

Ship effect Increase of the neutron background through cosmic ray interaction in the vicinity of large objects such as ships.

Signal-to-noise ratio Measure used to compare the level of a desired signal to the level of background noise.

Single channel analyzer (SCA) Part of DAQ that produces an output logic pulse if the peak amplitude of the pulse from the shaping amplifier falls within a preconfigured pulse-height window.

Special nuclear material (SNM) Plutonium, uranium enriched in the isotopes ^{233}U or ^{235}U, or any materials the U.S. Atomic Energy Commission determines to be SNM.

Spectroscopy Measurement of radiation energy distribution.

Spectrum unfolding Reconstruction of incident radiation spectrum from the measured detector response and the known detector response function.

Superconducting accelerator Device that uses superconducting magnets to allow significantly higher particle energies to be achieved at much reduced operational costs.

Technologically enhanced naturally occurring radioactive material (TENORM) Radioactive elements which have been further processed or concentrated.

Thermal neutrons Free neutrons with average energy of 0.0253 eV (corresponding to room temperature velocity) and large absorption and fission cross section.

Trigger Signal that tells the data acquisition electronics to start acquiring data.

Van de Graaff generator Electrostatic generator which uses a moving belt to accumulate electric charge on a hollow metal globe on the top of an insulated column.

Weapons of mass destruction Weapons that operate on the basis of nuclear, radiological, chemical, biological, or other principles that can lead to large loss of life, damage to infrastructure, or biosphere.

X ray Energetic photon produced in transition of an atomic electron transition or through electromagnetic interaction of energetic electrons.

X ray fluorescence Emission of characteristic secondary X rays from a material excited by bombarding with primary photons, usually X-ray source.

Index

I. Jovanovic, A. S. Erickson (eds.), *Active Interrogation in Nuclear Security*, Advanced Sciences and Technologies for Security Applications, https://doi.org/10.1007/978-3-319-74467-4

Printed by Printforce, the Netherlands